Electric Power Robots

Yaonan Wang • Yanjie Chen • Hang Zhong •
Jiacheng Liang

Electric Power Robots

Modeling, Control, and Applications

Springer

Yaonan Wang
College of Electrical and Information Engineering
Hunan University
Changsha, Hunan, China

Yanjie Chen
School of Mechanical Engineering and Automation
Fuzhou University
Fuzhou, Fujian, China

Hang Zhong
School of Robotics
Hunan University
Changsha, Hunan, China

Jiacheng Liang
College of Electrical and Information Engineering
Hunan University
Changsha, Hunan, China

ISBN 978-981-96-2511-6 ISBN 978-981-96-2512-3 (eBook)
https://doi.org/10.1007/978-981-96-2512-3

This Springer imprint is published by the registered company Springer Nature Singapore Pte Ltd.
The registered company address is: 152 Beach Road, #21-01/04 Gateway East, Singapore 189721, Singapore

Preface

The application of electric power robots is rapidly growing due to the immense power demands. This book is an attempt to address the topic of electric power robots in the electricity industry through a multidisciplinary approach. Unlike other publications that focus primarily on theoretical foundations, this one emphasizes practical applications within the power industry. Thus, it serves as an ideal companion to theoretical works, offering a comprehensive perspective on the subject matter.

This book introduces the typical operational robots utilized in the power industry, including condenser cleaning mobile manipulator robots, transmission line de-icing robot manipulators, and aerial manipulator robots. Each chapter of this book has been meticulously crafted by leading authorities in their respective fields, ranging from the intricacies of machinery mechanics to the complexities of control theory, informatics, and mechatronics.

The book is divided into eight chapters, with Chap. 1 covering the historical background of electric power operation robots. Chapter 2 introduces the background and structural modeling mathematical formulas of the condenser cleaning mobile manipulator robot. Following this, Chap. 3 discusses its motion control methods and applications within the power industry. Chapter 4 introduces the structural modeling and inverse kinematics algorithms of the transmission line de-icing robot manipulator. Following this, Chap. 5 discusses its motion control methods and applications within the power industry. Chapter 6 presents unmanned aerial vehicles designed for stable flight inspection in the power industry, with the design of adaptive controllers for unmanned aerial vehicles under disturbances and input saturation. Chapter 7 introduces aerial manipulator robots for operation in transmission lines and industrial pipelines. Chapter 8 summarizes the content of the book, analyzes the application challenges that electric power robots may encounter, and offers perspectives on future development directions and trends.

This book project can be foreseen as a reference for young professionals and researchers, providing an overview of the significant aspects of electric power robots in the power industry. Acknowledging the vast scope of the topic, this book is considered a first edition. As editors, we are genuinely receptive to feedback and are keen to incorporate suggestions for enriching future editions with additional insights and content.

We extend our heartfelt gratitude to all the authors and the expert blind reviewers for their invaluable contributions to this project. Their expertise and dedication have been instrumental in shaping its content. Additionally, we would like to express our sincere appreciation to the staff at Springer Nature for their unwavering professional support. Their guidance and counsel have been invaluable throughout the preparation of this book, ensuring its quality and coherence.

Finally, we extend our deepest appreciation to our families for their unwavering patience and understanding. Their support has been the bedrock that has enabled us to devote our time and energy to this book. Without their love and encouragement, this endeavor would not have been possible.

Changsha, Hunan, China — Yaonan Wang
Fuzhou, Fujian, China — Yanjie Chen
Changsha, Hunan, China — Hang Zhong
Changsha, Hunan, China — Jiacheng Liang

Contents

Chapter 1
Introduction to Electric Power Robots

1.1 Background and Motivation

In recent years, the demand for electrical energy has grown significantly due to the rapid expansion of the global population and economy, which has driven the development of power systems toward larger scale and more complexity. However, the expansions of power systems have raised new challenges in energy supply and placed increasing demands on the monitoring and maintenance of power systems. Take the accident that occurred in Brazil on March 21, 2018, for example [1]; a calibration error in the converter station circuit breaker triggered a chain reaction, resulting in a major blackout for a quarter of the population and a huge negative impact on the Brazilian economy. This accident highlights the challenges faced in operating and maintaining power systems, especially in such complex and large networks.

Effective monitoring and preventive maintenance are crucial for ensuring the stable operation of increasingly complex power systems and reducing the risk of faults occurring [2]. Although performing maintenance in the de-energized state can significantly reduce operational risks, it often results in considerable losses to users and economic benefits. On the contrary, conducting maintenance tasks in an energized state exposes the operator to greatly increased risk. Furthermore, electrical installations are generally located in harsh environments and mounted in inaccessible places, making inspection and maintenance tasks dangerous and inconvenient for technicians and limiting the efficiency and safety of the maintenance tasks. As shown in Fig. 1.1, operation workers typically need to climb high places to conduct high-risk high-voltage transmission inspection tasks, which are time-consuming, ineffective, and vulnerable to outdoor weather and other factors.

Against this background, the development and application of power operational robot (POR) technology is particularly critical in the field of power systems. These robotic systems can be remotely operated by humans or fully automated to replace

Y. Wang et al., *Electric Power Robots*,
https://doi.org/10.1007/978-981-96-2512-3_1

Fig. 1.1 Manual inspection and maintenance of power systems. (**a**) Manual ash removal and desulfurization in the generation plant. (**b**) Manual maintenance of high-voltage transmission lines. (**c**) Manual inspection of substation. (**d**) Manual maintenance of distribution

manual labor to perform high-risk and complex operational tasks in power environments [3–5]. The developed electric power robots are capable of performing tasks in harsh environments such as high temperatures and high voltages, which improves work efficiency and quality while reducing human resource requirements and operation risks.

Although electric power robots have exhibited potential application prospects in the field of power inspection and maintenance, they are still in the research and development stage. The successful implementations of the electric power robots are limited to the specific conditions of the tasks, and the current application framework is not yet mature enough to fully realize their application potential. For this purpose, this book will introduce the latest technological advancements in the field of electric power robots and explore their future application potential in the maintenance and operation of power systems. By analyzing existing research and practical examples, this book presents the current state of the art, research directions, and main contributions of electric power robots.

1.2 Research Status of Electric Power Robots

Power systems are composed of the generation of electric energy, transmission, and distribution to end consumers. Although the complexity and size of different power systems vary widely, they follow the same basic structure, including generation plant, transmission, substation, and distribution stages [6, 7]. Figure 1.2 provides a visual representation of a typical power system.

1.2.1 Electric Power Robots for Generation Plants

Generation plants convert other forms of energy into electricity, such as thermal, luminous, wind, and nuclear energies. In the following, the inspection and maintenance status of electric power robots in these generation plants will be comprehensively introduced.

1. **Thermal Generation Plant**

   Thermal power generation is a typical power generation method that uses heat energy generated when combustible materials are burned and converts it into electrical energy through a power generation device. The boiler header is one of the basic and key components of a thermal generation plant to generate electricity. Thus, the inspection and maintenance of the boiler headers are essential for the continuous operation of thermal generation plants. For this purpose, a slim and expandable header inspection robot was designed and developed for remote visual inspection of boiler headers in thermal power plants [8], as shown in Fig. 1.3a. The developed robotic device is very useful in the coverage of the internals of a header for detecting the presence of any foreign material in case of a newly erected header or for gross detection of internal deposits, blockages, degradation, and cracking on the stub-holes periphery. Further, boiler water wall in thermal generation plants is characterized by high-altitude detection requirements. The existing water wall-climbing robots are characterized by low

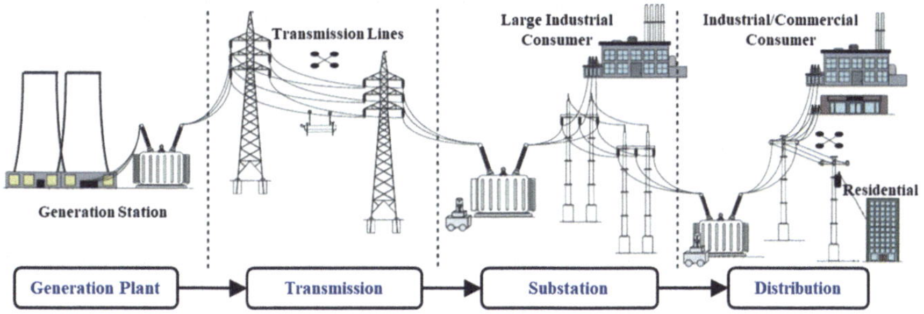

Fig. 1.2 Basic structure of a typical power system [6]

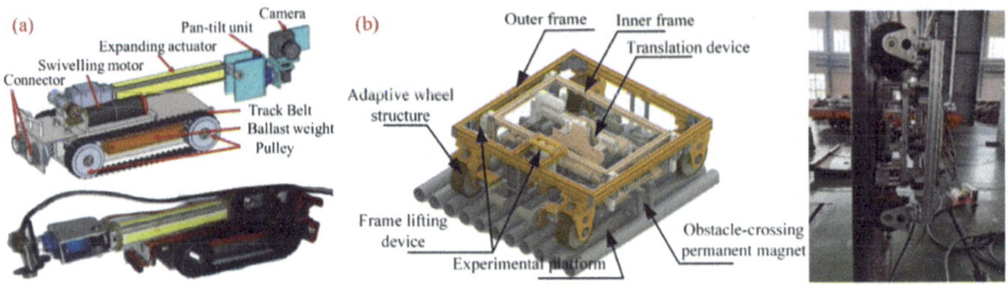

Fig. 1.3 Typical power operation robots for inspection and maintenance of thermal generation plants. (**a**) Slim and expandable header inspection robot [8]. (**b**) Wall-climbing robot [9]

obstacle-crossing performance, deviations, and a lack of autonomous crossing pipeline function. In view of this feature, a wall-climbing robot with permanent magnet and electromagnetic hybrid adsorption wheels was proposed for boiler water wall inspection [9], as shown in Fig. 1.3b. This robot can traverse independently and achieve transverse movement with adsorption, moving, and driving modes. Although thermal generation still occupies most of the electricity market, this method of power generation can lead to smoke pollution, dust pollution, and resource consumption.

2. **Photovoltaic Generation Plant**

   Photovoltaic power generation is a technology that uses the photovoltaic effect at the semiconductor interface to directly convert light energy into electrical energy. It is mainly composed of three parts: solar panels, controllers, and inverters. Solar cells connected in series are packaged and protected to form a large-area solar cell module and then combined with power controllers and other components to form a photovoltaic power generation device. To ensure the stable and safe operations of the photovoltaic generation system, the maintenance tasks involve photovoltaic module cleaning and sewage removal, inverter maintenance, and power system inspection. Regarding maintenance tasks, several wheeled mobile operational robots have been designed and developed for cleaning and sewage removal operations, as concluded in Table 1.1.

   Regarding inspection tasks, unmanned aerial vehicles integrated with inspection sensors have been employed for aerial inspection. The values of drones for solar farm inspections include quick and easy scanning of roofs and arrays, hotspot identification showing defective cells and geo-tagged images, diagnostic purposes and panel placement planning, infrared inspection to reveal defective panels, maintaining solar farms to run at peak efficiency, and saving time with aerial inspections of preselected waypoints. The typical examples of aerial inspection of photovoltaic generation plants are shown in Fig. 1.4.

3. **Wind Generation Plant**

   Wind energy is a clean and pollution-free renewable energy that has been used by people for a long time, such as pumping water and grinding flour through windmills. The use of wind power to generate electricity is very environmentally friendly and has attracted increasing attention from countries around the world

Table 1.1 Typical wheeled mobile operational robots for the maintenance of photovoltaic generation plants

Robot	Company	Salient features		Picture
GEKKO [10]	Serbot	Max. cleaning capacity	576 m^2/h	
		Max. speed	9.6 m^2/min	
		Size	1800 × 1250 × 432 mm	
		Weight	63 kg	
		Water consumption	0.5–3.0 L/min	
E4 [11]	Ecoppia	Dust removal rate	99%	
		Energy independent	100% solar	
		Cost-effectiveness	18 months return on investment	

(continued)

Table 1.1 (continued)

Robot	Company	Salient features		Picture
Solar cleaning robot [12]	Miraikikai	Weight	17 kg	
		Typical cost-effectiveness	80% reduction compared with manual cleaning	
		Power requirements	Li-ion battery	
RAYBOT [13]	Ecovacs	Cleaning speed	5 solar panels per hour	
		Steep roof tops	Up to 55°	
		Power requirements	Battery-powered	

(a) Drone-based automatic line inspection [14].

(b) Drone for infrared solar panel inspection [15].

Fig. 1.4 Drone-based inspection of photovoltaic generation plants. (**a**) Drone-based automatic line inspection [14]. (**b**) Drone for infrared solar panel inspection [15]

(a) Fenner wind turbine blade breakage accident in Madison County, New York, USA in 2016.

(b) Blade damage accident at Humber Gateway offshore wind farm in the UK in 2024.

Fig. 1.5 Typical wind turbine blade structural damage and fracture accident cases. (**a**) Fenner wind turbine blade breakage accident in Madison County, New York, USA, in 2016. (**b**) Blade damage accident at Humber Gateway offshore wind farm in the United Kingdom in 2024

due to the huge potential of wind energy in nature. Wind power plants convert the kinetic energy of the wind into mechanical kinetic energy and then convert the mechanical energy into electrical kinetic energy. Periodic inspection and maintenance are the key to ensuring the safe and stable operation of the wind power system. As the core component of the entire wind power generation equipment, the failure rate of wind turbine blades accounts for more than 40% of the entire wind power system failure. There have been many major accidents such as fires and large-scale power outages caused by wind turbine blade failures. For example, in 2016, a blade broke off and fell at the Fenner Wind Farm in Madison County, New York, USA (Fig. 1.5a), causing a fire. In January 2024, a wind turbine blade was damaged and fractured at the Humber Gateway offshore wind farm in the United Kingdom (Fig. 1.5b), which caused the wind farm to shut

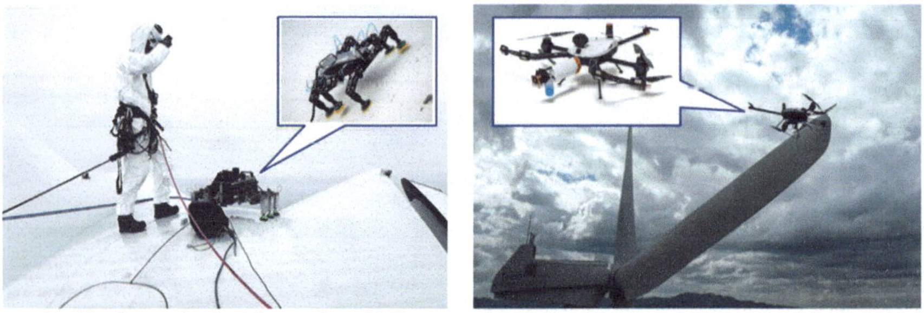

(a) Six-legged climbing robot [17]. (b) Voliro T robot for physical inspection [18].

Fig. 1.6 Electric power robots for the inspection and maintenance of wind turbines. (**a**) Six-legged climbing robot [17]. (**b**) Voliro T robot for physical inspection [18]

down for maintenance and caused 15,000 users to lose power, causing a direct economic loss of 32 million pounds. Therefore, periodic inspection and maintenance of wind turbine blades is very necessary, and it is also the key to ensuring the stable, safe, and efficient operation of the entire wind power generation system.

To cope with the above issues, several electric power robots have been designed and developed to implement periodic inspection and maintenance of wind generation plants. The British company BladeBUG has developed a six-legged climbing robot [16], shown in Fig. 1.6a. It can crawl stably on the surface of wind turbine blades through foot suction cups and detect the blades through the sensors it carries. In 2021, the world's first walking test was completed on a 7 MW offshore wind turbine blade in a Scottish offshore wind farm, which verified the feasibility of the solution. It has the advantages of flexible operation, wide coverage, and being able to operate on the blade surface. Further, unmanned aerial vehicles in CyberHawk company have been developed and employed to complete the inspection of up to six wind turbines per day and accurately detect minor defects at the level of ±5 mm under suitable meteorological conditions [17]. However, there is still room for improvement in the level of autonomy of these drones, which operate primarily through a human remote control mode and require operation within the operator's line of sight rather than operating in a fully autonomous manner. Furthermore, a novel drone "Voliro T" has been developed and equipped with a needle probe, which is used to create contact with the lightning protection system receptor [18], as shown in Fig. 1.6b. The payload is connected to a tether cable that connects the needle probe on the drone with a micro-ohmmeter on the ground. To close the loop, the ohmmeter is grounded to the wind turbine. The employment of "Voliro T" can reduce inspection time by 90% compared to traditional rope access techniques.

4. **Nuclear Generation Plant**

With the development and progress of nuclear technology, the heat energy released by nuclear fission in a nuclear reactor is utilized to generate electricity.

Absorbing the lessons from nuclear accidents, the inspection and maintenance of nuclear generation plants are crucial to ensure their safe operation. Reactor coolant system maintenance work conducted during a plant refueling outage involves complex tasks, and the environment is inherently hazardous because of the high radiation level and contamination. For this purpose, a series of mobile robots "KAEROT" have been developed to implement the maintenance of reactor coolant systems, which can be employed for the detection of loose parts in nuclear reactor vessels [19]. Further, a feeder pipe inspection robot was designed and developed for automobile movement and inspection on pressurized heavy water reactor tubes [19]. In addition, a remote robotic solution was proposed by Kinetrics company to implement inspection operations to minimize the impact of onsite exposure of workers [20]. As shown in Fig. 1.7, typical electric power robots for the inspection and maintenance of nuclear power plants are displayed. Regarding internal inspection operations in nuclear reactors, a robotic arm was designed to insert the tool through mesh sleeves in the reactor shielding wall to perform inspection and cleaning operations [21]. RICA robotic platform with a manipulator has been developed for fuel retreatment and waste storage in the nuclear field [22]. In [23], two robots were designed and used for investigation of the torus room in Units 2–3, where four-leg walking robot was used in Unit 2 and the Survey Runner was used in Unit 3. Moreover, JAEA-3 with gamma imaging system was developed and modified for undertaking gamma ray imaging on various floors of the reactor buildings and was deployed in Unit 2, 6 months after the accident [24]. Further, a modular snake robot was designed to attach to the MOTHERSHIP through an X-module base for waste storage, and its end effector may facilitate sensing instruments and manipulators [25].

1.2.2 Electric Power Robots for Transmissions

In order to generate high-quality electricity and deliver it to the places where electricity is consumed, it must rely on the transmission function of the power grid. High-voltage transmission lines are the main component of power systems, which are responsible for the transmission and distribution of power between power plants and substations and between substations. Due to the large transmission capacity of high-voltage transmission lines, once a safety accident occurs, it will have a great impact on the economy. At present, transmission line inspection and maintenance are still mainly manual, and it is necessary to rely on manual climbing to reach the transmission line for operation, which faces several challenges such as high-altitude and harsh working environment, low inspection efficiency, high labor costs, and high operating safety risks.

To solve the above problems, a LineMaster ground-based robotic arm was developed by the Quanta Services for inspection and maintenance of power transmission lines [26]. However, ground-based robotic arms are inconvenient for high-

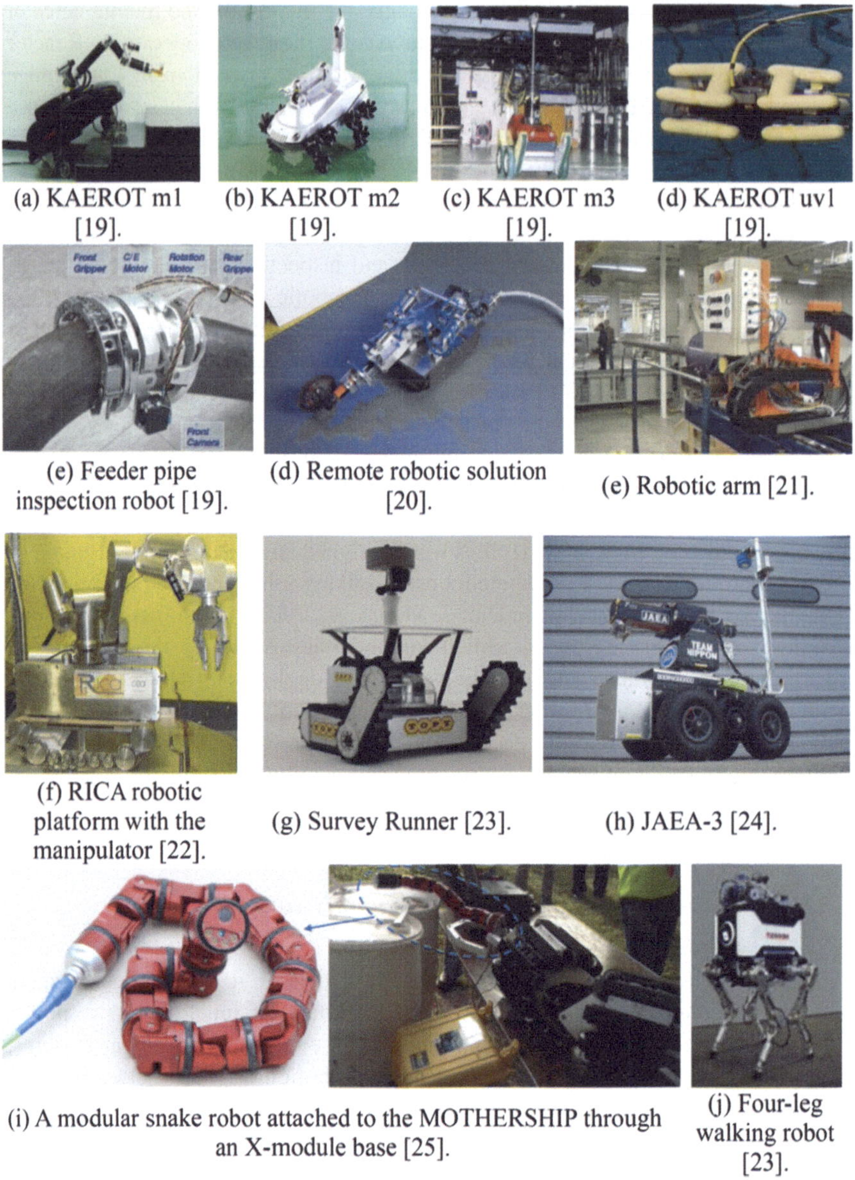

Fig. 1.7 Typical electric power robots for the inspection and maintenance of nuclear generation plants. (**a**) KAEROT m1 [19]. (**b**) KAEROT m2 [19]. (**c**) KAEROT m3 [19]. (**d**) KAEROT uv1 [19]. (**e**) Feeder pipe inspection robot [19]. (**f**) Remote robotic solution [20]. (**g**) Robotic arm [21]. (**h**) RICA robotic platform with the manipulator [22]. (**i**) Survey Runner [23]. (**j**) JAEA-3 [24]. (**k**) A modular snake robot attached to the MOTHERSHIP through an X-module base [25]. (**l**) Four-leg walking robot [23]

altitude operations, which are time-consuming and labor-intensive. The further solutions are that hanging line operational robots have been designed and developed for inspection and maintenance tasks, such as insulator inspection and replacement and bolt tightening [27–29]. A hanging line operational robot “Expliner” was designed and developed for the inspection of transmission lines [27]. The robot can stably run on double-split and quad-split wires through two sets of eight drive wheels and achieve obstacle-crossing functions through the coordinated movement of two vertical rods and counterweighted mechanical arms. However, the robot “Expliner” can only operate on two wires due to its structure, which makes this robot unable to perform inspection tasks on a single-split wire. Further, the robot “LineScout” was developed to temporarily repair damaged wires [28]. The robot “LineScout” can operate on a single line and cross obstacles, but its large weight and size make it inconvenient to transport and carry. Moreover, the robot “LinBot” has been developed to operate on the ground wire of transmission lines and negotiate all corresponding obstacles on the ground wire, including warning balls, suspended spires, and corner spires [29]. A noteworthy advantage of this robot is the use of active and passive mechanisms to enhance the robot’s stability while simplifying its control system. To cope with the problems of putting on and off transmission lines of hanging line operational robots, unmanned aerial vehicles equipped with camera or infrared sensors have been widely developed and employed for inspection of transmission lines [30] and insulators [31]. Further, an aerial vehicle equipped with an onboard operating mechanism has been designed to achieve hanging line inspection and maintenance operation tasks [32] (Fig. 1.8).

(a) LineMaster ground-based robotic arm [26]

(b) Expliner [27].

(c) LineScout [28].

(d) LinBot [29].

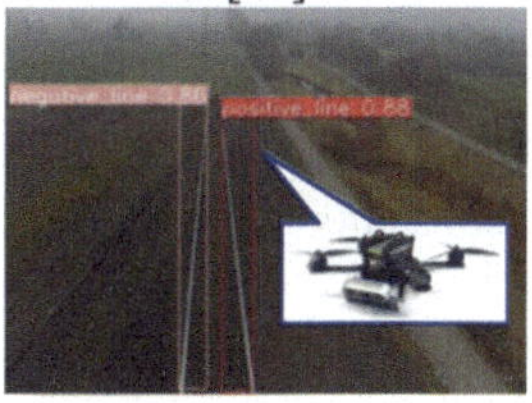

(e) Aerial line inspection [30].

(f) Insulator inspection [31].

(g) Aerial manipulator [32].

Fig. 1.8 Typical electric power robots for transmission stage. (**a**) LineMaster ground-based robotic arm [26]. (**b**) Expliner [27]. (**c**) LineScout [28]. (**d**) LinBot [29]. (**e**) Aerial line inspection [30]. (**f**) Insulator inspection [31]. (**g**) Aerial manipulator [32]

1.2.3 Electric Power Robots for Substations

A substation refers to a place in the power system that converts voltage and current as well as receives and distributes electrical energy. The inspection and maintenance tasks of substations involve real-world monitoring of the operating status of main primary equipment, such as main transformers, busbars, and switches in substations. The traditional inspection method of substations relies on manual recording. This method has several problems such as high labor intensity, low inspection efficiency, incomplete inspection, inconsistent inspection standards, and difficulty in inspection in harsh environments of rain and snow. Moreover, it is difficult to realize an objective, comprehensive, and accurate judgment by simply relying on the senses and experience of the inspection personnel, which creates potential safety hazards for the stable operation of the equipment.

To cope with the above issues, several inspection robots have been developed to replace humans to implement inspection of substations [33–36]. The fifth-generation substation inspection robot was developed by the State Grid Shandong Electric Power Research Institute [33], which combined with non-contact detection, multi-sensor fusion positioning and navigation, visual servo control, and other technologies to achieve all-weather and entire region autonomous operation in the outdoor environment of the substation. The robot system software was developed to realize equipment thermal defect analysis and early warning, switch and circuit breaker opening and closing status recognition, automatic instrument reading, equipment appearance abnormality and transformer sound abnormality detection and abnormal status alarm, etc. This robot is the first time in the world to realize autonomous inspection and application promotion in substations, which improves the automation and intelligence level of substation inspections and indoor inspections. Regarding inspection in extreme climates, an all-weather automatic inspection robot was developed for electrical substations to ensure reliable autonomous navigation and inspection [34]. This robot can accomplish two main types of inspections: visual and thermal. Further, as an important high-voltage equipment in a substation, gas-insulated switchgear is extremely important for the normal operation of the substation. However, since many instruments are located at a height of 4–5 m, it is not conducive to manual or ground robot inspection. Therefore, multi-rotor drones with strong maneuverability and high flexibility were employed to solve the problem of high-altitude inspection [35]. In addition, the collaboration of drones and intelligent ground robots forms a comprehensive three-dimensional inspection system, which can effectively improve the quality and efficiency of substation operation and maintenance and reduce the work intensity and operational risks of operation and maintenance personnel [36]. Moreover, a distribution network live operating robot was used for porcelain bottle replacement and removal, where operators can further enhance the adaptability and flexibility of the robot in different scenarios through local collaborative control of the robotic arm [37]. Further, a quadruped robot was designed and utilized for inspection in substations [38] (Fig. 1.9).

(a) Fifth-generation substation inspection robot [33].

(b) All-weather automatic inspection robot [34].

(c) Air-ground collaborative inspection robot system [36].

(d) Operating robot for porcelain bottle replacement and removal [37].

(e) Quadruped robot for inspection [38].

Fig. 1.9 Typical electric power robots for inspection of substations. (**a**) Fifth-generation substation inspection robot [33]. (**b**) All-weather automatic inspection robot [34]. (**c**) Air-ground collaborative inspection robot system [36]. (**d**) Operating robot for porcelain bottle replacement and removal [37]. (**e**) Quadruped robot for inspection [38]

1.2.4 Electric Power Robots for Distributions

The function of the distribution stage is to convert high-voltage and medium-voltage electric energy output from substations into low-voltage electric energy and then transmit it to industrial, commercial, and residential consumers. To ensure the normal operation of distributions, there exist several problems that need to be taken care of, such as daily inspection and maintenance of the distribution network, insulation aging, and line failures. For this purpose, an intelligent power distribution live-line operation robot system was proposed to replace frontline workers in completing maintenance tasks of distribution networks [39]. As shown in Fig. 1.10a, the system is mainly composed of an insulated lifting truck, a distribution network live operation robot with two arms as the main body, a vision system, and various operating tools. During operations, the operator remotely controls the robot through visual feedback and an interactive interface on the ground control platform to perform autonomous maintenance tasks. The robotic arm rises to the bottom of the overhead line along with the platform and cooperates with corresponding tools to perform various live operations. Moreover, as depicted in Fig. 1.10b, a track-type autonomous inspection robot system "LongSwords" was developed for the unattended inspection of electrical distribution rooms, which can provide a

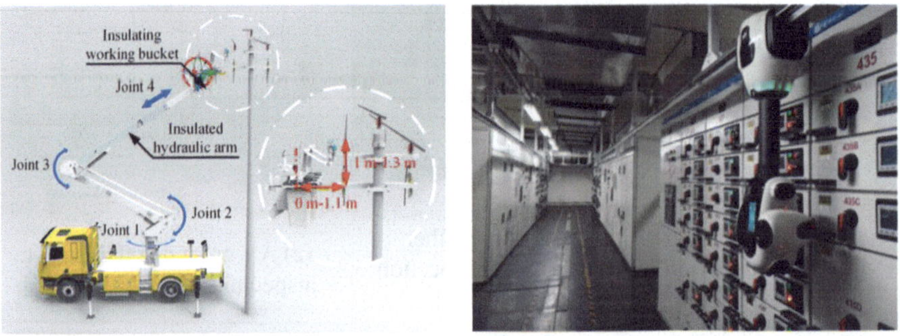

(a) Intelligent power distribution live-line operation robot system [39].

(b) Track-type autonomous inspection robot system "LongSwords" [40].

Fig. 1.10 Typical electric power robots for distribution stage. (**a**) Intelligent power distribution live-line operation robot system [39]. (**b**) Track-type autonomous inspection robot system "LongSwords" [40]

satisfying technical solution for equipment inspection with an optical zoom camera, a thermal imaging camera, or a partial discharge detector [40]. Further, with the development and progress of 5G wireless communication technology, an intelligent inspection enablement platform for distribution network fault detection was built based on 5G wireless communication connected drones, which can realize all-weather automated inspection of the distribution network and form a new type of autonomous periodic drone automated inspection system [41].

1.3 Challenges Faced by Electric Power Robots

Electric power robot systems have been widely developed and employed to replace human workers to perform monitoring, inspection, and maintenance tasks in complex and confined environments. However, there still exist several challenges for the applications of electric power robots.

1. **Environmental Impacts**

   Electric power robots are generally required to work in unpredictable and complex environments due to the inspection and maintenance requirements of power systems. These environments include extreme weather conditions such as strong winds, rain, snow, and even lightning, which may significantly impact the operation of electric power robots. For instance, the stability and safety of drones will be decreased in windy weather conditions. Further, camera images and infrared monitoring are affected by reduced visibility and sensor accuracy during rainy days, which poses a challenge to the perception capabilities of robots. Moreover, complex terrains may bring great challenges to the navigation and mapping effects of robots, and whether the robot can find the target location to complete operation tasks in the field is still a research topic worthy of study. In addition, the presence of strong magnetic fields near high-voltage live conductors will cause interference to circuit signals, which requires the robot to obtain effective signals under this circumstance. Therefore, power operational robot systems need to have strong adaptability and reliability to complex environments to maintain efficient and safe operations.

2. **Technical Constraints**

   Inspection and maintenance tasks in power systems mostly require high operation precision, even to the millimeter level. It means that for electric power robots, high precision ought to be pursued in all aspects, such as the optimized design of operating mechanisms, precise position and force control, and so on. However, the accuracy of electric power robots is determined by the precision of the positioning sensors and is limited by unavoidable perturbations. Therefore, electric power robots are expected to perform more precision control while performing tasks.

   Power operational robot systems involve the integration of multiple technologies such as mechanical design, intelligent control, robotics, and communication technology. Any one of these can make a huge impact on the final implementation. For example, advances in materials science could allow robots to achieve greater mechanical strength or lighter mass, which is especially important for payload-sensitive drones. Further, many electric power robots are integrated with built-in batteries, which often results in limited operation time due to limited battery storage capacity. Moreover, robots with enhanced computing powers could be a prerequisite for implementing fully autonomous and multi-task operations.

   Artificial intelligence technology has recently become a hot topic in all walks of life. With the improvement of artificial intelligence technology, electric power robots are driven to integrate the latest artificial intelligence technology to complete tasks autonomously. This requires robots to have a high degree of autonomy and intelligent decision-making capabilities and to be able to respond quickly according to environmental changes and task requirements.

3. **Cost**

   The development and deployment of advanced electric power robots require significant capital investment. However, the application of electric power robots is still in the preliminary stages, and most robots have not yet entered large-scale production. It means that the employment of electric power robots covers not only the development costs of robots but also the costs of system integration, testing, staff training, and necessary infrastructure modifications. Regarding power companies, effectively balancing these high initial investment costs with the long-term operational benefits is critical to ensuring the sustainability of electric power robots. This challenge requires companies to find the best balance between economic benefits and technological progress. The challenge is to find the optimal balance between economic efficiency and technological progress. In addition, the long-term operation and maintenance costs of power operational robot systems should be also considered.

4. **Safety and Security**

   Regarding inspection and maintenance tasks, electric power robots are typically required to physically interact with external environments, in particular with humans. Thus, it is necessary to develop safer robot platforms and new technologies to enhance safety. In addition, power operation robots rely on complex software to perform operation tasks. However, software flaws or vulnerabilities may cause unpredictable behavior or even cause the robot to lose control. Therefore, it is crucial to ensure that robot software is constantly updated and patched to protect against potential security threats.

1.4 Structure and Arrangement of this Book

This book aims to introduce the power operational robot system and related technologies in the power industry. This book is divided into eight chapters covering different areas and applications of electric power robots. The content and arrangement of each chapter is described below.

Chapter 1 provides a comprehensive overview of electric power robots. The background and motivation of electric power robots are introduced, and the current state of research on electric power robots is reviewed from the perspective of different power system stages and types of electric power robots. Further, the difficulties and challenges faced by electric power robots are discussed in their current commercial use and further maturation in the future.

Chapter 2 will focus on the condenser intelligent robot system and modeling. The design and modeling of condenser intelligent cleaning robots will be introduced in detail.

Chapter 3 will introduce the control approaches for the condenser cleaning mobile operation robot system, including adaptive sliding-mode control, fuzzy wavelet neural network-based control, and recurrent fuzzy CMAC neural network-based control approaches. Finally, experimental examples will be given.

Chapter 4 will introduce the system structure and modeling of the transmission line de-icing inspection robot system. The kinematic and dynamic modeling will be presented for the de-icing robot manipulator system, and inverse kinematic solutions using hybrid electromagnetism-like and modified MFP algorithms will be proposed for task-space control.

Chapter 5 will introduce the control methods and applications of the power line de-icing inspection robot system. Self-organizing single-input CMAC control and adaptive iterative learning control approaches will be employed for de-icing robot manipulator systems. Finally, real-world experiments and applications will be given.

Chapter 6 will introduce unmanned aerial vehicle in the power system. The kinematic and dynamic modeling will be presented for unmanned aerial vehicle. Regarding stable flight, the adaptive RBFBBs/integral sliding-mode control approach will be designed against external disturbances, and the adaptive and robust control method will be designed while considering input saturation.

Chapter 7 will introduce unmanned aerial robot manipulation in power systems. The kinematic and dynamic modeling will be presented for unmanned aerial manipulators. Regarding visual servo control, the kinematic-based and image dynamic-based visual servo control schemes will be proposed. Finally, the real-world experiments and applications of visual servo control of unmanned aerial manipulators will be given.

Chapter 8 summarizes the content of the book, analyzes the application challenges that electric power robots may encounter, and offers perspectives on future development directions and trends.

References

1. Liu Y. Analysis of Brazilian blackout on March 21st, 2018 and revelations to security for Hunan Grid. In: 2019 4th International Conference on Intelligent Green Building and Smart Grid. IEEE, 2019: 1–5.
2. De Michelis C, Rinaldi C, Sampietri C, et al. Condition monitoring and assessment of power plant components. In: Power Plant Life Management and Performance Improvement. Woodhead Publishing, 2011: 38–109.
3. Wang B, Guo R, Li B, et al. SmartGuard: An autonomous robotic system for inspecting substation equipment. Journal of Field Robotics, 2012, 29(1): 123–137.
4. Pouliot N, Montambault S. Field-oriented developments for LineScout Technology and its deployment on large water crossing transmission lines. Journal of Field Robotics, 2012, 29(1): 25–46.
5. Caprari G, Breitenmoser A, Fischer W, et al. Highly compact robots for inspection of power plants. Journal of Field Robotics, 2012, 29(1): 47–68.
6. Menendez O, Cheein F A A, Perez M, et al. Robotics in power systems: Enabling a more reliable and safe grid. IEEE Industrial Electronics Magazine, 2017, 11(2): 22–34.
7. Kothari D P, Nagrath I J. Modern power system analysis. Tata McGraw-Hill Publishing Company, 2003.
8. Ray D N. Slim and expandable header inspection robot for remote visual inspection of boiler headers in thermal power plants. In: Proceedings of the 2021 5th International Conference on Advances in Robotics. 2021: 1–8.
9. Jiang Z, Chen B, Ju Z, et al. Design and analysis of a wall-climbing robot for water wall inspection of thermal power plants. Journal of Field Robotics, 2023, 40(5): 1003–1013.
10. SERBOT. Mobile cleaning robot for PV installations on roof tops. [Online]. Available: http://www.serbot.ch/en/solar-panelscleaning/gekko-solar-robot
11. Ecoppia. Technology. [Online]. Available: http://www.ecoppia.com/technology/
12. MIRAI. Products: Solar cleaning robots. [Online]. Available: https://www.miraikikai.jp/products-e
13. RAYBOT. Products: Solar cleaning robots. [Online]. Available: http://www.china-nengyuan.com/product/135358.html
14. HUIBO ROBOTICS. Products: Inspection drone. [Online]. Available: https://ind.huiborobot.com/guangfuqingsaojiqiren/226.html
15. IG Drones. IG Drones - The best solution for solar panel inspection. [Online]. Available: http://igdrones.com/blog/ig-drones-the-best-solution-for-solar-panel-inspection
16. BladeBUG: Advanced robotics for turbine maintenance. [Online]. Available: https://www.bladebug.co.uk/media/blade-walk/
17. CyberHawk. Driving speed to decision. [Online]. Available: https://thecyberhawk.com/

18. VOLIRO. Drone inspection of wind turbine lightning protection systems. [Online]. Available: https://voliro.com/case-studies/wind-turbine-lps-drone-inspection
19. Kim S, Jung S H, Lee S U, et al. Application of robotics for the nuclear power plants in Korea. In: 2010 1st International Conference on Applied Robotics for the Power Industry. IEEE, 2010: 1–5.
20. Hamilton A, Burany S J, Peralta S B, et al. Robotic removal of high-activity debris from a nuclear primary heat transfer system. In: 2010 1st International Conference on Applied Robotics for the Power Industry. IEEE, 2010: 1–6.
21. Goldenberg A, Gryniewski M, Campbell T. AARM: A robot arm for internal operations in nuclear reactors. In: 2010 1st International Conference on Applied Robotics for the Power Industry. IEEE, 2010: 1–5.
22. Ducros C, Hauser G, Mahjoubi N, et al. RICA: A tracked robot for sampling and radiological characterization in the nuclear field. Journal of Field Robotics, 2017, 34(3): 583–599.
23. Tepco, 2017. [Online]. Available: https://www.tepco.co.jp/en/decommision/principles/robot/index-e.html
24. Kawatsuma S, Mimura R, Asama H. Unitization for portability of emergency response surveillance robot system: Experiences and lessons learned from the deployment of the JAEA-3 emergency response robot at the Fukushima Daiichi Nuclear Power Plants. Robomech Journal, 2017, 4: 1–7.
25. Voyles R, Abbaraju P, Choset H, et al. Novel serpentine robot combinations for inspection in hard-to-reach areas of damaged or decommissioned structures. In: Waste Manage. Symp. 2017: 1–15.
26. Quanta services. Line master robot arm. [Online]. Available: http://www.quantaenergized.com/proprietary-tools-and-techniques/
27. Debenest P, Guarnieri M, Takita K, et al. Expliner-robot for inspection of transmission lines. In: 2008 IEEE International Conference on Robotics and Automation. IEEE, 2008: 3978–3984.
28. Pouliot N, Richard P L, Montambault S. LineScout technology opens the way to robotic inspection and maintenance of high-voltage power lines. IEEE Power and Energy Technology Systems Journal, 2015, 2(1): 1–11.
29. Fakhari A, Mostashfi A. LinBot-design, analysis, and field test of a novel power transmission lines inspection robot. In: 2019 7th International Conference on Robotics and Mechatronics. IEEE, 2019: 132–137.
30. Xing J, Cioffi G, Hidalgo-Carrió J, et al. Autonomous power line inspection with drones via perception-aware MPC. In: 2023 IEEE/RSJ International Conference on Intelligent Robots and Systems. IEEE, 2023: 1086–1093.
31. Patsiouras E, Mygdalis V, Pitas I. Whitening transformation inspired self-attention for powerline element detection. In: 2022 26th International Conference on Pattern Recognition. IEEE, 2022: 4844–4849.
32. Ollero A, Suarez A, Marredo J M, et al. Application of intelligent aerial robots to the inspection and maintenance of electrical power lines. Robotics and Automation Solutions for Inspection and Maintenance in Critical Infrastructures, Delft, The Netherlands, 2024: 1–19.
33. Xu W, Yang S, Mu S, et al. Modularization design of inspection robot based on substation. In: 2018 IEEE 3rd Advanced Information Technology, Electronic and Automation Control Conference. IEEE, 2018: 2230–2233.
34. Dandurand P, Beaudry J, Hébert C, et al. All-weather autonomous inspection robot for electrical substations. In: 2022 IEEE/SICE International Symposium on System Integration. IEEE, 2022: 303–308.
35. Li Q, Dai Y, Han R, et al. GIS room autonomous inspection system based on multi-rotor UAV. In: 2021 International Conference on Electrical Materials and Power Equipment. IEEE, 2021: 1–4.
36. Zhang X, Qiu M, Feng J. Path optimization analysis of UAV-intelligent robot cooperative inspection of substation based on A* algorithm. In: 2022 IEEE 4th International Conference on Civil Aviation Safety and Information Technology. IEEE, 2022: 1461–1466.

37. SHENHAO, 2023. [Online]. Available: https://www.shenhaoinfo.com/
38. Hu X, He F, Xiao P, et al. Design of a quadruped inspection robot used in substation. In: 2021 IEEE 4th Advanced Information Management, Communicates, Electronic and Automation Control Conference (IMCEC). IEEE, 2021, 4: 766–769.
39. Chen Y, Wang Y, Tang X, et al. Intelligent power distribution live-line operation robot systems based on stereo camera. High Voltage, 2023, 8(6): 1306–1318.
40. Cheng M, Xiang D. The design and application of a track-type autonomous inspection robot for electrical distribution room. Robotica, 2020, 38(2): 185–206.
41. Wang J, Ye W, Xue T, et al. Intelligent inspection enabling platform for power distribution network based on 5G wireless communication connected unmanned aerial vehicle. In: 2022 5th International Conference on Energy, Electrical and Power Engineering. IEEE, 2022: 574–579.

Chapter 2
Intelligent Condenser Cleaning Mobile Manipulator Robot System

2.1 Background and Motivation

The large condenser is one of the key equipment in thermal power plants, nuclear power plants, chemical plants, etc. It is the cold source of the thermodynamic cycle of the large steam turbine set. It can decrease the exhaust stress and temperature of the steamer machine and improve the efficiency of the thermal cycle of the turbo-generator unit. As the larger steam condenser runs, the cooling water is sent to pass through the thermal tubes in the condenser, which may lead to the production of fouling due to the impurity of the cooling water and chemical reaction. The fouling is not favorable for heat exchange and will bring about a series of problems. It will decrease the heat exchange capacity, increase energy consumption, and lead to the blockages of condensing tubes.

In fact, the operating performance of the condenser is important to minimizing the operating costs of the turbo-generator units in industrial plants, in terms of both heat rate and generation capacity. Therefore, the problems of condenser cleaning have attracted attention from researchers. Currently, many cleaning techniques for industrial condensers have been presented, which can be divided into two main strategies: offline and online methods.

The popular offline condenser cleaning methods are the mechanical and chemical cleaning methods. Among the mechanical cleaning methods, the high-pressure water method is the most frequently chosen. The high-pressure water method is carried out by moving a jet along the tube slowly to clean the condensers. This method is simple, but the problem of longer operating time and the safety of tube sheet/tube coatings need to be paid careful attention to. The chemical method is implemented by using several mildly acidic products to remove deposits in the condenser tubes. By applying the chemical method, the performance of deposit cleaning is better than most other methods; but this method is expensive and takes longer for operating time to be completed. In addition, the subsequent disposal of the chemicals, an

Y. Wang et al., *Electric Power Robots*,
https://doi.org/10.1007/978-981-96-2512-3_2

Fig. 2.1 Workers with condenser cleaning

environmental hazard, is an important problem. Currently, a few online condenser cleaning methods are applied, but the best known method is the technique that uses recirculated sponge rubber balls as the cleaning vehicles, which are injected into the cooling water flow. The disadvantage of this method is the limitation of the operating cycle. Moreover, although the condenser tubes may become cleaner if the abrasive balls are applied, the wear can now become an important problem.

In the industrial sector, the majority of the condenser cleaning procedures are implemented offline, and the frequently chosen method is the mechanical cleaning method with supporting of cleaner chemicals. When this way is applied, workers have to clean condensers directly (in Fig. 2.1), and this work condition is unsafe in hazardous environments, such as nuclear power plants, chemical plants, and other plants. Therefore, there is an urgent need for an automated condenser cleaning robot system to improve labor efficiency, improve working conditions, and overcome operational limitations.

With the development of automation and technology, modern robotics has begun to appear. For instance, as shown in Fig. 2.2, KUKA offers a comprehensive range of industrial manipulator robots that meet various application needs with precision, flexibility, and efficiency, such as fluting and bending, tooling, and assembly [1]. KUKA robot manipulators perform precise control accuracy and repetition accuracy, and all components of KUKA robot manipulators are equipped with wear-resistant power trains that have long maintenance intervals due to their advanced and sturdy construction. FANUC robot manipulators have the characteristics of high precision, lightweight, and miniaturization [2]. The repeated positioning accuracy of FANUC's multifunctional six-axis small manipulator robot can reach ±0.02 mm. In addition, FANUC industrial robots are unique compared with other companies in that process control is more convenient, the base size of similar

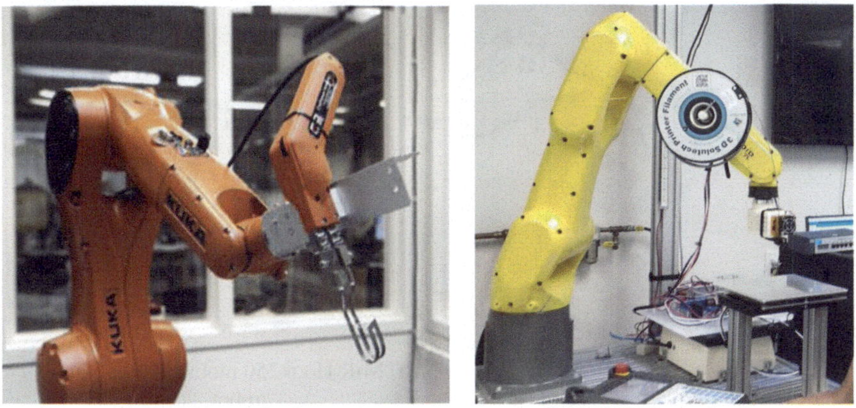

(a) KUKA manipulator. (b) FANUC manipulator.

Fig. 2.2 Robot manipulators. (**a**) KUKA manipulator. (**b**) FANUC manipulator

(a) ANYmal quadruped robot. (b) ADLINK ground mobile platform. (c) KUKA mobile platforms.

Fig. 2.3 Ground mobile robots. (**a**) ANYmal quadruped robot. (**b**) ADLINK ground mobile platform. (**c**) KUKA mobile platforms

robots is smaller, and they have unique arm designs. Although robot manipulators have many advantages in industry, their workspace is limited. Based on the development of mobile robotics, this drawback has been relaxed. Researching mobile robotics has attracted the attention of many researchers around the world, such as ANYmal quadruped robot [3], ADLINK ground mobile platform [4], and KUKA mobile platforms [5], as shown in Fig. 2.3. In the mobile robotics field, the mobile manipulator robot (MMR) has also been developed, such as the quadruped dexterous manipulation robot in Fig. 2.4a, MiR Hook250 mobile manipulator robot in Fig. 2.4b, and KUKA KMR QUANTEC mobile manipulator for inspection and maintenance in Fig. 2.4c.

Nowadays, mobile manipulator robots (MMRs) have been applied in a variety of different applications, such as construction, mining, waste management, outdoor exploration, and planetary sciences. It is actually a complex system that includes robotic arms and a mobile platform. In addition, the structure of the mobile

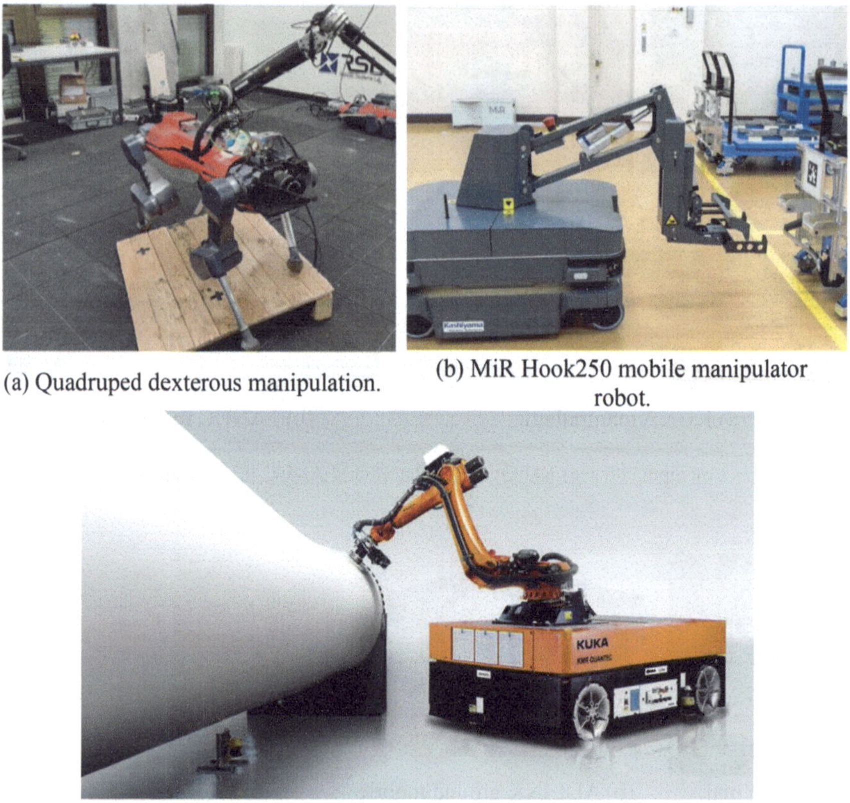

(a) Quadruped dexterous manipulation.

(b) MiR Hook250 mobile manipulator robot.

(c) KUKA KMR QUANTEC mobile manipulator for inspection and maintenance.

Fig. 2.4 Ground mobile manipulator robots. (**a**) Quadruped dexterous manipulation. (**b**) MiR Hook250 mobile manipulator robot. (**c**) KUKA KMR QUANTEC mobile manipulator for inspection and maintenance

manipulator robot is subject to kinematic and dynamic constraints, which make it a highly coupled dynamic nonlinear system. Therefore, the control of the mobile manipulator robot is a challenge and has attracted the attention of researchers.

Along with the development of robotics, control techniques have also been researched and developed quickly. Control theory is a widespread science that refers to many fields, such as mechanical, electrical, chemical, fluid, financial, and even biological. Therefore, robotic control is one of the aspects of the control theory. There are two major parts in the control theory, namely, classical and modern. Applications of the classical theory are usually limited to single-input and single-output (SISO) systems. Proportional-integral-derivative (PID) controller [6] can be considered a typical product of the classical theory, and it has been used in most industrial applications. Although the PID controllers are simply applied, the disadvantage of the conventional PID controllers is also presented in requiring knowledge/experiment of the controlled systems.

The modern control theory [7] can solve the limited problems of the classical control theory by implementing the control design in multi-input and multi-output (MIMO) systems with state variables. Based on the development of mathematical techniques, the control theory becomes flexible, and it can be applied in complex dynamic systems to achieve accurate control results. In the 1950s and 1960s, optimal control theory began to be explored, and it led to the development of many modern control methods in the 1970s and 1980s, such as robust, adaptive, optimal, and stochastic control methods [8–11]. As a result, the design of robot controllers is implemented more conveniently, and the control effectiveness is also improved. In addition, by inheriting the advantages of the robust/adaptive methods for system stability, many control techniques have been proposed, such as back-stepping, sliding mode, iterative learning, self-organizing, and hybrid control methods. The robust and adaptive controllers can handle the highly coupled nonlinearities of the robot dynamics, without prior knowledge of the robotic-controlled systems. Therefore, in the motion/force robotics control field, these methods play an important role in the controller design.

Since the 1940s, neural network (NN) theory has begun to be researched. Next, in 1956, the artificial intelligence concept was first introduced, and it was applied to control the behavior of the Shakey robot [12–14] from 1966 to 1972. The early onset of artificial intelligence and NNs made more favorable chances for the later development of intelligent control methods. Intelligent control is a class of control techniques that combines control theory and artificial intelligence to solve control problems. Nowadays, intelligent control has been widely applied in robotics control, which can be divided into some major subdomains, such as NN control, hybrid NN control, fuzzy control, expert systems, and genetic control. In the robotic control system-based intelligent control methods, the unknown highly coupled nonlinear dynamics of the robotic systems are usually eliminated by an estimating/approximating process that is implemented by the artificial intelligent methods [15–20]. Based on the requirement for condenser cleaning and the advantages of developments of robotics and control theory, a condenser cleaning mobile manipulator robot, after several improvement trials, has been created by Hunan University in recent years. For motion control of the condenser cleaning mobile manipulator robot (CCMMR), many researchers have proposed intelligent control strategies to improve the effectiveness of robots [21–26]. These methods use NNs/hybrid NN to deal with the dynamics of the CCMMR, but the kinematic constraints are still not concerned. In addition, the problems with the bounded parameters and stability conditions of the CCMMR control system also need more attention. Therefore, the control of CCMMR still has many challenges for research.

Considering the above analysis, we mainly introduce the structure of CCMMR control system in the next section.

2.2 Structure and Dynamics of Condenser Cleaning Mobile Manipulator Robot

2.2.1 *Structure and Operating Principle*

The CCMMR system, as shown in Fig. 2.5, is an intelligent robotic system combined with the auxiliary equipment for condenser cleaning. The CCMMR system is designed to guarantee the CCMMR can self-operate in various conditions under requirements for safety and cost-saving. The condenser cleaning is implemented by a high-pressure water jet and the support of cameras and sensor systems. The structure of the CCMMR system can basically be divided into the following main parts:

1. The mechanical and electrical structure of the mobile manipulator robot. The MMR includes one crawler mobile platform and three links (manipulators). The crawler mobile platform mechanical structure has many advantages in mobile operation and self-balancing, such as high obstacle-climbing capability and adaptability in steep topography environments. The structure of manipulators consists of two revolute joints and one prismatic joint. The CCMMR's manipulators are movable in the condenser water chamber. With this structure, the CCMMR has a large workspace (two rotary links) and high accuracy for its end-effector (one transitional link). The manipulators and crawler mobile platform are driven by servomotors combined with tachometers/encoders. In

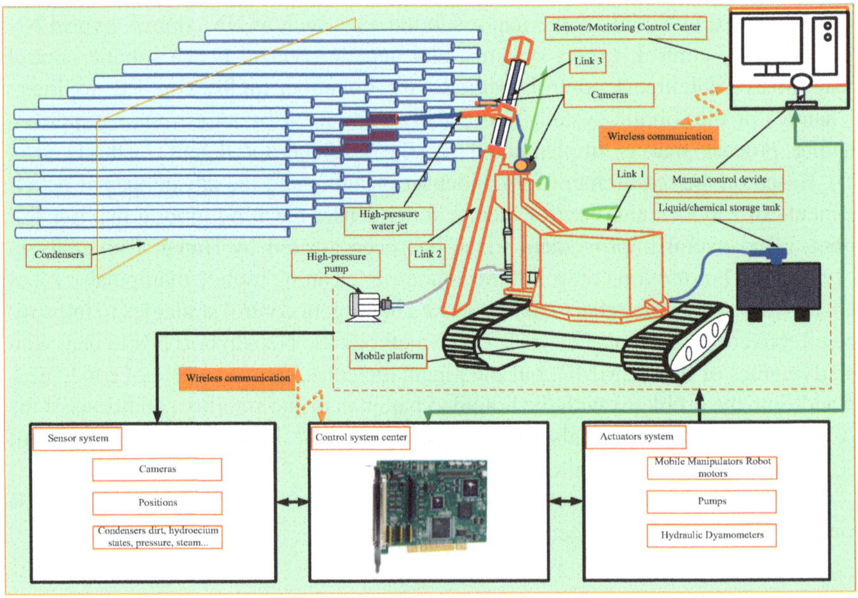

Fig. 2.5 Mechanical structure and control system of the CCMMR system

addition, the moving operation of the CCMMR is also further improved by the underwater cameras and location compass.

2. The auxiliary equipment for condenser cleaning includes a high-pressure water jet integrated into the end-effector (link 3), high-pressure pumps, chemical tanks, pumps/motor (impurity hydraulic, gas, chemicals), and sensors (cooling water hardness, temperature, steam flow, pressure, chemicals).
3. The CCMMR control system. The control system includes a center control system, sub-controllers, and a remote control system. The sub-controllers are the separate control modules (robot positions, condenser cleaning). The CCMMR control system is designed based on the distribution control structure. The communication of the control modules is processed through the control area network (CAN) bus protocol. In addition, a wireless network system is also connected to the CCMMR and control system module, guaranteeing that remote control and monitoring operations are implemented easily. Therefore, the design of the CCMMR control system makes it possess the ability for high-speed information processing, real-time communication, and so on.

The condenser cleaning procedures of the CCMMR system are implemented offline, and the condenser cleaning method is utilized by using a high-pressure water jet with supporting cleaner chemicals. The condenser cleaning can be controlled by both manual and automatic methods. With the manual method, a worker will control and supervise the activities of the CCMMR via the remote control system and cameras. With the automatic method, the operating principle of the CCMMR system can be presented in the following steps:

Step 1: At the workspace, the CCMMR control system will check all operating conditions to guarantee the CCMMR can work. Next, the CCMMR will automatically search the exact position of each condenser tube. The information on positions can be determined by the pre-set trajectories or image processing from the cameras.

Step 2: The CCMMR will automatically determine the optimal pressure washing, cleaning time, or cleaning cycles. This determination is implemented by supporting the condenser dirt sensors and cameras.

Step 3: The high-pressure water jet on the CCMMR will be utilized for the condenser cleaning by high-pressure pumps from water/chemical tanks.

Step 4: After the cleaning cycles, the CCMMR system will repeat step 1.

Based on the structure and operating behavior of the CCMMR system, the condenser cleaning will not only be resolved effectively but also bring safety and cost-saving significantly.

2.2.2 Dynamical Model

In this section, the proposed intelligent control methods will be applied to improve the tracking performances of the CCMMR. Here, the tracking problems of robot positions and constraint force (motion and force) will be considered. To achieve the best results, this section first reviews the system modeling of the CCMMR via its dynamics structure. In general, the dynamics of the CCMMR can be considered as the dynamics of the mobile manipulator robot (MMR). To take advantage of the design of the proposed controllers, the dynamics of the CCMMR can be examined as the dynamics of multiple degrees of freedom (DOF) manipulator mounted on a mobile platform that can be expressed as a Lagrange function of the following form [27]:

$$\boldsymbol{M}(\boldsymbol{q})\ddot{\boldsymbol{q}} + C(\boldsymbol{q},\dot{\boldsymbol{q}})\dot{\boldsymbol{q}} + \boldsymbol{G}(\boldsymbol{q}) + \tau_d = \boldsymbol{B}(\boldsymbol{q})\tau + f \tag{2.1}$$

The kinematic constraints are described by

$$A(\boldsymbol{q})\dot{\boldsymbol{q}} = 0 \tag{2.2}$$

where $\boldsymbol{q}, \dot{\boldsymbol{q}}, \ddot{\boldsymbol{q}} \in \mathbb{R}^{n\times 1}$ are the joint position, velocity, and acceleration vectors, respectively. $\boldsymbol{M(q)} \in \mathbb{R}^{n \times n}$ is the inertia matrix. $\boldsymbol{C}(\boldsymbol{q}, \dot{\boldsymbol{q}})\dot{\boldsymbol{q}} \in \mathbb{R}^{n\times 1}$ expresses the centripetal and Coriolis term. $\boldsymbol{G(q)} \in \mathbb{R}^{n \times 1}$ is the gravity vector. $\boldsymbol{\tau}_d \in \mathbb{R}^{n \times 1}$ denotes unknown disturbances. $\boldsymbol{\tau} \in \mathbb{R}^{r_c \times 1}$ is the input torque vector. $r_c = n - m$, and $\boldsymbol{B}(\boldsymbol{q}) \in \mathbb{R}^{n\times r_c}$ is the input transformation matrix. $f = \boldsymbol{A(q)}^{\top}\gamma$, where $\boldsymbol{A(q)} \in \mathbb{R}^{m \times n}$ is the full rank matrix. $\gamma \in \mathbb{R}^{m \times 1}$ denotes the Lagrangian multiplier. For convenience, a 3-DOF MMR model, as shown in Fig. 2.6, is applied to verify the dynamics properties that are given in the next sections.

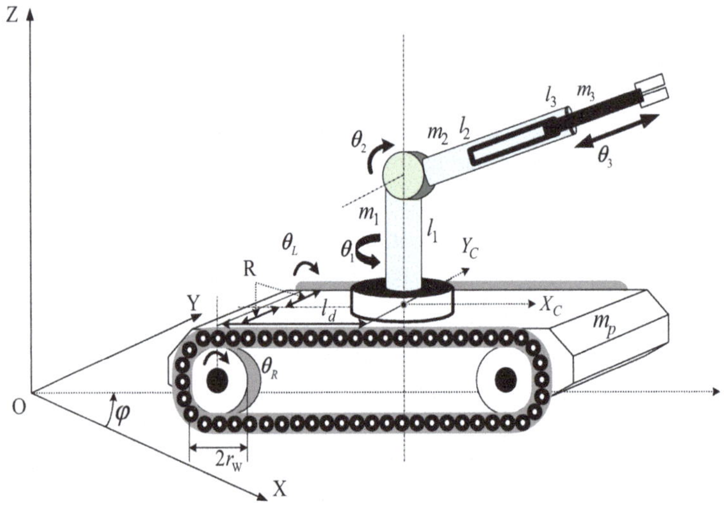

Fig. 2.6 3-DOF mobile manipulator robot model

Assume that the MMR is subject to nonholonomic constraints. Thus, the parameters in (2.1) can be expressed in the following forms [27]:

$$\begin{cases} \boldsymbol{M}(\boldsymbol{q}) = \begin{bmatrix} \boldsymbol{M}_n & \boldsymbol{M}_{nh} \\ \boldsymbol{M}_{hn} & \boldsymbol{M}_h \end{bmatrix} \\ \boldsymbol{C}(\boldsymbol{q}, \dot{\boldsymbol{q}}) = \begin{bmatrix} C_n & C_{nh} \\ C_{hn} & C_h \end{bmatrix} \\ \boldsymbol{G}(\boldsymbol{q}) = \begin{bmatrix} \boldsymbol{G}_{\mathbf{n}} \\ \boldsymbol{G}_{\mathbf{h}} \end{bmatrix} \\ \boldsymbol{\tau}_d = \begin{bmatrix} \boldsymbol{\tau}_{dn} \\ \boldsymbol{\tau}_{dh} \end{bmatrix} \\ \boldsymbol{\tau} = \begin{bmatrix} \boldsymbol{\tau}_n \\ \boldsymbol{\tau}_h \end{bmatrix} \\ \boldsymbol{B}(\boldsymbol{q}) = \begin{bmatrix} \boldsymbol{B}_n & 0 \\ 0 & \boldsymbol{B}_h \end{bmatrix} \\ f = \begin{bmatrix} \boldsymbol{A}_n^\top(\boldsymbol{q}_n)\gamma_n \\ 0 \end{bmatrix} = \begin{bmatrix} f_n \\ 0 \end{bmatrix} = \boldsymbol{A}^\top \gamma \end{cases} \tag{2.3}$$

where $\boldsymbol{M}_n$, $\boldsymbol{M}_h$ are the inertia matrices of the mobile platform and robot manipulators, respectively. $\boldsymbol{M}_{nh}$, $\boldsymbol{M}_{hn}$ are the coupling inertia matrices of the mobile platform and robot manipulators, respectively. C_n, C_h are the centripetal and Coriolis torques of the mobile platform and robot manipulators, respectively. C_{nh}, C_{hn} are the coupling centripetal and Coriolis torques of the mobile platform and robot manipulators, respectively. $\boldsymbol{G}_n$, $\boldsymbol{G}_h$ are the gravitational torque vectors of the mobile platform and robot manipulators, respectively. τ_{dn}, τ_{dh} are the disturbances on the mobile platform and robot manipulators, respectively. γ_n is the associated Lagrangian multiplier with generalized nonholonomic constraints. $\boldsymbol{A}_n$ is the kinematic nonholonomic constraint matrix of the mobile platform. f_n is the nonholonomic constraint force. The joint position is expressed as $q = [\boldsymbol{q}_n, \boldsymbol{q}_h]^\top$, where $\boldsymbol{q}_n \in \mathbb{R}^{n_n \times 1}, \boldsymbol{q}_h \in \mathbb{R}^{n_h \times 1}$ are the joint position vectors of the mobile platform and robot manipulators, respectively.

Based on the above analysis, (2.2) can be represented as follows:

$$\boldsymbol{A}_n(\boldsymbol{q}_n)\dot{\boldsymbol{q}}_n = 0 \tag{2.4}$$

where $\boldsymbol{A}_n(\boldsymbol{q}_n)$ is the nonholonomic constraint matrix of the mobile platform, which is described as $\boldsymbol{A}_n(\boldsymbol{q}_n) = [\boldsymbol{A}_{n1}^\top(\boldsymbol{q}_n), \boldsymbol{A}_{n2}^\top(\boldsymbol{q}_n), \ldots, \boldsymbol{A}_{nm}^\top(\boldsymbol{q}_n)] \in \mathbb{R}^{m \times n_n}$, and assume that it has the full rank of m. The effect of the constraints can be viewed as a restriction of the dynamics on the manifold [27, 28] satisfying

$\mathrm{E}_n = \{(\boldsymbol{q}_n, \dot{\boldsymbol{q}}_n) | \boldsymbol{A}_n(\boldsymbol{q}_n)\dot{\boldsymbol{q}}_n = 0\}$. Assume that there exists a full-rank matrix $\boldsymbol{Z}(\boldsymbol{q}_n) = [\boldsymbol{Z}_1(\boldsymbol{q}_n), \boldsymbol{Z}_2(\boldsymbol{q}_n), \ldots, \boldsymbol{Z}_{(n_n-m)}(\boldsymbol{q}_n)] \in \mathbb{R}^{n_n \times (n_n - m)}$, and the $n_n - m$ columns of this matrix span the null space of $A_n(q_n)$

$$\boldsymbol{Z}^\top(\boldsymbol{q}_n)\boldsymbol{A}_n^\top(\boldsymbol{q}_n) = 0 \tag{2.5}$$

where $\boldsymbol{Z}^\top(\boldsymbol{q}_n)\boldsymbol{Z}(\boldsymbol{q}_n)$ is also a full-rank matrix.

From (2.4) and (2.5), there exists a vector $\boldsymbol{\rho}$ and its derivation $\dot{\boldsymbol{\rho}}$ satisfies

$$\dot{\boldsymbol{q}}_n = \boldsymbol{Z}(\boldsymbol{q}_n)\dot{\boldsymbol{\rho}} \tag{2.6}$$

where $\dot{\boldsymbol{\rho}} \in \mathbb{R}^{(n_n - m) \times 1}$.

Defining $\xi = [\dot{\boldsymbol{\rho}}^\top, \dot{\boldsymbol{q}}_h^\top]^\top$, the derivation of joint position can be expressed as

$$\dot{\boldsymbol{q}} = \boldsymbol{Z}_q(\boldsymbol{q})\dot{\xi} \tag{2.7}$$

where $\boldsymbol{Z}_q(\boldsymbol{q}) = \begin{bmatrix} \boldsymbol{Z}(\boldsymbol{q}_n) & 0 \\ 0 & \mathrm{I} \end{bmatrix}$.

Differentiating (2.7) yields

$$\ddot{\boldsymbol{q}} = \dot{\boldsymbol{Z}}_q(\boldsymbol{q})\dot{\xi} + \boldsymbol{Z}_q(\boldsymbol{q})\ddot{\xi} \tag{2.8}$$

According to (2.8), the dynamics of the MMR (2.1) can be rewritten as the following form:

$$\boldsymbol{M}(\boldsymbol{q})\boldsymbol{Z}_q(\boldsymbol{q})\ddot{\xi} + [\boldsymbol{M}(\boldsymbol{q})\dot{\boldsymbol{Z}}_q(\boldsymbol{q}) + C(\boldsymbol{q}, \dot{\boldsymbol{q}})\boldsymbol{Z}_q(\boldsymbol{q})]\dot{\xi} + \boldsymbol{G}(\boldsymbol{q}) + \boldsymbol{\tau}_d = u + f \tag{2.9}$$

where $u = \boldsymbol{B}(\boldsymbol{q})\boldsymbol{\tau}$.

By multiplying two sides of (2.9) by $\boldsymbol{Z}_q^\top(\boldsymbol{q})$, one can obtain

$$\begin{aligned} &\boldsymbol{Z}_q^\top(\boldsymbol{q})\boldsymbol{M}(\boldsymbol{q})\boldsymbol{Z}_q(\boldsymbol{q})\ddot{\xi} + \boldsymbol{Z}_q^\top[\boldsymbol{M}(\boldsymbol{q})\dot{\boldsymbol{Z}}_q(\boldsymbol{q}) + C(\boldsymbol{q}, \dot{\boldsymbol{q}})\boldsymbol{Z}_q(\boldsymbol{q})]\dot{\xi} \\ &+\boldsymbol{Z}_q^\top\boldsymbol{G}(\boldsymbol{q}) + \boldsymbol{Z}_q^\top\boldsymbol{\tau}_d = \boldsymbol{Z}_q^\top u \end{aligned} \tag{2.10}$$

From (2.10), the dynamics of the MMR can be represented as

$$\boldsymbol{M}_\xi\ddot{\xi} + \boldsymbol{C}_\xi\dot{\xi} + \boldsymbol{G}_\xi + \boldsymbol{\tau}_{\xi d} = \boldsymbol{Z}_q^\top u \tag{2.11}$$

where $\boldsymbol{M}_\xi = \begin{bmatrix} \boldsymbol{Z}^\top\boldsymbol{M}_n\boldsymbol{Z} & \boldsymbol{Z}^\top\boldsymbol{M}_{nh} \\ \boldsymbol{M}_{hn}\boldsymbol{Z} & \boldsymbol{M}_h \end{bmatrix}$, $\boldsymbol{C}_\xi = \begin{bmatrix} \boldsymbol{Z}^\top\boldsymbol{C}_n\boldsymbol{Z} + \boldsymbol{Z}^\top\boldsymbol{C}_n\dot{\boldsymbol{Z}} & \boldsymbol{Z}^\top\boldsymbol{C}_{nh} \\ \boldsymbol{C}_{hn}\boldsymbol{Z} + \boldsymbol{C}_{hn}\dot{\boldsymbol{Z}} & \boldsymbol{C}_n \end{bmatrix}$, $\boldsymbol{G}_\xi = \begin{bmatrix} \boldsymbol{Z}^\top\boldsymbol{G}_n \\ \boldsymbol{G}_h \end{bmatrix}$, and $\boldsymbol{\tau}_{\xi d} = \begin{bmatrix} \boldsymbol{Z}^\top\boldsymbol{\tau}_{dn} \\ \boldsymbol{\tau}_{dh} \end{bmatrix}$.

The nonholonomic force multiplier γ_n can be obtained by (2.9)

$$\gamma_n = \Upsilon\left[\left(\boldsymbol{M}\dot{\boldsymbol{Z}}_q + \boldsymbol{C}\boldsymbol{Z}_q\right)\dot{\xi} + \boldsymbol{G} + \boldsymbol{\tau}_d - u\right] \tag{2.12}$$

where $\Upsilon = (\boldsymbol{A}\boldsymbol{M}^{-1}\boldsymbol{A}^{\top})^{-1}\boldsymbol{A}\boldsymbol{M}^{-1}$.

Based on the above analysis, the dynamics of CCMMR are influenced by three main terms, which are the torques, constraint force input, and disturbances. It is well known that these terms can interact with each other. The result in (2.11) is a reduced formulation of the nonholonomic CCMMR/MMR system that is denoted by (2.1), (2.2), and (2.3). This new model allows many methods proposed for robot manipulators that can directly be utilized for MMR motion control. In addition, the dynamics of this new model is considered to satisfy the following properties [28]:

Property 2.1 $\boldsymbol{M}$ is uniformly bounded and continuous.

Property 2.2 $\boldsymbol{M}_\xi$ is a positive definite symmetric matrix and $\boldsymbol{M}_\xi$ is uniformly bounded: $\lambda_{\min}(\boldsymbol{M}_\xi)\|x\|^2 \leq x^{\top}\boldsymbol{M}_\xi x \leq \lambda_{\max}(\boldsymbol{M}_\xi)\|x\|^2$, $\forall x \in \mathbb{R}^{(n-m)\times 1}$.

Property 2.3 Skew symmetric matrix. $\boldsymbol{S}_\xi \equiv \dot{\boldsymbol{M}}_\xi - 2\boldsymbol{C}_\xi$, where S_ξ is a skew symmetric matrix.

References

1. Willette A, Brell-Cokcan S, Braumann J. Robotic Fabrication in Architecture, Art and Design 2014, Springer Science & Business Media, 2014.
2. Urbina P. 3D-Printing with fanuc robot arm final report. Central Connecticut State University, 2019.
3. Hutter M, Gehring C, Jud D, et al. Anymal-a highly mobile and dynamic quadrupedal robot. In: IEEE/RSJ International Conference on Intelligent Robots and Systems (IROS). IEEE, 2016: 38–44.
4. ADLINK. Products: SMR250 Series. [Online]. Available: https://www.adlinktech.com/Products/Smart_Manufacturing/Autonomous_Mobile_Robot/SMR250_series?lang=zh-cn
5. KUKA. Products: Autonomous mobile robots. [Online]. Available: https://www.kuka.com/zh-cn/products/autonomous-mobile-robots-amr/mobile-platforms
6. Bennett S A. A history of control engineering. IET, 1993.
7. Ogata K. Modern control engineering. Prentice-Hall, Inc, 1996.
8. Lim K, Eslami M. Robust adaptive controller designs for robot manipulator systems. IEEE Journal on Robotics and Automation, 1987, 3(1): 54–66.
9. Slotine J J E, Weiping L. Adaptive manipulator control: a case study. IEEE Transactions on Automatic Control, 1988, 33(11): 995–1003.
10. Spong M W. Adaptive control of flexible joint manipulators. Systems & Control Letters, 1989, 13(1): 15–21.
11. Reed J S, Ioannou P A. Instability analysis and robust adaptive control of robotic manipulators. IEEE Transactions on Robotics and Automation, 1989, 5(3): 381–386.
12. Nilsson N J, Rosen C A, Raphael B. Application of intelligent automata to reconnaissance. Rome Air Development Center, 1969.
13. Nilsson N J, Automaton A M. An application of artificial intelligence techniques. In: Proc. of IJCAI, 1969, 509.

14. Nilsson N J. Shakey the robot. Menlo Park, California: Sri International, 1984.
15. Ge S S, Lee T H, Zhu G. Genetic algorithm tuning of lyapunov-based controllers: an application to a single-link flexible robot system. IEEE Transactions on Industrial Electronics, 1996, 43(5): 567–574.
16. Watanabe K, Tang J, Nakamura M, et al. A fuzzy-gaussian neural network and its application to mobile robot control. IEEE Transactions on Control Systems Technology, 1996, 4(2): 193–199.
17. Huang S J, Lian R. A hybrid fuzzy logic and neural network algorithm for robot motion control. IEEE Transactions on Industrial Electronics, 1997, 44(3): 408–17.
18. Mbede J B, Huang X, Wang M. Robust neuro-fuzzy sensor-based motion control among dynamic obstacles for robot manipulators. IEEE Transactions on Fuzzy Systems, 2003, 11(2): 249–261.
19. Zhou F, Wang J, Li Y, et al. Control of an inspection robot for 110KV power transmission lines based on expert system design methods. In: Proceedings of 2005 IEEE Conference on Control Applications, 2005, CCA 2005. IEEE, 2005.
20. Chang Y, Chang C, Chen C, et al. Fuzzy sliding-mode formation control for multirobot systems: design and implementation. IEEE Transactions on Systems, Man, and Cybernetics, Part B (Cybernetics), 2011, 42(2): 444–457.
21. Fan S. Research on approaches of intelligent measurement and control for fouling cleaning incondenser. Hunan University, 2006.
22. Peng J. Research on methods of intelligent robust control of cleaning mobile manipulator of large condenser. Hunan University, 2008.
23. Zhang Y. Research on visual control methods for cleaning robot of large condenser. Hunan University, 2010.
24. Wu X. The control theory research of neural network for the condenser cleaning robot. Hunan University, 2013.
25. Wang Y, Wu X. Neural networks-based adaptive robust control of crawler-type mobile manipulators using sliding mode approach. Industrial Robot: An International Journal, 2012, 39(3): 260–270.
26. Wu X, Wang Y, Dang X. Robust adaptive sliding-mode control of condenser-cleaning mobile manipulator using fuzzy wavelet neural network. Fuzzy Sets and Systems, 2014, 235: 62–82.
27. Li Z, Ge S S, Ming A. Adaptive robust motion/force control of holonomic-constrained nonholonomic mobile manipulators. IEEE Transactions on Systems, Man, and Cybernetics, Part B (Cybernetics), 2007, 37(3): 607–616.
28. Li Z, Ge S S, Adams M, et al. Robust adaptive control of uncertain force/motion constrained nonholonomic mobile manipulators. Automatica, 2008, 44(3): 776–784.

Chapter 3
Control Methods for Condensing Cleaning Mobile Manipulator Robot

3.1 Background and Motivation

The tracking control of mobile robots has received increasing attention owing to its wide applicability in various fields, such as mining, waste management, forestry, planetary exploration, and military [1, 2]. It is well known that mobile manipulators possess strongly coupled dynamics of mobile platforms and manipulators. The mobile platform extends the manipulators' workspace, whereas the manipulators offer much operational functionality. Therefore, the control design for these systems has attracted much attention and is a challenging problem in the nonholonomic mobile robots [3].

Generally, the condenser cleaning mobile platform has closed chain mechanism with redundant actuation. The redundant actuation provides an effective means for eliminating singularities and improving the performance such as Cartesian stiffness and homogeneous output forces. At the same time, a major difficulty, which prevents vast control strategies from applying in redundant actuated systems, is the lack of an efficient dynamic model and control method. However, without dynamic control, it is difficult to perform coordinated motion of the mobile platform and the dynamically controlled manipulator. Hence, there are many researches about dynamic modeling and control of omnidirectional wheeled mobile manipulators with redundant or nonredundant actuation [4, 5]. Recently, several results about dynamic modeling and control of holonomic systems were published [6, 7], which provided effective means for such problems of CCMMs. The major components of well-known control methods are robust control, adaptive control, sliding-mode control, robust Lyapunov stability-based control, and a combination of adaptive and sliding-mode control. For example, in [8], the authors consider multiple mobile manipulators grasping a rigid object in contact with a deformable working surface, whose geometric and real physical parameters are unknown but boundedness of physical parameters is known. They proposed a neuro-adaptive robust force/motion tracking

Y. Wang et al., *Electric Power Robots*,
https://doi.org/10.1007/978-981-96-2512-3_3

controller for coordinated mobile manipulators. Tsai et al. [9] present methodologies for dynamic modeling and trajectory tracking of a nonholonomic wheeled mobile manipulator with dual arms. In [10], the authors propose a novel control law based on sliding-mode theory in order to drive mobile robots to a target location, which is specified by a previously acquired reference image. However, the robust control schemes for these uncertain system dynamics are not fully explored. In practical applications, environmental uncertainties arise in mobile manipulator applications, which can affect the system stability and performance. It is well known that the main advantage of using sliding-mode control is strong robustness with respect to system uncertainties and external disturbances. In this section, to handle unknown dynamics of mobile robot systems, robust adaptive sliding-mode controls have been extensively investigated for mobile robot manipulator systems.

FWNN, one of the most popular intelligent computation approaches, has an inherent learning ability and can approximate a nonlinear continuous function to arbitrary accuracy. This property is crucial in the controller design for complex model identifying and unstructured uncertainties compensating. Therefore, the development of FWNN control in mobile manipulators has attracted considerable interest [11–13]. The FWNN can be represented by a linear combination of fuzzy wavelet basis functions. Different from wavelet neural networks, fuzzy wavelet basis functions can be specified by experts as traditional fuzzy systems. Furthermore, the architecture of wavelet networks and fuzzy wavelet networks can provide at least the same order of approximation error as neural networks.

Intelligent control method based on the adaptive control and NNs is also a useful method to deal with the unknown/uncertain dynamics of the MMR system. The NNs have self-learning characteristic and good approximation capabilities [14, 15] and have been applied successfully in robot control systems. Lin and Goldenberg [16] proposed an adaptive online NN controller to identify the dynamics of the MMR and eliminate disturbances. But the parameters of Gaussian functions were all chosen a priori or kept fixed. Li et al. in [17] introduced a neuro-adaptive force/motion controller for the MMR in the presence of unknown dynamic model and uncertain constraints. In [18], a robust neural network based on sliding-mode controller, which used NNs to identify the unstructured system dynamics directly, was proposed. This controller is capable of disturbance rejection in the presence of unknown disturbance bounds, but the assumption of the bounded signals, fixed parameter of the NNs, and approximation errors still need to be improved. Intelligent control methods based on FNNs have also been widely applied in robotics. For the MMR control, Mbede et al. [19] have developed a modular fuzzy navigation method in changing of dynamic unstructured environments. In this method, the integration of robust controller and modified Elman NNs is to deal with uncertainties. But the bounded parameters (disturbances, Taylor parameters) are assumed to be known.

The CMAC system has been researched and developed rapidly in recent years. The CMAC, presented by Albus [20], is an auto-associative memory feedforward NN with overlapping receptive fields. The input-output relationship is similar to the model of the human memory, where the local receptive region is expected to perform the corresponding input-output mapping. The CMAC has demonstrated favorable features, such as fast learning, good generalization capability, and simplicity of computation, when compared with the NNs. This assertion was evident in control applications of the CMAC both in theory and practical applications [21, 22]. In the CMAC application, the partition of the input variables directly impacts the control performance. Hence, it can imply that the finer partitions can achieve a better control performance. However, the size of the networks and the memories of the CMAC will be increased. Based on the combination of fuzzy techniques and the CMAC, this drawback can be eliminated. The difference in the feature between the traditional CMAC and the fuzzy CMAC (FCMAC) is that the FCMAC builds membership functions in its mapping process such that the control performance can be enhanced by programming the appropriate membership functions rather than modifying the partition size to increase the computation burden. The advantages of the FCMAC have been demonstrated in some applications [23, 24].

By inheriting the merits of the previous research results for the CCMMR, this section continues developing intelligent control methods to further improve the operating effectiveness of the CCMMR. Some new intelligent control methods for the motion/force control of the CCMMR that combine the adaptive robust control methods and the universal approximation capability of the hybrid NNs are proposed in this section. The improvements of these methods focus on tracking position errors and force constraint control in the presence of time-varying uncertain parameters (disturbances) without the knowledge of the CCMMR control system. In addition, the proposed control method was designed based on the Lyapunov stability theorem, and the stability of the controlled system was verified through simulations and experimental results of performing the condenser cleaning.

3.2 Robust Adaptive Sliding-Mode Control with a Fuzzy Wavelet Neural Network

Consider a n-DOF CCMMR mounted on a nonholonomic mobile platform, as shown in Fig. 3.1. The dynamic equations of the mechanical system with external disturbances can be described by

$$\boldsymbol{M}(\boldsymbol{q})\ddot{\boldsymbol{q}} + \boldsymbol{V}(\boldsymbol{q},\dot{\boldsymbol{q}})\dot{\boldsymbol{q}} + F(\dot{\boldsymbol{q}}) + \boldsymbol{G}(\boldsymbol{q}) + \tau_d = \boldsymbol{B}(\boldsymbol{q})\boldsymbol{\tau} + \boldsymbol{A}^{\mathrm{T}}(\boldsymbol{q})\boldsymbol{\lambda} \tag{3.1}$$

where $\dot{\boldsymbol{q}} = \left[\dot{x}, \dot{y}, \dot{\varphi}, \dot{\theta}_R, \dot{\theta}_L\right]^{\mathrm{T}}$, $\boldsymbol{M}(\boldsymbol{q}) \in R^{n \times n}$ is a symmetric positive definite inertia matrix, $\boldsymbol{V}(\boldsymbol{q},\dot{\boldsymbol{q}}) \in R^n$ represents the vector of centripetal and Coriolis forces, $F(\dot{\boldsymbol{q}}) \in R^n$ is the surface friction and $\boldsymbol{G}(\boldsymbol{q}) \in R^n$ denotes the gravitational torque vector, $\tau_d \in R^{n \times 1}$ denotes bounded unknown disturbances including unstructured unmodeled dynamics, $\boldsymbol{B}(\boldsymbol{q}) \in \mathbb{R}^{n \times n}$ is the input transformation matrix, $\boldsymbol{\tau} \in \mathbb{R}^{n \times 1}$ is the input vector, $\boldsymbol{A}(\boldsymbol{q}) \in \mathbb{R}^{m \times n}$ is the matrix associated with the constraints, and $\boldsymbol{\lambda} \in \mathbb{R}^{m \times 1}$ is the vector of constraint forces.

The kinematic constraints are considered independent of time and can be expressed as

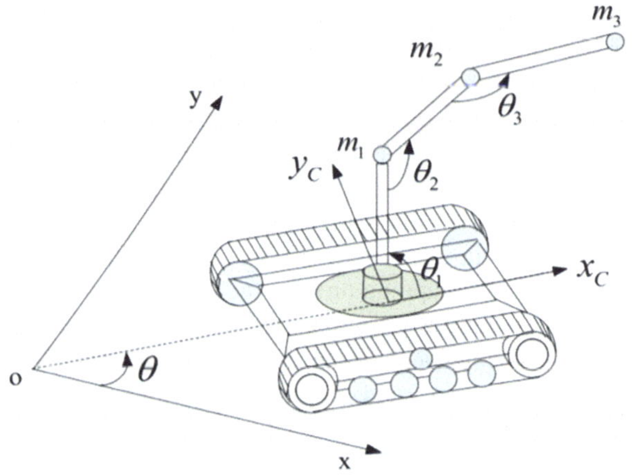

Fig. 3.1 The 3-DOF CCMMR

$$A(q)\dot{q} = 0 \tag{3.2}$$

Let $S(q)$ be a full-rank matrix formed by a set of smooth and linearly independent vector fields spanning the null space of $A(q)$, i.e.,

$$S^{\mathrm{T}}(q)A^{\mathrm{T}}(q) = 0 \tag{3.3}$$

In order to reduce complexity of the whole system, applying the projection of the dynamic model into the null space of the constraints, we can obtain

$$\dot{q} = S(q)\dot{z} \tag{3.4}$$

$$\overline{M}\ddot{z} + \overline{V}\dot{z} + \overline{G} + \overline{F} + \overline{\tau}_d = \overline{\tau} \tag{3.5}$$

where $\overline{M} = S^{\mathrm{T}}MS$, $\overline{V} = S^{\mathrm{T}}\left(M\dot{S} + VS\right)$, $\overline{G} = S^{\mathrm{T}}G$, $\overline{F} = S^{\mathrm{T}}F$, $\overline{\tau}_d = S^{\mathrm{T}}\tau_d$, and $\overline{\tau} = S^{\mathrm{T}}B(q)\tau$. For (3.5), some fundamental properties are addressed as follows:

Property 3.1 The matrix $\overline{M}(q)$ is symmetric and positive definite.

Property 3.2 The matrix $\dot{\overline{M}}(q) - 2\overline{V}(q,\dot{q})$ is skew-symmetric, i.e.,$\xi^{\mathrm{T}}\left(\dot{\overline{M}}(q) - 2\overline{V}(q,\dot{q})\right)\xi = 0$, $\forall \xi \in \mathbb{R}^{n-m}$.

3.2.1 *Fuzzy Wavelet Neural Network*

This section describes the fuzzy wavelet neural network, which has been proven to be an accurate function approximator. This function approximation plays an important role in the mobile robot control. Thus, a fuzzy wavelet neural network is

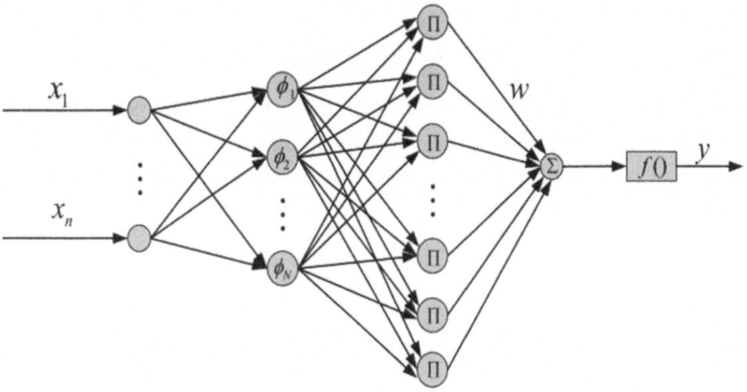

Fig. 3.2 Fuzzy wavelet neural network structure

employed to be a universal approximator in controller design. A general n-input-m-output fuzzy system with N rules can be described as follows:

R_j (j-th rule): *if* x_1 is A_{j1} and x_2 is A_{j2}...and x_n is A_{jn}, *then* f_1 is B_{1j} and f_2 is B_{2j}... and f_m is B_{mj}

where x_i is the i-th state variable, $i = 1, 2, \ldots, n$; f_k is the k-th output variable, $k = 1, 2, \ldots, m$; A_j is a fuzzy term set of x_i for the j-th fuzzy rule R_j, $j = 1, 2, \ldots, N$; and B_{kj} is a fuzzy term set of f_i of the j-th fuzzy rule R_j.

Generally, all fuzzy sets in the antecedent part are characterized by their corresponding fuzzy membership functions $\mu_{Aji}(x_i)$. In this section, for ease of computation, the fuzzy wavelet neural network is employed as shown in Fig. 3.2.

The n - input - m - output fuzzy system with N rules is expressed in the following form:

R_j(j-th rule): IF x_1 is A_{j1} and x_2 is A_{j2}...and x_n is A_{jn}, THEN $\varphi_1 = \omega_1 g_{j1}(x_1) \cdots g_{jn}(x_n)$ and $\varphi_2 = \omega_2 g_{j1}(x_1) \cdots g_{jn}(x_n)$, ..., and $\varphi_m = \omega_m g_{j1}(x_1) \cdots g_{jn}(x_n)$,

where $x_1, x_2, \ldots, x_n$ are the input variables, $\varphi_1, \varphi_2, \ldots, \varphi_n$ are the output variables, and $A_{j1}, A_{j2}, \ldots, A_{jn}$ are the same as in the above fuzzy system.

Using a singleton fuzzifier, product inference, and weighted average defuzzifier, the output of the fuzzy wavelet model can be written as follows:

$$\varphi_k = \sum_{j=1}^{N} \omega_{kj} \psi_j(x) \tag{3.6}$$

The fuzzy wavelet basis function for the j-th rule is defined as

$$\psi_j = \frac{\left(\cap_{i=1}^{n} g_{ji}(x_i)\right)\phi_j}{\sum_{j=1}^{n} \phi_j} \tag{3.7}$$

where $g_{ji}(x_i)=1-\varpi_{ji}^2(x_i-c_{ji})^2$ and the firing strength of the j-th rule is $g_{ji}(x_i)=1-\varpi_{ji}^2(x_i-c_{ji})^2$ and the membership function μ_{Aji} is chosen as $\mu_{Aji}(x_i)=e^{-\varpi_{ji}^2(x_i-c_{ji})^2}$. The parameters ϖ_{ji} and c_{ji} of the fuzzy wavelet basis function can be viewed as dilation and translation parameters.

For ease of notation, the following vector form is used to represent outputs of the fuzzy wavelet model:

$$y=f=\mathbf{W}^{\mathrm{T}}\psi(x,c,\varpi) \tag{3.8}$$

where $\psi(x,c,\varpi)=[\psi_1(x),\psi_2(x),\ldots,\psi_N(x)]^{\mathrm{T}}$ is the fuzzy wavelet basis function vector, $\boldsymbol{x}=[\boldsymbol{x}_1,\boldsymbol{x}_2,\ldots,\boldsymbol{x}_n]^{\mathrm{T}}\in\mathbb{R}^n$ is the input vector, $\varpi=[\varpi_1,\varpi_2,\ldots,\varpi_n]^{\mathrm{T}}\in\mathbb{R}^n$ is the dilation parameter vector, $\boldsymbol{c}^{\mathrm{T}}=\left[\boldsymbol{c}_1^{\mathrm{T}},\boldsymbol{c}_2^{\mathrm{T}},\ldots,\boldsymbol{c}_N^{\mathrm{T}}\right]^{\mathrm{T}}\in\mathbb{R}^N$ is the translation parameter vector, and $\boldsymbol{W}^{\mathrm{T}}=[\boldsymbol{\omega}_{kj}]_{m\times N}$ is the weight matrix.

3.2.2 Controller Design

In this section, the objective of mobile robot control is to determine the control torque τ so that the error between the joint position vector z and the desired joint position z_d can decay to zero within finite time. Thus, the robust adaptive sliding-mode control scheme with the fuzzy wavelet neural network is employed to estimate the unknown functions of mobile robot control system, so that the stability of control system can be guaranteed. The architecture of the mobile robot system is shown in Fig. 3.3.

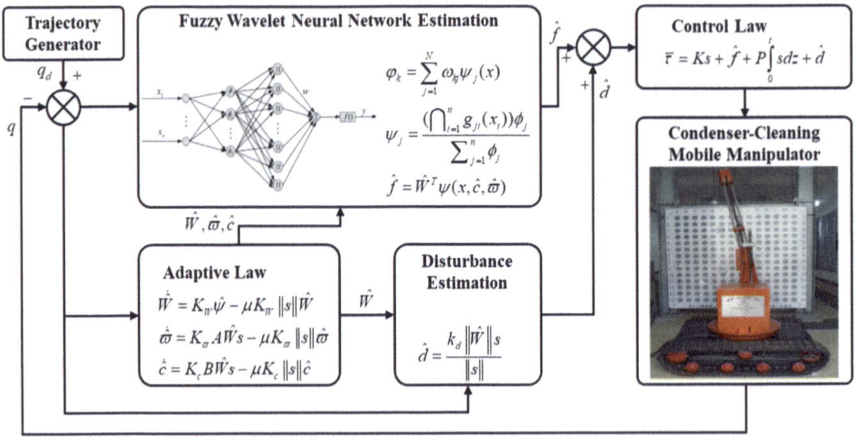

Fig. 3.3 The diagram of the mobile robot system

Denote the tracking error vector $\boldsymbol{e}(t)$ and a sliding variable $\boldsymbol{s}(t)$ as

$$\boldsymbol{e}(t) = z_d(t) - z(t) \tag{3.9}$$

$$\boldsymbol{s}(t) = \dot{\boldsymbol{e}}(t) - \Lambda \boldsymbol{e}(t) \tag{3.10}$$

where $\Lambda = \Lambda^{\mathrm{T}} > 0$. It is typical to define an error metric $s(t)$ to be a performance measure. When the sliding surfaces $s = 0$, the sliding mode is governed by the following differential equation according to the theory of sliding-mode control:

$$\dot{\boldsymbol{e}}(t) = -\Lambda \boldsymbol{e}(t) \tag{3.11}$$

Obviously, the behavior of the system on the sliding surface is determined by the structure of the matrix Λ. In other words, when the error $s(t)$ is smaller, the system performance is better. Therefore, the mobile robot system (3.5) in terms of the sliding variable s can be expressed by

$$\overline{\boldsymbol{M}}\dot{s} = -\overline{\boldsymbol{V}}s + f + \overline{\boldsymbol{\tau}}_d - \overline{\boldsymbol{\tau}} \tag{3.12}$$

where the unknown function $f = \overline{\boldsymbol{M}}(\ddot{z}_d + \Lambda\dot{e}) + \overline{\boldsymbol{V}}(\dot{z}_d + \Lambda e) + \overline{\boldsymbol{G}} + \overline{\boldsymbol{F}}$.

Since the unknown function f existed in the mobile robot system and the fuzzy wavelet neural network (FWNN) can provide accurate function approximation, the FWNN can be chosen to estimate it. Compared with neural networks, the FWNN can achieve the same quality of approximation with a smaller size of the network, especially for higher-dimensional functions. In addition, the FWNN can take advantage of the rigorous approximation theory of wavelet basis function expansions. Therefore, the FWNN can be employed as a universal approximator in our controller design. If we would like to approximate the unknown function f, then there exist an ideal matrix $\boldsymbol{W}^*$ and ideal vectors ϖ^* and $\boldsymbol{c}^*$ of an FWNN such that

$$f = \boldsymbol{W}^* \boldsymbol{\psi}(x, \boldsymbol{c}^*, \varpi^*) + \boldsymbol{\varepsilon}_f(\boldsymbol{x}) \tag{3.13}$$

where $\boldsymbol{\varepsilon}_f(\boldsymbol{x})$ is a small and bounded error vector. The weights W, ϖ, and c are constrained within specified regions: $\Omega_W \equiv \{W \in \mathbb{R}^{N \times m} | \, \|W\| \le b_W, b_W > 0\}$, $\Omega_\varpi \equiv \{\varpi \in \mathbb{R}^{nN} | \, \|\varpi\| \le b_\varpi, b_\varpi > 0\}$, and $\Omega_c \equiv \{c \in \mathbb{R}^{nN} | \, \|c\| \le b_c, b_c > 0\}$, respectively. The ideal weights W^*, ϖ^*, and c^* are denoted as

$$(\boldsymbol{W}^*, \varpi^*, \boldsymbol{c}^*) = \underset{W \in \Omega_W,\, \varpi \in \Omega_\varpi,\, c \in \Omega_c}{\arg\min} \Phi(x)$$

where $\Phi(x) = (\sup \| f(x) - \boldsymbol{W}^{\mathrm{T}}(x) \boldsymbol{\psi}(x, c, \varpi) \|)$.

The corresponding FWNN estimate $\widehat{f}$ can be described as

$$\widehat{f} = \widehat{W}^{\mathrm{T}} \psi(x, \hat{c}, \hat{\varpi}) \tag{3.14}$$

where $\widehat{W}, \widehat{c}, \widehat{\varpi}$ of the FWNN are used to estimate W, c, ϖ. It is convenient to represent ideal and estimated fuzzy wavelet basis function as $\psi^* = \psi(x(t), c^*, \varpi^*)$ and $\widehat{\psi} = \psi(x(t), \widehat{c}, \widehat{\varpi})$, respectively.

According to (3.13) and (3.14), the approximation error f is denoted as

$$\begin{aligned} \widetilde{f} &= f - \widehat{f} \\ &= \boldsymbol{W}^{*\mathrm{T}} \boldsymbol{\psi}(x(t), \boldsymbol{c}^*, \varpi^*) - \widehat{W}^{\mathrm{T}} \boldsymbol{\psi}(x(t), \widehat{c}, \widehat{\varpi}) \\ &= \boldsymbol{W}^{*\mathrm{T}} \widetilde{\psi} + \widetilde{W} \widehat{\psi} + \boldsymbol{\varepsilon}_f(\boldsymbol{x}) \end{aligned} \tag{3.15}$$

where $\widetilde{W} = \boldsymbol{W}^* - \widehat{W}$ and $\widetilde{\psi} = \psi^* - \widehat{\psi}$.

In this section, our goal is to design a sliding-mode controller such that the control system is stable. For achieving the goal, the linearization technique is employed to transform the nonlinear fuzzy wavelet basis functions into partially linear form so that the Lyapunov theorem extension can be applied. Therefore, taking the expansion of $\widetilde{\psi}$ in a Taylor series, we obtain

$$\widetilde{\psi} = A^{\mathrm{T}} \varpi + B^{\mathrm{T}} \widetilde{c} + h \tag{3.16}$$

where $\widetilde{\varpi} = \varpi^* - \widehat{\varpi}$, $A = \left[\frac{\partial \psi_1}{\partial \varpi}, \cdots, \frac{\partial \psi_r}{\partial \varpi}\right]\Big|_{\varpi = \widehat{\varpi}}$, $B = \left[\frac{\partial \psi_1}{\partial c}, \cdots, \frac{\partial \psi_r}{\partial c}\right]\Big|_{c = \widehat{c}}$ and $\boldsymbol{h}$ is a vector of higher-order terms.

Substituting (3.16) into (3.15), we obtain

$$\begin{aligned} \widetilde{f} &= \widetilde{W}^{\mathrm{T}} \widetilde{\psi} + \widehat{W}^{\mathrm{T}} \widetilde{\psi} + \widetilde{W} \widehat{\psi} + \varepsilon_f(x) \\ &= \widetilde{W} \widehat{\psi} + \widehat{W}^{\mathrm{T}} \left(A^{\mathrm{T}} \widetilde{\varpi} + B^{\mathrm{T}} \widetilde{c}\right) + d \end{aligned} \tag{3.17}$$

where $d = \widetilde{W}^{\mathrm{T}} \widetilde{\psi} + \widetilde{W}^{\mathrm{T}} h + \varepsilon_f(x)$.

For convenience, we define an optimal matrix $\boldsymbol{\Theta}^*$ including all optimal weights and an estimating matrix $\widehat{\boldsymbol{\Theta}}$ as

$$\boldsymbol{\Theta}^* = \begin{bmatrix} \boldsymbol{W}^* & 0 & 0 \\ 0 & \varpi^* & 0 \\ 0 & 0 & \boldsymbol{c}^* \end{bmatrix}, \widehat{\boldsymbol{\Theta}} = \begin{bmatrix} \widehat{W} & 0 & 0 \\ 0 & \widehat{\varpi} & 0 \\ 0 & 0 & \widehat{c} \end{bmatrix} \tag{3.18}$$

In the following, we can make the following assumptions:

Assumption 3.1 Suppose that the desired vector z_d is continuous and differentiable and the vectors z_d, $\dot{z}_d$, $\ddot{z}_d$ are uniformly bounded.

Assumption 3.2 The norms of optimal weights W^*, ϖ^*, and c^* are bounded by known positive real values, i.e., $\|W^*\| \leq W_m$, $\|\varpi^*\| \leq \varpi_m$, $\|c^*\| \leq c_m$, with some known W_m, ϖ_m, c_m.

Assumption 3.3 The approximation error $\varepsilon_f(x)$, disturbance $\overline{\tau}_d$, and higher-order term h are bounded, i.e., specified b_ε and b_τ satisfying $\|\varepsilon_f(x)\| \leq b_\varepsilon$ and $\|\tau_d\| \leq b_\tau$, respectively.

Assumption 3.4 Since the values of fuzzy basis functions are positive and not greater than one, $\widetilde{\psi}$ is bounded by $\|\widetilde{\psi}\| \leq 1$. Therefore, $\widetilde{W}^{\mathrm{T}}\widetilde{\psi}$ is bounded by $\left\|\widetilde{W}^{\mathrm{T}}\widetilde{\psi}\right\| \leq \left\|\widetilde{W}^{\mathrm{T}}\right\|\widetilde{\psi} \leq \left\|\widetilde{W}^{\mathrm{T}}\right\| \leq \left\|\widetilde{\Theta}^{\mathrm{T}}\right\|$.

Based on the above assumptions, we know $\boldsymbol{\Theta}^*$ is bounded by a known positive real value $\boldsymbol{\Theta}_m$, i.e., $\|\boldsymbol{\Theta}^*\| \leq \boldsymbol{\Theta}_m$. Moreover. we can find d satisfying the following formula:

$$\begin{aligned} \|d\| &= \left\|\widetilde{W}^{\mathrm{T}}\widetilde{\psi} + \widetilde{W}^{\mathrm{T}}h + \varepsilon_f(x)\right\| \\ &\leq \left\|\widetilde{\Theta}\right\| + \left\|\widetilde{W}^{\mathrm{T}}\right\|k_d + b_\varepsilon \end{aligned} \tag{3.19}$$

Define the control law as

$$\overline{\tau} = \boldsymbol{K}s + \widehat{f} + P\int_0^t s dz + \widehat{d} \tag{3.20}$$

where the positive matrix $\boldsymbol{K} = \boldsymbol{K}^{\mathrm{T}} > 0$, P is positive definite, and $\widehat{d}$ is the robustifying term to attenuate disturbances given by

$$\widehat{d} = \frac{k_d\left\|\widehat{W}\right\|s}{\|s\|} \tag{3.21}$$

where $k_d \geq \|h\|$ is assumed to be satisfied. The parameters are updated by the following learning rules:

$$\begin{aligned} \dot{\widehat{W}} &= \boldsymbol{K}_{\boldsymbol{W}}\widehat{\psi} - \mu\boldsymbol{K}_{\boldsymbol{W}}\|s\|\widehat{W} \\ \dot{\widehat{\varpi}} &= \boldsymbol{K}_{\varpi}A\widehat{W}s - \mu\boldsymbol{K}_{\varpi}\|s\|\widehat{\varpi} \\ \dot{\widehat{c}} &= \boldsymbol{K}_{\boldsymbol{c}}B\widehat{W}s - \mu\boldsymbol{K}_{\boldsymbol{c}}\|s\|\widehat{c} \end{aligned} \tag{3.22}$$

where $\boldsymbol{K}_W$, $\boldsymbol{K}_\varpi$, and $\boldsymbol{K}_c$ are diagonal positive square matrices and $\mu > 0$ is a positive real value. The first terms of (3.22) are similar to the modified backpropagation algorithm that can tune weights and gains of nodes (neurons) [25]. As to the last terms of (3.22), they are similar to the e-modification of adaptive control theory [26, 27].

Remark 3.1 The FWNN law (3.20) consists of three components. The first component $\left[\text{i.e.}, Ks + P\int_0^t s\mathrm{d}z\right]$ is the sliding-mode controller, which is used to guarantee the stability of the system and achieve the uniformly ultimately bounded performance. The second component $\left[\text{i.e.}, \widehat{f}\right]$ is the adaptive FWNN to approximate the uncertain term, and the third component $\left[\text{i.e.}, \widehat{d}(x)\right]$ is the robust term to attenuate disturbances. Thus, in practice, the FWNN law is a hybrid sliding-mode robust controller.

Using the control input (3.20), the mobile robot control system (3.12) can be expressed as follows:

$$\overline{\boldsymbol{M}}\dot{s} = -\left(\overline{\boldsymbol{V}} + \boldsymbol{K}\right)s + \widetilde{f} + \overline{\tau}_d - P\int_0^t s\mathrm{d}z - \widehat{d} \tag{3.23}$$

Theorem 3.1 Under Assumptions 3.1–3.4, if we select the control law (3.20) and learning rules (3.22) to the mobile robot system (3.23) with the robust term (3.21), then the following performance can be achieved:

1. Along with increasing the gain matrix $\boldsymbol{K}$, the tracking error e of the uncertain mobile robot system (3.23) will be more and more small.
2. With suitable control input t, the error metric $s(t)$ and the weights $\widehat{W}, \widehat{\varpi}, \widehat{c}$ are uniformly ultimately bounded.
3. The proposed control law can guarantee the global stability of the mobile robot system (3.23).

Proof Consider the Lyapunov function candidate

$$\begin{aligned} V = \frac{1}{2}s^{\mathrm{T}}\overline{\boldsymbol{M}}s + \frac{1}{2}\left(\int_0^t s\mathrm{d}z\right)^{\mathrm{T}} P\left(\int_0^t s\mathrm{d}z\right) + \frac{1}{2}\mathrm{tr}\left\{\widetilde{W}^{\mathrm{T}}\boldsymbol{K}_W^{-1}\widetilde{W}\right\} \\ + \frac{1}{2}\mathrm{tr}\left\{\widetilde{\varpi}^{\mathrm{T}}\boldsymbol{K}_\varpi^{-1}\widetilde{\varpi}\right\} + \frac{1}{2}\mathrm{tr}\left\{\widetilde{c}^{\mathrm{T}}\boldsymbol{K}_c^{-1}\widetilde{c}\right\} \end{aligned} \tag{3.24}$$

According to Property 3.2 and formulas (3.8)–(3.16), the time derivative of V along the trajectory of (3.23) is

$$
\begin{aligned}
\dot{V}=&\frac{1}{2}s^{\mathrm{T}}\dot{\overline{\boldsymbol{M}}}s+s^{\mathrm{T}}\overline{\boldsymbol{M}}\dot{s}+s^{\mathrm{T}}P\int_{0}^{t}s\mathrm{d}z+\mathrm{tr}\left\{\widetilde{W}^{\mathrm{T}}\boldsymbol{K}_{W}^{-1}\dot{\widetilde{W}}\right\}+\mathrm{tr}\left\{\widetilde{\varpi}^{\mathrm{T}}\boldsymbol{K}_{\varpi}^{-1}\dot{\widetilde{\varpi}}\right\}+\mathrm{tr}\left\{\widetilde{c}^{\mathrm{T}}\boldsymbol{K}_{\varpi}^{-1}\dot{\widetilde{c}}\right\}\\
=&-s^{\mathrm{T}}\boldsymbol{K}s+\frac{1}{2}s^{\mathrm{T}}\left(\dot{\overline{\boldsymbol{M}}}-2\overline{\boldsymbol{V}}\right)s+s^{\mathrm{T}}\widetilde{f}+s^{\mathrm{T}}\overline{\boldsymbol{\tau}}_{d}-s^{\mathrm{T}}\widehat{d}-\mathrm{tr}\left\{\widetilde{W}^{\mathrm{T}}\boldsymbol{K}_{W}^{-1}\dot{\widehat{W}}\right\}\\
&-\mathrm{tr}\left\{\widetilde{\varpi}^{\mathrm{T}}\boldsymbol{K}_{\varpi}^{-1}\dot{\widehat{\varpi}}\right\}-\mathrm{tr}\left\{\widetilde{c}^{\mathrm{T}}\boldsymbol{K}_{\varpi}^{-1}\dot{\widehat{c}}\right\}\\
=&-s^{\mathrm{T}}\boldsymbol{K}s+-s^{\mathrm{T}}\left[\widetilde{W}^{\mathrm{T}}\widehat{\psi}+\widehat{W}^{\mathrm{T}}\left(A^{\mathrm{T}}\widetilde{\varpi}+\boldsymbol{B}^{\mathrm{T}}\widetilde{c}\right)+\widetilde{d}\right]+s^{\mathrm{T}}\overline{\boldsymbol{\tau}}_{d}-s^{\mathrm{T}}\widehat{d}\\
&-\mathrm{tr}\left\{\widetilde{W}^{\mathrm{T}}\left(\widehat{\psi}s^{\mathrm{T}}-\mu\|s\|\widehat{W}\right)\right\}-\mathrm{tr}\left\{\widetilde{\varpi}^{\mathrm{T}}\left(A\widehat{W}s-\mu\|s\|\widehat{\varpi}\right)\right\}-\mathrm{tr}\left\{\widetilde{c}^{\mathrm{T}}\left(\boldsymbol{B}\widehat{W}s-\mu\|s\|\widehat{c}\right)\right\}
\end{aligned}
\tag{3.25}
$$

Based on the matrix trace, inner product, and Frobenius norm theory, the following property holds:

$$
\begin{aligned}
\mathrm{tr}\left(\widetilde{\Theta}\widehat{\Theta}\right)&=\mathrm{tr}\left(\widetilde{\Theta}^{\mathrm{T}}\left(\boldsymbol{\Theta}^{*}-\widetilde{\Theta}\right)\right)\\
&=\left\langle\widetilde{\Theta},\boldsymbol{\Theta}^{*}\right\rangle-\left\|\widetilde{\Theta}\right\|^{2}\\
&\leq\left\|\widetilde{\Theta}\right\|\left\|\boldsymbol{\Theta}^{*}\right\|-\left\|\widetilde{\Theta}\right\|^{2}
\end{aligned}
\tag{3.26}
$$

In view of (3.19), we obtain

$$
\begin{aligned}
\dot{V}\leqslant&-s^{\mathrm{T}}\boldsymbol{K}s+\left\|s\right\|\left\|\widetilde{\Theta}\right\|+\left\|s\right\|\left\|\widetilde{W}^{\mathrm{T}}\right\|\boldsymbol{K}_{d}+\left\|s\right\|b_{\varepsilon}+\left\|s\right\|b_{\tau}-\boldsymbol{K}_{d}\left\|\widehat{W}\right\|\left\|s\right\|\\
&+\mu\left\|s\right\|\mathrm{tr}\left\{\widetilde{\Theta}^{\mathrm{T}}\left(\boldsymbol{\Theta}^{*}-\widetilde{\Theta}\right)\right\}\\
\leqslant&-s^{\mathrm{T}}\boldsymbol{K}s+\left\|s\right\|\left\|\widetilde{\Theta}\right\|+\left\|s\right\|b_{\varepsilon}+\left\|s\right\|b_{\tau}+\mu\left\|s\right\|\left(\left\|\widetilde{\Theta}\right\|\left\|\widetilde{\Theta}\right\|-\left\|\widetilde{\Theta}\right\|^{2}\right)\\
\leqslant&-\left\|s\right\|\left\{K_{\min}\left\|s\right\|+\mu\left[\left\|\widetilde{\Theta}\right\|-\left(\frac{\Theta_{m}}{2}+\frac{1}{2\mu}\right)\right]^{2}-D\right\}
\end{aligned}
\tag{3.27}
$$

where $K_{\min}$ is the minimum singular value of K and $D=b_{\varepsilon}+b_{\tau}+\mu\left(\frac{\Theta_m}{2}+\frac{1}{2\mu}\right)^2$.

Thus, $\dot{V}\leq 0$. This implies that the Lyapunov derivative $\dot{V}$ is negative outside the compact set ($\|s\| < D/K_{\min}$). In other words, outside the compact set, the tracking errors and parameter errors will be decreased. As for inside the compact region around the origin, the tracking errors and parameter errors are bounded. Therefore, according to a standard Lyapunov theorem extension [28], we can prove that the error $s(t)$ and the weights $\widehat{W}, \widehat{\varpi}, \widehat{c}$ are uniformly ultimately bounded. The tracking error e can be kept as small as desired by increasing $K_{\min}$. This completes the proof.

3.2.3 Simulation Results

To verify the feasibility and effectiveness of the proposed controller, let us consider the 3-DOF CCMMR system shown in Fig. 3.1. The physical parameters of the CCMMR system are shown in Table 3.1.

Let $z = [\theta_l, \theta_r, \theta_1, \theta_2, \theta_3]^{\mathrm{T}}$, and θ_l and θ_r are the rotation angle of the left caterpillar and the right caterpillar, respectively. Let $q_v = [x_c, y_c, \theta]^{\mathrm{T}}$ be the position and direction of mobile platform and $q_r = [\theta_1, \theta_2, \theta_3]^{\mathrm{T}}$ be the link angle of manipulators. The generalized coordinate is chosen as $q = [q_v, q_r]^{\mathrm{T}}$, where the parameters are shown in Fig. 3.1. The mobile manipulator is subjected to the following constraints:

$$\dot{x}_c \sin\theta - \dot{y}_c \cos\theta + d\dot{\theta} + R\theta_r = 0$$

$$-\dot{x}_c \sin\theta + \dot{y}_c \cos\theta + d\dot{\theta} + R\theta_l = 0$$

$$\dot{x}\cos\theta + \dot{y}_c \sin\theta = 0$$

Then, the constraint matrix according to formula (3.2) is

$$\boldsymbol{A}(\boldsymbol{q}) = [-\sin\theta,\ \cos\theta,\ -1, 0, 0, 0]^{\mathrm{T}}$$

Let $\boldsymbol{S}(\boldsymbol{q})$ be one of the bases of the null space of $\boldsymbol{A}(\boldsymbol{q})$:

$$\boldsymbol{S}(\boldsymbol{q}) = \begin{bmatrix} \frac{r}{2}\cos\theta + \frac{r}{R}\sin\theta & \frac{r}{2}\cos\theta - \frac{r}{R}\sin\theta & 0 & 0 & 0 \\ \frac{r}{2}\sin\theta - \frac{r}{R}\cos\theta & \frac{r}{2}\sin\theta + \frac{r}{R}\cos\theta & 0 & 0 & 0 \\ -\frac{r}{R} & \frac{r}{R} & 0 & 0 & 0 \\ 0 & 0 & 1 & 0 & 0 \\ 0 & 0 & 0 & 1 & 0 \\ 0 & 0 & 0 & 0 & 1 \end{bmatrix}$$

Table 3.1 Parameters of the mobile manipulator

Parameter	Value	Unit
Mass of the platform m_0	10	kg
Mass of link 1 m_w	1	kg
Mass of link 2 m_w	3	kg
Mass of link 3 m_w	3	kg
Inertia of the platform I_0	0.3	$\mathrm{kg \cdot m^2}$
Inertia of the platform I_1	0.1	$\mathrm{kg \cdot m^2}$
Inertia of the platform I_2	0.2	$\mathrm{kg \cdot m^2}$
Inertia of the platform I_3	0.3	$\mathrm{kg \cdot m^2}$
Radius of the wheel r	0.05	m
Length of link 1 l_1	0.05	m
Length of link 2 l_2	0.15	m
Length of link 3 l_3	0.2	m
Distance between the two caterpillars	0.2	m

Using the Lagrangian approach, we can obtain the standard form with

$$\boldsymbol{M}(\boldsymbol{q}) = \begin{bmatrix} \boldsymbol{M}_{11} & \boldsymbol{M}_{12} & \boldsymbol{M}_{13} & \boldsymbol{M}_{14} & \boldsymbol{M}_{15} & \boldsymbol{M}_{16} \\ \boldsymbol{M}_{21} & \boldsymbol{M}_{22} & \boldsymbol{M}_{23} & \boldsymbol{M}_{24} & \boldsymbol{M}_{25} & \boldsymbol{M}_{26} \\ \boldsymbol{M}_{31} & \boldsymbol{M}_{32} & \boldsymbol{M}_{33} & \boldsymbol{M}_{34} & \boldsymbol{M}_{35} & \boldsymbol{M}_{36} \\ \boldsymbol{M}_{41} & \boldsymbol{M}_{42} & \boldsymbol{M}_{43} & \boldsymbol{M}_{44} & \boldsymbol{M}_{45} & \boldsymbol{M}_{46} \\ \boldsymbol{M}_{51} & \boldsymbol{M}_{52} & \boldsymbol{M}_{53} & \boldsymbol{M}_{54} & \boldsymbol{M}_{55} & \boldsymbol{M}_{56} \\ \boldsymbol{M}_{61} & \boldsymbol{M}_{62} & \boldsymbol{M}_{63} & \boldsymbol{M}_{64} & \boldsymbol{M}_{65} & \boldsymbol{M}_{66} \end{bmatrix}$$

$$\boldsymbol{V}(\boldsymbol{q}, \dot{\boldsymbol{q}}) = \begin{bmatrix} \boldsymbol{V}_{11} & \boldsymbol{V}_{12} & \boldsymbol{V}_{13} & \boldsymbol{V}_{14} & \boldsymbol{V}_{15} & \boldsymbol{V}_{16} \\ \boldsymbol{V}_{21} & \boldsymbol{V}_{22} & \boldsymbol{V}_{23} & \boldsymbol{V}_{24} & \boldsymbol{V}_{25} & \boldsymbol{V}_{26} \\ \boldsymbol{V}_{31} & \boldsymbol{V}_{32} & \boldsymbol{V}_{33} & \boldsymbol{V}_{34} & \boldsymbol{V}_{35} & \boldsymbol{V}_{36} \\ \boldsymbol{V}_{41} & \boldsymbol{V}_{42} & \boldsymbol{V}_{43} & \boldsymbol{V}_{44} & \boldsymbol{V}_{45} & \boldsymbol{V}_{46} \\ \boldsymbol{V}_{51} & \boldsymbol{V}_{52} & \boldsymbol{V}_{53} & \boldsymbol{V}_{54} & \boldsymbol{V}_{55} & \boldsymbol{V}_{56} \\ \boldsymbol{V}_{61} & \boldsymbol{V}_{62} & \boldsymbol{V}_{63} & \boldsymbol{V}_{64} & \boldsymbol{V}_{65} & \boldsymbol{V}_{66} \end{bmatrix}$$

$$\boldsymbol{F}(\dot{\boldsymbol{q}}) = [0 \;\; 0 \;\; 0 \;\; 0 \;\; m_2 g l_2 \sin(\theta_2) \;\; 0]$$

$$\boldsymbol{B}(\boldsymbol{q}) = \frac{1}{r} \begin{bmatrix} \cos\theta & \cos\theta & 0 & 0 & 0 \\ \sin\theta & \sin\theta & 0 & 0 & 0 \\ R & -R & 0 & 0 & 0 \\ 0 & 0 & r & 0 & 0 \\ 0 & 0 & 0 & r & 0 \\ 0 & 0 & 0 & 0 & r \end{bmatrix}$$

Let the desired output trajectory z_d be

$$z_d = [\,2\cos t \quad 2\sin t \cos(2t) \sin(2t) - \cos(2t)]$$

Assume that the disturbance to the system is

$$\tau_d = [\,-\sin t \quad 3\cos t \quad -2\sin t \quad \sin t \quad -\cos t \quad 2\cos t\,]$$

For the convenience of simulations, the control parameters are chosen as $\mu = 0.01$, $k_d = 100.0$, $K =$ diag $(1.6, 1.8, 1.0, 1.4, 0.9)$, and $P =$ diag $(1.2, 1.2, 1.2, 1.2, 1.2)$. The adaptation laws are specified as $\boldsymbol{K}_W = 100I$, $\boldsymbol{K}_\varpi = 50I$, and $\boldsymbol{K}_c = 50I$ (I is the identity matrix with compatible dimensions). The initial values of centers $\widehat{c}(0)$ and $\widehat{W}(0)$ are set to be small random numbers, and the inverse radius $\widehat{\varpi}(0)$ is specified to be 1. The FWNN is constructed with 25 input neurons and 5 output neurons to online estimate the unknown nonlinear function f of the mobile robot system. The input pattern of the neural network controller is chosen as $(z_d, \dot{z}_d, \ddot{z}_d, e, s)$. The outputs of the FWNN represent the compensation torques for nonlinear dynamics of all joints. The weights of the FWNN are simply initialized at zero and updated timely at each control cycle using the proposed learning law (3.22).

The proposed RASMC is applied to the CCMMR, and the simulation results are shown in Fig. 3.4. The actual computed trajectory of the platform is shown in Fig. 3.4a. Position tracking performances are shown in Fig. 3.4b–d. Figure 3.5 shows the control torques of two caterpillars and three joints. From the aforementioned explanations, it is clear that the system converges to the desired trajectory quickly and achieves good tracking performance. Therefore, the proposed control scheme is effective and robust to both the plant uncertainties and external disturbances.

Remark 3.2 In this section, our approach can take advantage of the strong robustness performance of adaptive robust control, nonlinear mapping properties, and the self-learning ability of neural networks, to deal with the unstructured uncertainties. Moreover, FWNN can alleviate the chattering to some extent and the learning process is online. In this section, the center vector and the width vector are chosen as equal interval distances, and no optimization method is applied to neural networks. However, the system dynamics identification and trajectory tracking performance are still quite effective. Therefore, we may expect the controller output and the simulation results to be more perfect if some standard optimization methods are adopted.

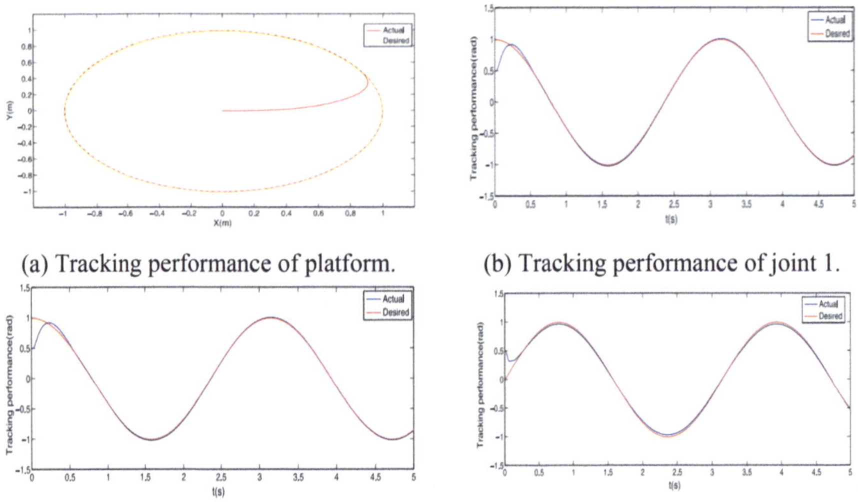

(a) Tracking performance of platform. (b) Tracking performance of joint 1.

(c) Tracking performance of joint 2. (d) Tracking performance of joint 3.

Fig. 3.4 Tracking performance of the mobile platform and three joints. (**a**) Tracking performance of platform. (**b**) Tracking performance of joint 1. (**c**) Tracking performance of joint 2. (**d**) Tracking performance of joint 3

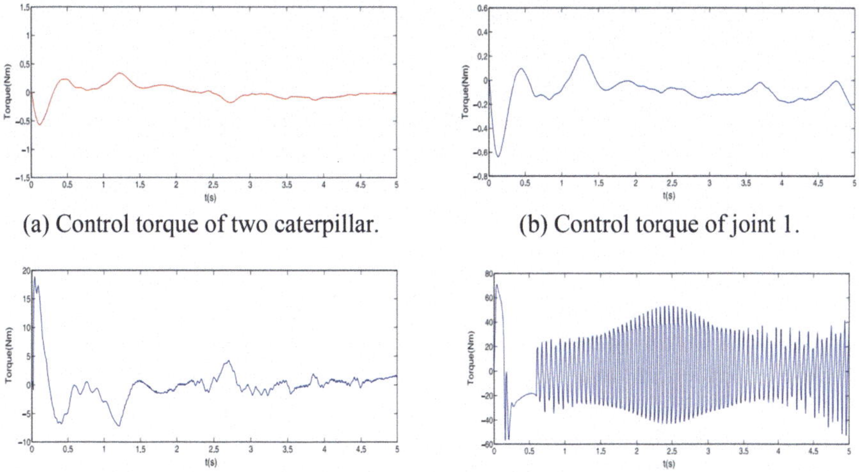

(a) Control torque of two caterpillar. (b) Control torque of joint 1.

(c) Control torque of joint 2. (d) Control torque of joint 3.

Fig. 3.5 Control torques of two caterpillars and three joints. (**a**) Control torque of two caterpillars. (**b**) Control torque of joint 1. (**c**) Control torque of joint 2. (**d**) Control torque of joint 3

3.3 Intelligent Force/Motion Tracking Control Using Fuzzy Wavelet Neural Networks

In this section, by inheriting the advantages of the FWNNs and the recurrent FWNNs (RFWNNs), an intelligent adaptive position tracking system and a force control strategy based on the RFWNNs are presented for the CCMMR/MMR. In general, it is not easy to adopt a model-based method to achieve this control object due to the uncertainties of the CCMMR/MMR, such as unknown dynamics, disturbances, and parameter variations. To solve this problem, an adaptive RFWNN control scheme with the online learning ability is utilized to approximate the unknown dynamics without the requirement of prior system information. In addition, an adaptive robust compensator is proposed to eliminate uncertainties that consist of approximation errors, disturbances, optimal parameters, and higher-order terms in Taylor series. According to adaptive position tracking control design, an adaptive robust control strategy is also considered for the nonholonomic constraint force. The design of adaptive online learning algorithms is derived using Lyapunov stability theorem. Therefore, the proposed intelligent control system based on the RFWNNs proves that it not only can guarantee the stability of mobile robot manipulator control system but also guarantee tracking performance. The effectiveness and robustness of the proposed method are demonstrated by comparing simulation results.

3.3.1 Fuzzy Wavelet Neural Networks

The proposed FWNN structure consists of the Takagi-Sugeno-Kang (TSK) fuzzy system [29, 30] and the WNNs. The TSK FWNN model is constructed on the base of a set of fuzzy rules. Each fuzzy rule is considered a sub-WNN space. In this TSK FWNN structure, each sub-WNN contains wavelets with a specified dilation value (resolution). Thus, the sub-WNNs at different resolution levels can capture different behaviors of the approximated function. Here, the role of the fuzzy set is to determine the contribution of the sub-FWNN space to the input/output of the FWNNs. Therefore, each output of the proposed FWNNs contains a wavelet basis function with multiresolution capability. With the main feature stated above, the proposed FWNNs can improve the accuracy of the approximation process in terms of dilation and translation parameters of the wavelets without increasing the number of wavelet bases [31] or it can achieve the same approximation accuracy with a smaller size in comparison with the FNNs/NNs [32]. Figure 3.6 shows the structure of the proposed FWNNs, which is explained as follows:

Layer 1: The layer of the inputs x_1, x_2, ..., x_n. The input signals are moved to layer 2 by neurons.

Layer 2: The fuzzification layer. Each input linguistic term in this layer is represented by the corresponding fuzzy membership function $\mu_{A_i^j}^{fw}(x_i)$ that is calculated by the following formula:

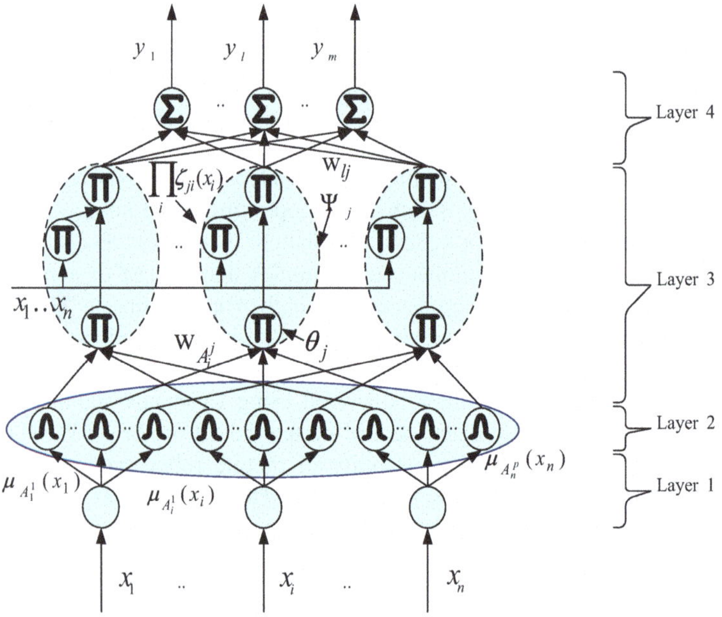

Fig. 3.6 The structure of FWNNs

$$\mu_{A_i^j}^{fw}(x_i) = e^{-\left(d_{ji}^{fw}\right)^2\left(x_i - c_{ji}^{fw}\right)^2} \tag{3.28}$$

where d_{ji}^{fw} is the dilation parameter and c_{ji}^{fw} is the translation parameter, $j = 1, \ldots, p$, $i = 1, \ldots, n$, $p, n \in \mathbb{N}$. p is the number of rules. The outputs of this layer are the values of the membership functions.

Layer 3: The fuzzy rule layer. Each neuron in this layer is represented as a rule, and it performs precondition matching of a rule. The AND operator is used to calculate the outputs of this layer:

$$\theta_j^{fw} = \prod_i w_{A_i^j}^{fw} \mu_{A_i^j}^{fw}(x_i) \tag{3.29}$$

The operation of the product $\prod$ can be referred to fuzzy inference mechanism, and it is used to determine the firing strength. $w_{A_i^j}^{fw}$ presents the weight between the fuzzification layer and the rule layer. It is assumed to be unity. Then, we simplify the firing strength of rule j by the combination of θ_j^{fw} and $\prod_i \zeta_{ji}^{fw}(x_{ji})$ to constitute the fuzzy wavelet basis function:

$$\Psi_j^{fw}(x) = \theta_j^{fw} \prod_i \zeta_{ji}^{fw}(x_i) \tag{3.30}$$

where $\zeta_{ji}^{fw}(x_i) = 1 - \left(d_{ji}^{fw}\right)^2\left(x_i - c_{ji}^{fw}\right)^2$ and $x = [x_1, \ldots, x_n]^{\mathrm{T}} \in \mathbb{R}^{n \times 1}$. From (3.30), it is easy to see that the function $\Psi_j^{fw}(x)$ is the wavelet basis function, so it can be expressed as the multidimensional wavelet function:

$$\Psi_j^{fw}(x) = \prod_i \psi_j^{fw}(x_i) \tag{3.31}$$

where $\psi_j^{fw}(x_i) = \left[1 - \left(d_{ji}^{fw}\right)^2\left(x_i - c_{ji}^{fw}\right)^2\right] e^{-\left(d_{ji}^{fw}\right)^2\left(x_i - c_{ji}^{fw}\right)^2}$ and it is the Mexican hat wavelet function.

Layer 4: The fuzzy output layer. Each node in this layer expresses the output linguistic variable, and it is computed by the summation of all input signals:

$$y_l^{fw} = \sum_j \mathrm{w}_{lj}^{fw} \Psi_j^{fw}(x) \tag{3.32}$$

where w_{lj}^{fw} is the weight between the rule layer and output layer, $l = 1, \ldots, m$, $m \in \mathbb{N}$, and m is the number of the FWNN outputs. The output node can be referred to as the normal defuzzification procedure. In addition, it can be denoted as the following vector form:

$$\boldsymbol{y}_{fw}\left(\boldsymbol{x}, \boldsymbol{d}_{fw}, \boldsymbol{c}_{fw}, \boldsymbol{W}_{fw}\right) = \boldsymbol{W}_{fw}{}^{T}\boldsymbol{\Psi}_{fw}\left(\boldsymbol{x}, \boldsymbol{d}_{fw}, \boldsymbol{c}_{fw}\right) \tag{3.33}$$

where $\boldsymbol{x} = [x_1, \ldots, x_n]^{\mathrm{T}} \in \mathbb{R}^{n \times 1}$, $\boldsymbol{y}_{fw} = \left[y_1^{fw}, \ldots, y_m^{fw}\right]^{\mathrm{T}} \in \mathbb{R}^{m \times 1}$, $\boldsymbol{d}_{fw} = \left[d_{11}^{fw}, \ldots, d_{1n}^{fw}, \ldots, d_{p1}^{fw}, \ldots, d_{pn}^{fw}\right]^{\mathrm{T}} \in \mathbb{R}^{pn \times 1}$, $\boldsymbol{c}_{fw} = \left[c_{11}^{fw}, \ldots, c_{1n}^{fw}, \ldots, c_{p1}^{fw}, \ldots, c_{pn}^{fw}\right]^{\mathrm{T}} \in \mathbb{R}^{pn \times 1}$, $\boldsymbol{\Psi}_{fw}\left(x, d_{fw}, c_{fw}\right) = \left[\Psi_1^{fw}(x), \ldots, \Psi_p^{fw}(x)\right]^{\mathrm{T}} \in \mathbb{R}^{p \times 1}$, and

$$\boldsymbol{W}_{fw}^{\mathrm{T}} = \begin{bmatrix} w_{11}^{fw} & w_{12}^{fw} & \cdots & w_{1p}^{fw} \\ w_{21}^{fw} & w_{22}^{fw} & \cdots & w_{2p}^{fw} \\ \vdots & \vdots & \ddots & \vdots \\ w_{m1}^{fw} & w_{m2}^{fw} & \cdots & w_{mp}^{fw} \end{bmatrix} \in \mathbb{R}^{m \times p}.$$

This FWNN structure is different from the previous FWNN structures [33–36]. The difference is shown in the conclusion part of fuzzy rules; the proposed FWNN rules contain a part of the wavelet function. The wavelet basis functions only appear at the outputs of the FWNNs. Here, the simple firing strength of rules is created to increase the speed of computational capability of the FWNNs, which is necessary when the FWNNs are in complex dynamic systems.

This section has no knowledge of the dynamic robotic system. Therefore, the proposed FWNNs are applied to deal with the unknown/uncertain dynamics of the robot manipulator system. Based on the universal approximation error analysis, there exists an ideal FWNN structure with its optimal parameters to approximate the unknown function y_d in (3.23), such that

$$y_d(x(t)) = \mathrm{W}_{fw}^{*\mathrm{T}}\Psi_{fw}\left(x(t), d_{fw}^*, c_{fw}^*\right) + \Delta_{fw}(x(t)) \tag{3.34}$$

where W_{fw}^*, d_{fw}^*, c_{fw}^* are the optimal parameters of the W_{fw}, d_{fw}, c_{fw}, respectively, and $\boldsymbol{\Delta}_{fw}(\boldsymbol{x}(t))$ is the approximation error vector.

Assumption 3.5 The norms of the FWNN-optimal weights are bounded; $\left\|\mathrm{W}_{fw}^*\right\| \le b_{\mathrm{W}}^{fw}, \left\|d_{fw}^*\right\| \le b_d^{fw}, \left\|c_{fw}^*\right\| \le b_c^{fw}$, where b_{W}^{fw}, b_d^{fw}, b_c^{fw} are the positive real values.

Assumption 3.6 Error of FWNN approximation process and disturbances of robotic system are bounded: $\left\|\Delta_{fw}\right\| \le b_{\Delta}^{fw}$, $\|\tau_d\| \le b_\tau$, b_{Δ}^{fw}, b_τ are the positive real values.

3.3.2 Recurrent Fuzzy Wavelet Neural Networks

The proposed RFWNN structure is the combination of the recurrent structure and the proposed FWNNs. In this RFWNN structure, the appearance of the feedback loop with the time delay at the FWNN fuzzification layer will make the RFWNNs become

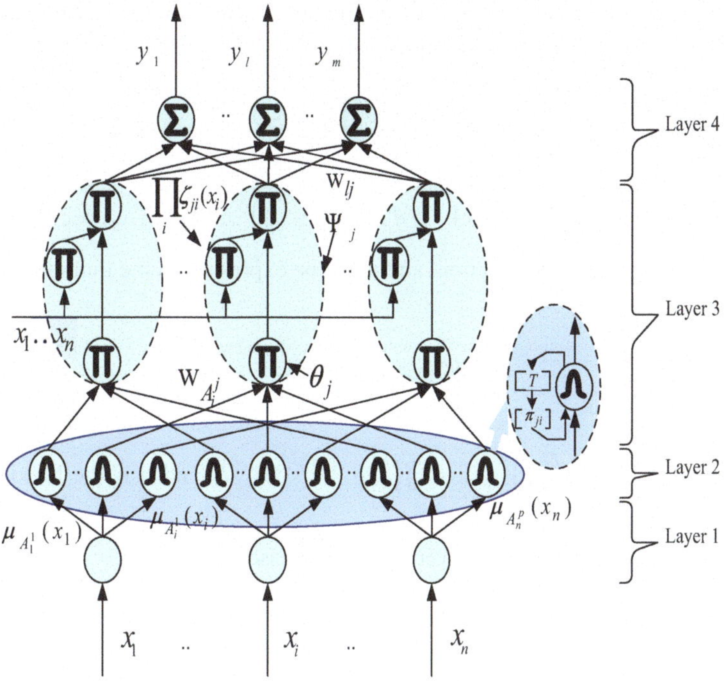

Fig. 3.7 The structure of RFWNNs

a dynamic structure that can improve the ability to deal with the high dynamics of the robot manipulator control system. Basically, the structure of the RFWNNs is only different from the FWNN structure from layer 2. Figure 3.7 shows the structure of the proposed RFWNNs, which is explained as follows:

Layer 2: The local feedback unit with the real time delay part is added into this layer. Therefore, the input of this layer will be represented as follows:

$$x_{ri}(t) = x_i(t) + \pi_{ri}^R \mu_{A_i^j}^R(x_i(t-T)) \tag{3.35}$$

where the membership function $\mu_{A_i^j}^R(x_i(t-T))$ expresses the time delay value of $\mu_{A_i^j}^R(x_i(t))$ via an interval T. $\mu_{A_i^j}^R(x_i(t))$ and $\mu_{A_i^j}^R(x_i(t-T))$ are computed based on formula (3.28). $\boldsymbol{\pi}_{ri}^R$ is the link weight vector (recurrent weight) of the feedback unit. $r = 1, \ldots, p$. Here, the dynamic of the RFWNNs is existent because each input of the membership layer contains the past information of the network. The output of this layer is expressed as

$$\mu_{A_i^j}^R(x_{ri}) = e^{-\left(d_{ji}^R\right)^2\left(x_{ri} - c_{ji}^R\right)^2} \tag{3.36}$$

where d_{ji}^R is the dilation parameter and c_{ji}^R is the translation parameter.

Layer 3: The structure of this layer is similar to layer 2 in the FWNN structure. The outputs of the layer are represented as

$$\Psi_j^R(x_r) = \theta_j^R \prod_i \zeta_{ji}^R(x_{ri}) \tag{3.37}$$

where $\theta_j^R = \prod_i \mu_{A_i^j}^R(x_{ri})$, $\zeta_{ji}^R(x_{ri}) = 1 - \left(d_{ji}^R\right)^2 \left(x_{ri} - c_{ji}^R\right)^2$, $x_r = [x_{r1}, \ldots, x_{rn}]^{\mathrm{T}} \in \mathbb{R}^{pn \times 1}$, $\Psi_j(x_r)$ is the wavelet basis function, and it can be expressed as the multidimensional wavelet function:

$$\Psi_j^R(x_r) = \prod_i \psi_j^R(x_{ri}) \tag{3.38}$$

where $\psi_j^R(x_{ri}) = \left[1 - \left(d_{ji}^R\right)^2 \left(x_{ri} - c_{ji}^R\right)^2\right] e^{-\left(d_{ji}^R\right)^2 \left(x_{ri} - c_{ji}^R\right)^2}$ and it is the Mexican hat wavelet function.

Layer 4: The output of this layer can be expressed as

$$y_l^R = \sum_j w_{lj}^R \Psi_j^R(x_r) \tag{3.39}$$

where w_{lj}^R is the weight between the rule layer and output layer. The vector form of the output of RFWNNs is denoted as

$$\boldsymbol{y_R}(\boldsymbol{x_r}, \boldsymbol{d_R}, \boldsymbol{c_R}, \boldsymbol{\pi_R}, \boldsymbol{W_R}) = \boldsymbol{W_R^T} \boldsymbol{\Psi_R}(\boldsymbol{x_R}, \boldsymbol{d_R}, \boldsymbol{c_R}, \boldsymbol{\pi_R}) \tag{3.40}$$

where $\boldsymbol{x} = [x_1, \ldots, x_n]^{\mathrm{T}} \in \mathbb{R}^{n \times 1}$, $\boldsymbol{y_R} = \left[y_1^R, \ldots, y_m^R\right]^{\mathrm{T}} \in \mathbb{R}^{m \times 1}$, $\boldsymbol{d_R} = \left[d_{11}^R, \ldots, d_{1n}^R, \ldots, d_{p1}^R, \ldots, d_{pn}^R\right]^{\mathrm{T}} \in \mathbb{R}^{pn \times 1}$,
$\boldsymbol{c_R} = \left[c_{11}^R, \ldots, c_{1n}^R, \ldots, c_{p1}^R, \ldots, c_{pn}^R\right]^{\mathrm{T}} \in \mathbb{R}^{pn \times 1}$,
$\boldsymbol{\pi_R} = \left[\pi_{11}^R, \ldots, \pi_{1n}^R, \ldots, \pi_{p1}^R, \ldots, \pi_{pn}^R\right]^{\mathrm{T}} \in \mathbb{R}^{pn \times 1}$,
$\boldsymbol{\Psi_R}(\boldsymbol{x_r}, \boldsymbol{d_R}, \boldsymbol{c_R}, \boldsymbol{\pi_R}) = \left[\Psi_1^R(x_r), \ldots, \Psi_p^R(x_r)\right]^{\mathrm{T}} \in \mathbb{R}^{p \times 1}$, and

$$\boldsymbol{W_R^T} = \begin{bmatrix} w_{11}^R & w_{12}^R & \cdots & w_{1p}^R \\ w_{21}^R & w_{22}^R & \cdots & w_{2p}^R \\ \vdots & \vdots & \ddots & \vdots \\ w_{m1}^R & w_{m2}^R & \cdots & w_{mp}^R \end{bmatrix} \in \mathbb{R}^{m \times p}.$$

The proposed RFWNNs are also used to deal with the unknown/uncertain dynamics of the robot manipulator system. Based on the universal approximation error analysis, there exists an ideal RFWNN structure with its optimal parameters to approximate the unknown function y_d, such that

$$y_d(x(t)) = \mathrm{W}_R^{*\top}\Psi_R\left(x_r(t), d_R^*, c_R^*, \pi_R^*\right) + \Delta_R(x_r(t)) \tag{3.41}$$

where $\mathrm{W}_R^*, d_R^*, c_R^*, \pi_R^*$ are the optimal parameters of the W_R, d_R, c_R, π_R, respectively, and $\boldsymbol{\Delta}_R(\boldsymbol{x}_r(t))$ is the approximation error vector.

Assumption 3.7 The norms of the RFWNN-optimal weights are bounded: $\|\mathrm{W}_R^*\| \le b_{\mathrm{W}}^R$, $\|d_R^*\| \le b_d^R$, $\|c_R^*\| \le b_c^R$, and $\|\pi_R^*\| \le b_\pi^R$, where $b_{\mathrm{W}}^R, b_d^R, b_c^R, \pi_c^R$ are the positive real values.

Assumption 3.8 Error of FWNN approximation process is bounded: $\|\Delta_{fw}\| \le b_\Delta^{fw}$, where b_Δ^{fw}, b_τ are the positive real values.

The proposed RFWNN and FWNN structures are a combination of fuzzy concepts, the theory of multiresolution analysis, and the recurrent scheme. Hence, besides the advantages of universal approximation, a few drawbacks still exist. The multiresolution of the RFWNNs/FWNNs may create the size of a huge network, and the selection of wavelons becomes difficult and important. The fuzzy controller with the linguistic rules alone may fail to control a system under some conditions. To overcome these difficulties, the rule bases of the RFWNNs/FWNNs need to be adjusted adaptively. Therefore, an approach for an adaptive algorithm for the dilation and translation parameters, the recurrent parameters, and the weights of the RFWNNs and FWNNs will be developed in the following parts.

3.3.3 *Adaptive Force/Motion Control Design on RFWNNs*

Since the mobile robot manipulator system is subjected to the nonholonomic constraint in (2.5), the states q_n, q_h are not independent. By a proper partition of q_n, ρ is uniquely determined by $\xi = \left[\rho^\top, q_h^\top\right]^\top$. Therefore, it is not necessary to consider the control of q_n.

The control goal is to find the adaptive control laws such that the joint position of the CCMMR/MMR $\xi = \left[\rho^\top, q_h^\top\right]^\top$ tracks the desired joint position $\xi_d = \left[\rho_d{}^\top, q_{hd}^\top\right]^\top$ and the nonholonomic constraint force f_n tracks the desired constraint force f_{nd} or, equivalently, the multiplier γ_n will converge to the desired multiplier γ_{nd}, with expectation that performance control achieves the highest possible accuracy. The desired reference trajectory ξ_d is assumed to be bounded and uniformly continuous. It also has bounded and uniformly continuous derivatives up to the second orders. γ_{nd} is also bounded and uniformly continuous. The structure of the proposed controllers is described in Fig. 3.8.

Define the tracking error vectors $\boldsymbol{e}_\xi(t)$, $\boldsymbol{e}_\gamma(t)$ and the filtered tracking error $\boldsymbol{s}_\xi$ as follows:

$$\boldsymbol{e}_\xi(t) = \xi_d - \xi \tag{3.42}$$

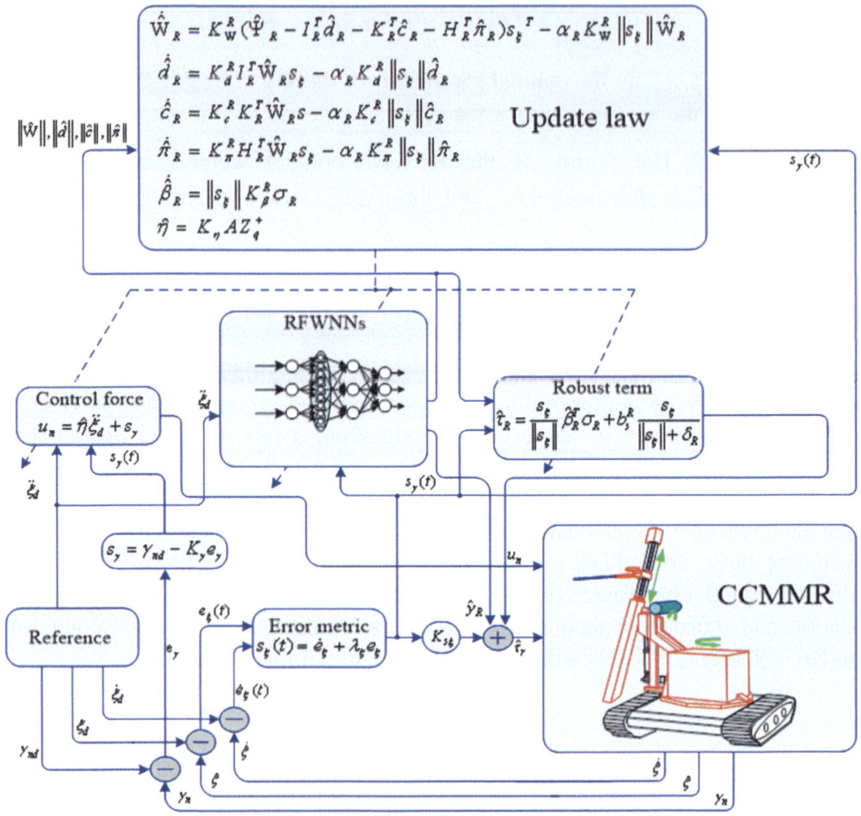

Fig. 3.8 Diagram of the proposed adaptive position/force controller based on RFWNNs

$$e_{\gamma}(t) = \gamma_n - \gamma_{nd} \tag{3.43}$$

$$s_{\xi}(t) = \dot{e}_{\xi} + \lambda_{\xi} e_{\xi} \tag{3.44}$$

where $\lambda_{\xi} = \mathrm{diag}\left(\lambda_1^{\xi}, \lambda_2^{\xi}, \ldots, \lambda_{n-m}^{\xi}\right)$ are positive constants.

In this section, a decoupling control scheme is introduced to control the generalized position and the constraint force, separately. Therefore, this section will consider the control input of the CCMMR/MMR control system in the following form:

$$u = u_{\tau} - A^{\mathrm{T}} u_n \tag{3.45}$$

where two terms on the right side of (3.45) are the control inputs for torque and nonholonomic constrained force.

According to (3.45), the dynamic model of the CCMMR/MMR can be rewritten as follows:

$$\boldsymbol{M}_{\xi}\ddot{\xi} + \boldsymbol{C}_{\xi}\dot{\xi} + \boldsymbol{G}_{\xi} + \tau_{\xi d} = \boldsymbol{Z}_q^{\mathrm{T}} u_{\tau} \tag{3.46}$$

$$\gamma_n = \Upsilon\left[\left(\boldsymbol{M}\dot{\boldsymbol{Z}}_q + \boldsymbol{C}\boldsymbol{Z}_q\right)\dot{\xi} + \boldsymbol{G} + \boldsymbol{\tau}_d - u_{\tau}\right] + u_n = -\Upsilon\boldsymbol{Z}_q^{\dagger^{\mathrm{T}}}\boldsymbol{M}_{\xi}\ddot{\xi} + u_n \tag{3.47}$$

where $\boldsymbol{Z}_q^{\dagger} = \left(\boldsymbol{Z}_q^{\mathrm{T}}\boldsymbol{Z}_q\right)^{-1}\boldsymbol{Z}_q^{\mathrm{T}}$.

From definitions (3.42)–(3.44), the dynamic model of the CCMMR/MMR in (3.46) and (3.47) can be represented as the following forms:

$$\boldsymbol{M}_{\xi}\dot{s}_{\xi} + (\boldsymbol{K}_{s\xi} + \boldsymbol{C}_{\xi})\boldsymbol{s}_{\xi} = y_{\xi} + \tau_{\xi d} - \boldsymbol{Z}_q^{\mathrm{T}} u_{\tau} + \boldsymbol{K}_{s\xi}\boldsymbol{s}_{\xi} \tag{3.48}$$

$$\boldsymbol{e}_{\gamma}\left(1 + \boldsymbol{K}_{\gamma}\right) = -\Upsilon\boldsymbol{Z}_q^{\dagger^{\mathrm{T}}}\boldsymbol{M}_{\xi}\ddot{\xi} + u_n - \gamma_{nd} + \boldsymbol{K}_{\gamma}\boldsymbol{e}_{\gamma} \tag{3.49}$$

where y_{ξ} is the function of $\xi(t), \dot{\xi}(t), \xi_d(t), \dot{\xi}_d(t), \ddot{\xi}_d(t)$ and it is defined as $y_{\xi} = M_{\xi}\left(\ddot{\xi}_d + \lambda_{\xi}\dot{e}_{\xi}\right) + C_{\xi}\left(\dot{\xi}_d + \lambda_{\xi}e_{\xi}\right) + G_{\xi}$, $K_{s\xi} = \boldsymbol{diag}\left\{k_1^{s\xi}, \ldots, k_{n-m}^{s\xi}\right\}$ is the positive constant matrix, and K_{γ} is the positive constant. If this section designs the controllers that can force the right hand of (3.48) and (3.49) to be zero, then the closed-loop control system ($\boldsymbol{M}_{\xi}\dot{s}_{\xi} + (\boldsymbol{K}_{s\xi} + \boldsymbol{C}_{\xi})\boldsymbol{s}_{\xi} = 0; \boldsymbol{e}_{\gamma}\left(1 + \boldsymbol{K}_{\gamma}\right) = 0$) can not only guarantee the stability of the CCMMR/MMR controlled system and the exponential of position tracking errors but also the tracking performance of γ_n. Hence, the ideal control laws for position/force tracking can be designed as the following forms:

$$\boldsymbol{Z}_q^{\mathrm{T}} u_{\tau}^{*} = \boldsymbol{y}_{\xi} + \boldsymbol{\tau}_{\xi d} + \boldsymbol{K}_{s\xi}\boldsymbol{s}_{\xi} \tag{3.50}$$

$$u_n^{*} = \Upsilon\boldsymbol{Z}_q^{+^{\mathrm{T}}}\boldsymbol{M}_{\xi}\ddot{\xi} + \gamma_{nd} - \boldsymbol{K}_{\gamma}\boldsymbol{e}_{\gamma} \tag{3.51}$$

Unfortunately, this section cannot implement these ideal feedback controllers, (3.50) and (3.51), for torque and constraint force control because of various kinds of uncertainties and the unknown knowledge of robot control systems. Therefore, this section needs adaptive/robust controllers that can online identify the unknown/uncertain parameters of the robot control system. Based on this analysis, an adaptive/robust controller based on RFWNNs, with adaptive online learning algorithms, will be proposed to approximate the unknown/uncertain parameters of the robot control system. In addition, with the existence of inevitable approximation errors and disturbances of the mobile manipulator system, a robust compensation term will be proposed to compensate for the existing problems. For the constraint force control problem, an adaptive controller will be subsequently proposed to address the problem of force control.

3.3.4 Tracking Position Controller Design

The ideal control law (3.50) depends on the unknown dynamics of the CCMMR/MMR model. The part y_ξ will be approximated by the proposed RFWNNs. In addition, a robust term is proposed to estimate and compensate for the $\tau_{\xi d}$ and the approximation errors. An actual torque control law is proposed as follows:

$$Z_q^{\mathrm{T}} u_\tau = \widehat{y}_R + K_{s\xi} s_\xi + \widehat{\tau}_R \tag{3.52}$$

where $\widehat{\tau}_R$ also is the robust term. The part $\widehat{y}_R$ is the RFWNN approximation function of the function y_ξ. The robust term $\widehat{\tau}_R$ and the online learning algorithms for the proposed RFWNN position controllers are proposed as follows:

$$\widehat{\tau}_R = \frac{s_\xi}{\|s_\xi\|} \widehat{\beta}_R^{\mathrm{T}} \sigma_R + b_{s\xi}^R \frac{s_\xi}{\|s_\xi\| + \delta_R} \tag{3.53}$$

$$\begin{cases} \dot{\widehat{\mathrm{W}}}_R = K_{\mathrm{W}}^R \left(\widehat{\Psi}_R - I_R^{\mathrm{T}} \widehat{d}_R - K_R^{\mathrm{T}} \widehat{c}_R - H_R^{\mathrm{T}} \widehat{\pi}_R \right) s_\xi^{\mathrm{T}} - \alpha_R K_{\mathrm{W}}^R \|s_\xi\| \widehat{\mathrm{W}}_R \\ \dot{\widehat{d}}_R = K_d^R I_R^{\mathrm{T}} \widehat{\mathrm{W}}_R s_\xi - \alpha_R K_d^R \|s_\xi\| \widehat{d}_R \\ \dot{\widehat{c}}_R = K_c^R K_R^{\mathrm{T}} \widehat{\mathrm{W}}_R s - \alpha_R K_c^R \|s_\xi\| \widehat{c}_R \\ \dot{\widehat{\pi}}_R = K_\pi^R H_R^{\mathrm{T}} \widehat{\mathrm{W}}_R s_\xi - \alpha_R K_\pi^R \|s_\xi\| \widehat{\pi}_R \\ \dot{\widehat{\beta}}_R = \|s_\xi\| K_\beta^R \sigma_R \end{cases} \tag{3.54}$$

From the analysis about the ideal controller for the constraint force in (3.51), $\Upsilon Z_q^{\dagger\mathrm{T}} M_\xi \ddot{\xi}$ is unknown. Therefore, an adaptive robust controller is developed to estimate $\Upsilon Z_q^{\dagger\mathrm{T}} M_\xi \ddot{\xi}$. Consider the adaptive robust control law as

$$u_n = \widehat{\eta} \ddot{\xi}_d + s_\gamma \tag{3.55}$$

where $s_\gamma = \gamma_{nd} - K_\gamma e_\gamma$ and $K_\gamma e_\gamma$ is the linear proportional derivative-type compensator. Then, $\widehat{\eta}$ is defined as

$$\widehat{\eta} = K_\eta A Z_q^\dagger \tag{3.56}$$

where K_η is the positive constant.

3.3.5 Stability Analysis

Theorem 3.2 By considering the CCMMR/MMR model in (2.11) and (2.12), all the assumptions for RFWNNs and CCMMR/MMR hold. If the control law for position tracking is (3.52), the control law for the constraint force is (3.55), the adaptive online learning algorithms for the RFWNN-based controller are designed as (3.54), and the adaptive robust update algorithm for the constraint force control is designed as (3.56), then:

1. The parameters of the RFWNNs and the approximation errors are bounded, and the filtered tracking error s_ξ converges to zero.
2. The variable e_γ and the control inputs are bounded for $t \geq 0$, and the stability of the proposed control system is guaranteed.

Proof (Theorem 3.2(1)) $s_\xi(t), \widetilde{\mathrm{W}}_R, \widetilde{d}_R, \widetilde{c}_R, \widetilde{\pi}_R, \widetilde{\beta}_R$ are bounded for $t \geq 0$. The filtered tracking error $s_\xi(t)$ will converge to zero as time tends to infinity.

Proof (Theorem 3.2(2)) By substituting (3.55) into (3.51), one can obtain

$$(I + K_\gamma)e_\gamma = -\Upsilon Z_q^{\dagger^{\mathrm{T}}} M_\xi \ddot{\xi} + \widehat{\eta}\ddot{\xi}_d \tag{3.57}$$

Based on the results of proof (Theorem 3.2(1)), $\Upsilon, \boldsymbol{Z}_q, \boldsymbol{M}_\xi, \ddot{\xi}_d$ and $-\Upsilon \boldsymbol{Z}_q^{\dagger^{\mathrm{T}}} \boldsymbol{M}_\xi \ddot{\xi} + \widehat{\eta}\ddot{\xi}_d$ are bounded. Therefore, the size of $\boldsymbol{e}_\gamma$ can be adjusted by selecting the proper gain $\boldsymbol{K}_\gamma$, $\boldsymbol{K}_\eta$. Here, it is easy to conclude that the control inputs are bounded from (3.52) and (3.55). Based on proof (Theorem 3.2(1)) and proof (Theorem 3.2(2)) results, this section can conclude that the stability of the proposed control system is guaranteed.

In the above analysis, the requirement for bounded uncertainties in the proposed control robot system is relaxed. Here, the drawbacks of the previous controller-based NNs for the MMR, the problems of bounded parameters, have been eliminated. The assumptions for the bounds of parameters are used to analyze the stability of the control system, and these assumptions disappear in this designed controller. In addition, according to [37], it is easy to show that the proposed control system, with adaptive online update laws, is strictly state-passive. This advantageous feature can be used to conclude that some internal boundlessness properties of the controlled system without the assumptions of observability and stability.

3.3.6 Simulation Results

For the sake of simplicity in the simulation process, this section considers the CCMMR dynamics as a 2-DOF MMR model, as shown in Fig. 3.9, to verify the effectiveness of the proposed method. In this MMR model, two links of the manipulator part are considered as link 1 and link 2 of the CCMMR's manipulators. The detailed mechanical parameters of this MMR are similar to those of the CCMMR. The dynamics of this MMR are given as follows:

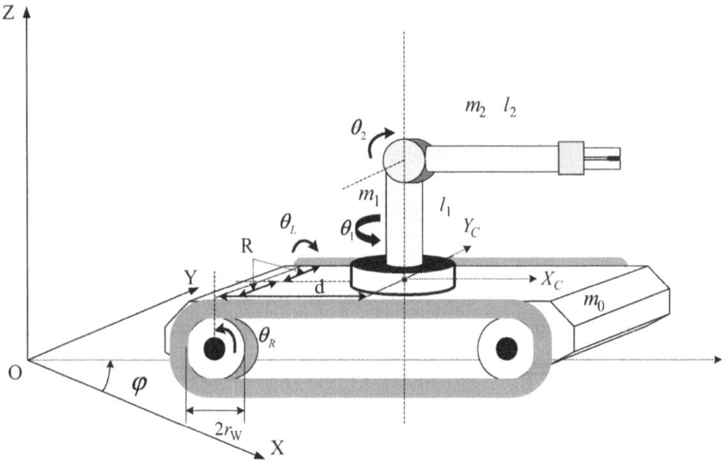

Fig. 3.9 The 2-DOF MMR model

$$\boldsymbol{q} = [\boldsymbol{q}_n, \boldsymbol{q}_h]^{\mathrm{T}}, \boldsymbol{q}_n = [\boldsymbol{X}_C, \boldsymbol{Y}_C, \boldsymbol{\varphi}]^{\mathrm{T}}, \boldsymbol{q}_h = [\theta_1, \theta_2]^{\mathrm{T}}$$

$$\boldsymbol{M}_n = \begin{bmatrix} m_{n11} & m_{n12} & m_{n13} \\ m_{n21} & m_{n22} & m_{n23} \\ m_{n31} & m_{n32} & m_{n33} \end{bmatrix}, \boldsymbol{M}_{nh} = \begin{bmatrix} 0 & 0 \\ 0 & 0 \\ I_1 + I_2 & 0 \end{bmatrix}, \boldsymbol{M}_{hn} = \begin{bmatrix} 0 & 0 & I_1 + I_2 \\ 0 & 0 & 0 \end{bmatrix},$$

$$\boldsymbol{M}_h = \begin{bmatrix} I_1 & 0 \\ 0 & I_2 \end{bmatrix}$$

$$\boldsymbol{C}_n = \begin{bmatrix} c_{n11} & c_{n12} & c_{n13} \\ c_{n21} & c_{n22} & c_{n23} \\ 0 & 0 & 0 \end{bmatrix}, \boldsymbol{C}_{nh} = \begin{bmatrix} 0 & 0 \\ 0 & 0 \\ 0 & 0 \end{bmatrix}, \boldsymbol{C}_h = \begin{bmatrix} 0 & 0 \\ 0 & 0 \end{bmatrix}, \boldsymbol{C}_{hn} = \begin{bmatrix} 0 & 0 & 0 \\ 0 & 0 & 0 \end{bmatrix}$$

$$\boldsymbol{B} = \frac{1}{r} \begin{bmatrix} \cos(\varphi) & \cos(\varphi) & 0 & 0 \\ \sin(\varphi) & \sin(\varphi) & 0 & 0 \\ R & -R & 0 & 0 \\ 0 & 0 & r_{\mathrm{W}} & 0 \\ 0 & 0 & 0 & r_{\mathrm{W}} \end{bmatrix}, \boldsymbol{G}_n = [0 \quad 0 \quad 0]^{\mathrm{T}}, \boldsymbol{G}_h = [0 \quad 0.5 g m_2 l_2 \sin(\theta_2)]^{\mathrm{T}}$$

where $m_{n11} = m_0 + m_1 + m_2 + \frac{2I_p \sin^2(\varphi)}{r_{\mathrm{W}}^2}$, $m_{n12} = -\frac{2I_p \sin^2(\varphi)}{r_{\mathrm{W}}^2}$, $m_{n13} = -(m_1 + m_2) d \sin(\varphi)$, $m_{n21} = m_{n12}$, $m_{n22} = m_{n11}$, $m_{n23} = m_{n13}$, $m_{n31} = m_{n13}$, $m_{n32} = m_{n23}$, $m_{n33} = I_p + I_1 + I_2 + (m_1 + m_2) d^2 + \frac{2I_b R^2}{r_{\mathrm{W}}^2}$, $c_{n11} = \frac{2I_p \dot{\varphi} \sin(\varphi)\cos(\varphi)}{r_{\mathrm{W}}^2}$, $c_{n12} = \frac{2I_p \dot{\varphi} \sin^2(\varphi)}{r_{\mathrm{W}}^2}$, $c_{n21} = c_{n12}$, $c_{n13} = -(m_1 + m_2) d\dot{\varphi}\cos(\varphi)$, $c_{n22} = m_0 + m_1 + m_2 + \frac{2I_p \sin^2(\varphi)}{r_{\mathrm{W}}^2}$, $c_{n23} = c_{n13}$, $m_p \equiv m_0$, and $g = 9.8$.

This 2-DOF MMR is subjected to nonholonomic constraints that are given as

$$\begin{cases} -\dot{X}_C \sin(\varphi) + \dot{Y}_C \cos(\varphi) = 0 \\ \dot{X}_C \cos(\varphi) + \dot{Y}_C \cos(\varphi) - R\dot{\varphi} - r_W\dot{\theta}_L = 0 \\ \dot{X}_C \cos(\varphi) + \dot{Y}_C \cos(\varphi) + R\dot{\varphi} - r_W\dot{\theta}_R = 0 \end{cases}$$

Therefore, the nonholonomic constraint matrix is described as

$$\boldsymbol{A}_n = [-\sin(\varphi), \cos(\varphi), 0]$$

The vector $\boldsymbol{\rho}$ is defined as $\boldsymbol{\rho} = [\theta_L, \theta_R]$ and its derivation $\dot{\boldsymbol{\rho}}$ is $\dot{\boldsymbol{\rho}} = \left[\dot{\theta}_L, \dot{\theta}_R\right]$. $\boldsymbol{Z}(\boldsymbol{q}_n)$ is determined as

$$\boldsymbol{Z}(\boldsymbol{q}_n) = \begin{bmatrix} \frac{r_W}{2}\cos(\varphi) & \frac{r_W}{2}\cos(\varphi) \\ \frac{r_W}{2}\sin(\varphi) & \frac{r_W}{2}\sin(\varphi) \\ -\frac{r_W}{R} & \frac{r_W}{R} \end{bmatrix}$$

In order to exhibit the superior control performance and effectiveness of the proposed scheme, the PID controller (PIDC), the NN controller, and the FWNN controller are examined in the meanwhile. The PIDC law can be expressed as

$$u_\tau = K_{p\tau} e_\xi(t) + K_{i\tau} \int_0^t e_\xi(t)\mathrm{dt} + K_{d\tau}\dot{e}_\xi(t)$$

where $\boldsymbol{K}_{p\tau}$, $\boldsymbol{K}_{i\tau}$, $\boldsymbol{K}_{d\tau}$ denote proportional, integral, and differential gain matrices, respectively, and they are designed by a compromise between the superiority of control performance and the magnitude of control effort. $K_{p\tau}$, $K_{i\tau}$, $K_{d\tau}$ are given as 1.

The structure of the proposed RFWNNs can be characterized by $n_i = 20, p = 5$, and $n_o = 4$ nodes; n_i and n_o are the numbers of the inputs and outputs of the RFWNNs, respectively. The control parameters of the proposed RFWNNs are given as follows:

$$\begin{cases} K_{s\xi} = \mathrm{diag}(300, 300, 300, 300), \lambda_\xi = \mathrm{diag}(10, 10, 20, 20) \\ K_{\mathrm{W}}^R = \mathrm{diag}(30, 30, 30, 30, 30), K_{\mathrm{d}}^R = K_{\mathrm{c}}^R = K_\pi^R = \mathrm{diag}(20, \ldots, 20, \ldots, 20) \in \mathbb{R}^{n_ip \times n_ip} \\ K_\beta^R = \mathrm{diag}(0.001, 0.001, 0.001, 0.001, 0.001), \alpha_R = 0.001, K_\eta = 0.01, K_\gamma = 0.02 \\ \sigma_R = 0.001, \delta_R = 0.001, b_{s\xi}^R = 0.01 \end{cases}$$

The structure of the proposed FWNNs can be characterized by $n_i = 20, p = 7$, and $n_o = 4$nodes. Similarly, the control parameters of the FWNNs of the CCMMR/MMR are given as follows:

$$\begin{cases} K_{s\xi} = \mathrm{diag}(300, 300, 300, 300), \lambda_\xi = \mathrm{diag}(10, 10, 20, 20) \\ K_{\mathrm{W}}^{fw} = \mathrm{diag}(30, 30, 30, 30, 30, 30, 30), K_{\mathrm{d}}^{fw} = K_{\mathrm{c}}^{fw} = \mathrm{diag}(20, \ldots, 20, \ldots, 20) \in \mathbb{R}^{n_i p \times n_i p} \\ K_\beta^{fw} = \mathrm{diag}(0.001, 0.001, 0.001, 0.001, 0.001), \alpha_{fw} = 0.001, K_\eta = 0.01, K_\gamma = 0.02 \\ \sigma_{fw} = 0.001, \delta_{fw} = 0.001, b_{s\xi}^{fw} = 0.01 \end{cases}$$

For recording the respective control performance, the mean square error (MSE) of the position tracking response is defined as

$$MSE = \frac{1}{T} \sum_{k=1}^{T} [\xi(k) - \xi_d(k)]^2$$

where T is the total sampling instant. Based on this definition, the normalized MSE (NMSE) value of the position tracking response using a per-unit value with 1 rad is used to examine the control performance. The desired joint positions and the nonholonomic constraint are considered as

$$\begin{cases} \xi_d = [4t, 4.5t, \sin(2\pi t/4), \sin(2\pi t/5)](\mathrm{rad}) \\ \gamma_{nd} = 10(\mathrm{N}) \end{cases}$$

To investigate the robustness of the proposed scheme, two simulation cases including the parameter variations and disturbances are considered as follows:

Case 1
With $\tau_d = [0.1\sin(2t); 0.2\cos(2t); 0.4\sin(2t); 0.1\sin(2t); 0.2\cos(2t)]$, there are no tip loads on the links.

Case 2
With $\tau_d = 30[\sin(20t); \cos(20t); \sin(20t); \sin(20t); \cos(20t)]$, tip loads (4.5 kg, 2 kg, and 50 kg on link 1, link 2, and the mobile platform, respectively) will occur at 8 s of the simulation process.

In addition, the friction term is also considered in this simulation process:

$$F_\xi = \left[10\dot{\xi}_1 + 0.2\mathrm{sign}(\dot{\xi}_1); 10\dot{\xi}_2 + 0.5\mathrm{sign}(\dot{\xi}_2); 5\dot{\xi}_3 + 0.2\mathrm{sign}(\dot{\xi}_3); 5\dot{\xi}_4 + 0.2\mathrm{sign}(\dot{\xi}_4)\right]$$

The simulation is carried out by using the MATLAB package. The time sample for the simulation process is 0.001 s.

The simulated comparisons of PIDC, NNs, FWNNs, and the proposed scheme are depicted in Figs. 3.10 and 3.12 (for case 1) and Figs. 3.11 and 3.13 (for case 2). The simulated comparison NMSE values of each method are also presented in Table 3.2. In cases 1 and 2, good tracking position can be obtained with the RFWNN, FWNN, PIDC, and NN methods. However, the tracking errors of the proposed RFWNN and FWNN strategies converge faster than that of the NN and PIDC methods. In addition, based on the NMSE measures in Table 3.2, the proposed RFWNN strategy

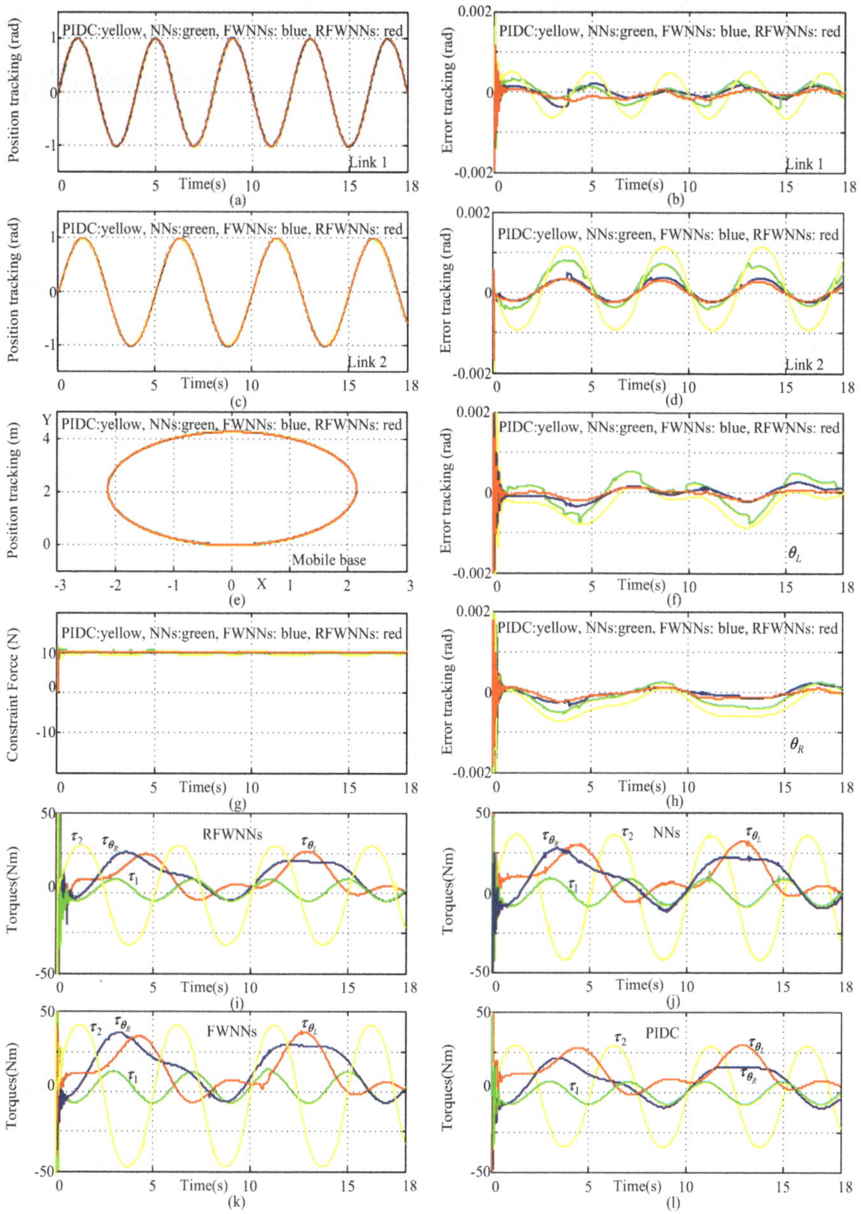

Fig. 3.10 Simulation results with case 1: the tracking positions, the position tracking errors, and the control inputs. (**a**, **c**, and **e**) Tracking position performances. (**b**, **d**, **f**, and **h**) Tracking error performances. (**g**) Constraint force input. (**i–l**) Torque control inputs of RFWNNs, FWNNs, FNNs, and PIDC methods

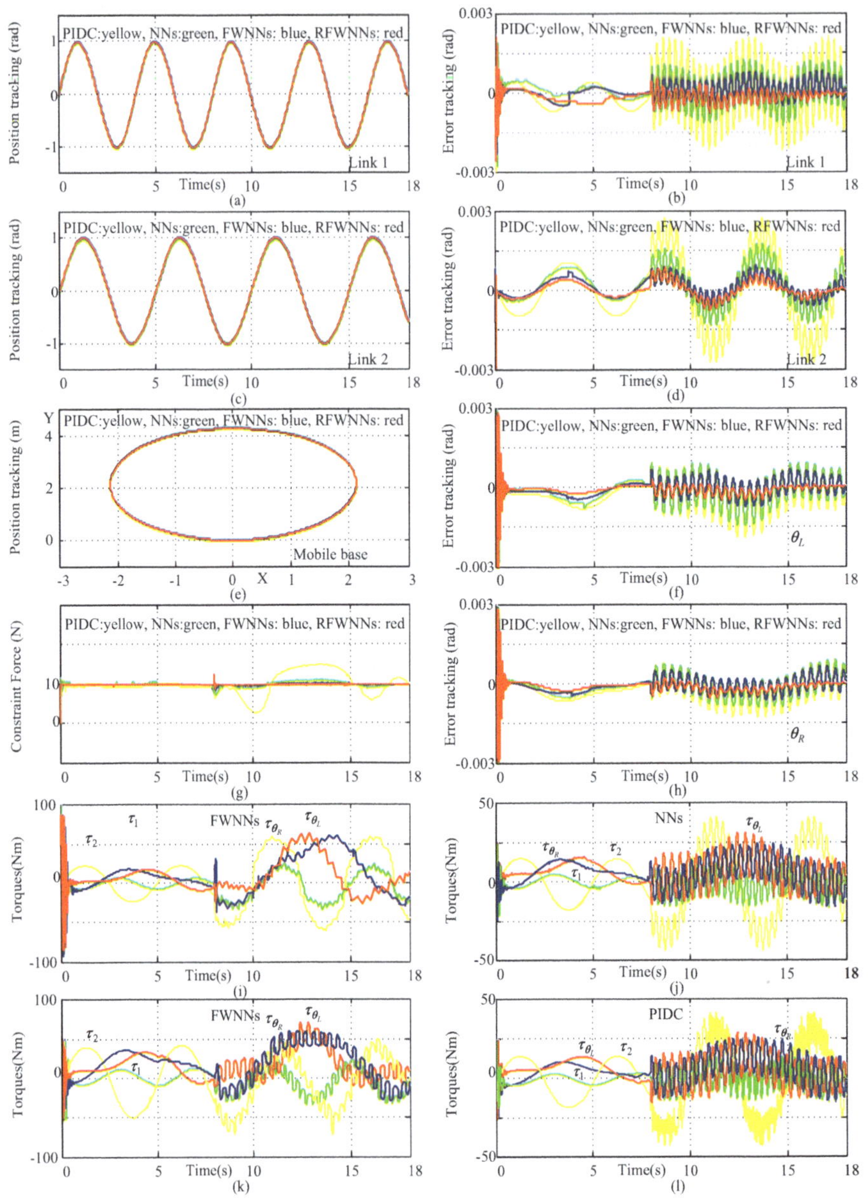

Fig. 3.11 Simulation results with case 2: the tracking positions, the position tracking errors, and the control inputs. (**a**, **c**, and **e**) Tracking position performances. (**b**, **d**, **f**, and **h**) Tracking error performances. (**g**) Constraint force input. (**i**–**l**) Torque control inputs of RFWNNs, FWNNs, FNNs, and PIDC methods

Table 3.2 Simulation performance comparisons of RFWNN, FWNN, NN, and PIDC schemes

Simulation NMSE ($\times 10^{-4}$)	Case 1 (unit: rad)				Case 2 (unit: rad)			
	θ_L	θ_R	Link 1	Link 2	θ_L	θ_R	Link 1	Link 2
RFWNNs	1.641	1.837	0.664	0.750	1.753	1.880	0.894	0.925
FWNNs	1.772	1.937	0.778	0.849	1.879	2.021	1.094	0.972
NNs	2.327	2.631	1.775	1.241	2.712	2.837	1.935	1.365
PIDC	2.442	2.708	1.925	2.416	2.928	2.984	3.306	4.230

has position tracking improvements better than that of the FWNN, PIDC, and NN schemes. Figure 3.10i–l (for case 1) and Fig. 3.11i–l (for case 2) represent the torque control inputs of the RFWNN, FWNN, FNN, and PIDC methods. In case 1, the performances of control torque inputs of all the methods are good. In case 2, while the proposed RFWNN, the FWNN, and the NN strategies can show good torque input performance at the parameter variation conditions (higher disturbance frequency, changing load on links), the PIDC torque input performance has occurred chattering phenomena. The performances of the PIDC, shown in these figures, are the best performance in terms of handling the heuristic process, but the chattering phenomena still exist when the frequency of the external disturbances increases or a load of manipulators is changed suddenly. In the simulation process of NNs, some control parameters, such as the constraint bounds of tracking errors and the number of nodes of NNs, are not easy to determine to achieve better performances.

The simulation comparisons with the constraint force of all the methods are presented in Figs. 3.10g and 3.11g. The performance of constraint force control depends on the performance of tracking position control. The design for the constraint force control in the NN controller is the same as the RFWNNs/FWNNs. However, the design for the constraint force in the PIDC scheme is different; one more PIDC for the constraint force control is considered. In case 1, the performance of the constraint force control of the RFWNN, FWNN, and NN methods is good, but the simulation result of the PIDC is poor in the steady state. In case 2, while the proposed and FWNN methods can show good constraint force performance at the parameter variation conditions, the simulation results of the NN and PIDC methods are not good, and the undesired response appears when the time increases. The robust characteristic of the proposed RFWNN and FWNN controllers can be set under the parameter variation and external disturbances.

The simulated results of the robust term, approximating functions, and estimated parameters of the RFWNN and FWNN methods (for cases 1 and 2) are depicted in Figs. 3.12 and 3.13. The performances of the approximating functions and estimated parameters are good, and they have proved the correctness of the proposed method including the boundedness of the control system parameters. In case 2, this section can observe that the approximating capability of the proposed method (the estimated parameters and approximating functions), as shown in Fig. 3.13c–f, is better than the FWNN method under variation conditions. Based on the comparison simulations, the proposed RFWNN controller is more suitable to be implemented to control the

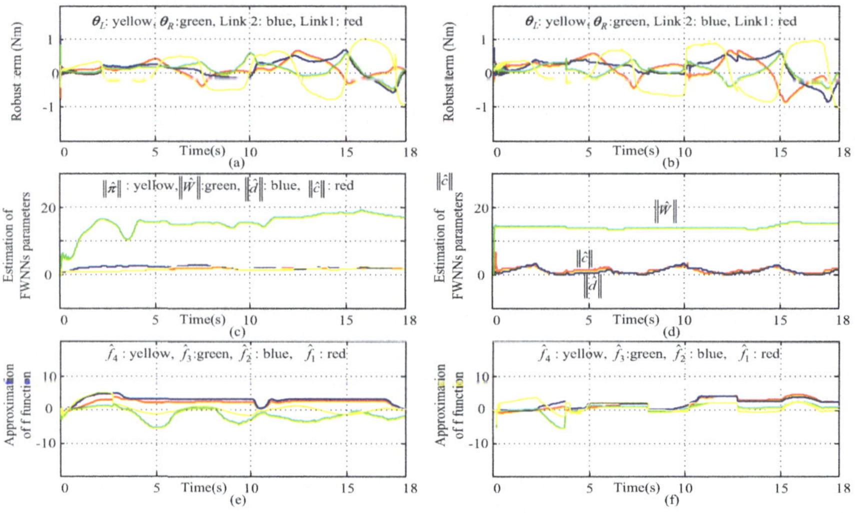

Fig. 3.12 Simulation results with case 1: the estimated parameters, the approximation functions, and the robust terms. (**a**) and (**b**) Robust term performances of RFWNN and FWNN methods. (**c**) and (**d**) Estimated parameter performances of RFWNN and FWNN methods. (**e**) and (**f**) Approximation functions of RFWNN and FWNN methods

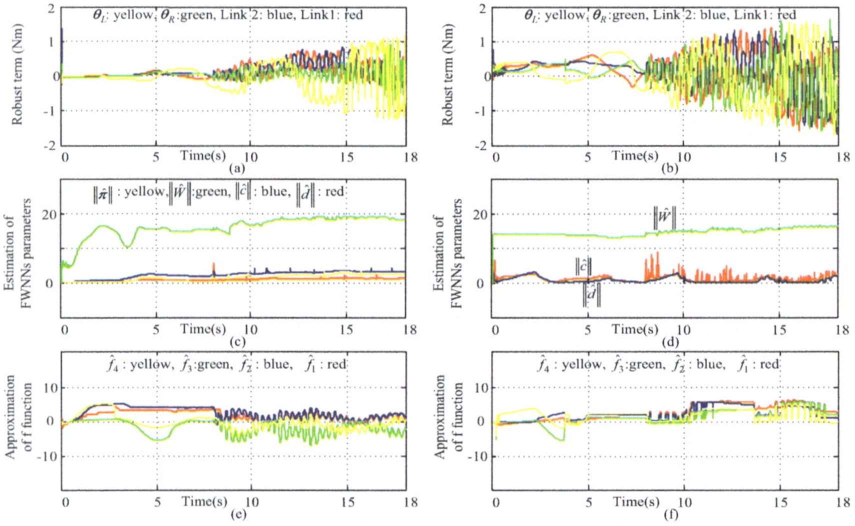

Fig. 3.13 Simulation results with case 2: the estimated parameters, the approximation functions, and the robust terms. (**a**) and (**b**) Robust term performances of RFWNN and FWNN methods. (**c**) and (**d**) Estimated parameter performances of RFWNN and FWNN methods. (**e**) and (**f**) Approximation functions of RFWNN and FWNN methods

CCMMR/MMR under the occurrence of parameter variation and external disturbances.

3.4 Intelligent Force/Motion Tracking Control with Recurrent Fuzzy Wavelet CMAC Neural Networks

In this section, an intelligent adaptive motion/force control strategy based on the recurrent fuzzy wavelet cerebellar model articulation control neural networks (RFWCMACNNs) is proposed for the nonholonomic CCMMR/MMR system. Based on the advantages of the RFWCMACNNs, the unknown dynamics and parameter variations of the CCMMR/MMR control system are relaxed by an approximation process. An adaptive robust compensator is also used to eliminate uncertainties that consist of unknown approximation errors and uncertain disturbances. According to the adaptive position tracking control design, an adaptive robust control scheme is also developed for the nonholonomic constraint force control. The design of the adaptive online learning algorithms for the proposed control system is obtained by using the Lyapunov stability theorem. Therefore, the proposed controllers prove that they not only can guarantee stability and robustness but also the tracking performance of the CCMMR/MMR. In addition, this section also summarizes the control structure of controller-based FWCMACNNs to compare and enhance the proposed RFWCMACNN control structure. The effectiveness and robustness of the proposed methods are verified by comparative simulation and experimental results that are examined in the CCMMR control system.

The proposed TSK RFWCMACNNs' strategy is to combine the TSK fuzzy system with the WNNs, the recurrent units, and CMAC. The TSK fuzzy wavelet CMAC neural network (WCMACNN) model is constructed on the base of a set of TSK fuzzy rules. Each fuzzy rule is considered as a WCMACNN subspace. In this TSK RFWCMACNN model, each WCMACNN subspace consists of wavelets with a specified dilation value (resolution). Hence, the WCMACNN subspace at different resolution levels is able to capture different behaviors of the approximated functions. Here, the role of the fuzzy set is to determine the contribution of the WCMACNN subspace to the input and output space of the RFWCMACNNs. In this proposed RFWCMACNN structure, each output of the RFWCMACNNs contains a fuzzy wavelet basis function with multiresolution capability. In addition, to overcome the static problem of conventional CMAC and improve the adaptive ability of the FWCMACNN structure, the recurrent method is also applied. With the main features stated above, the proposed RFWCMACNNs first can improve the approximation function's accuracy in terms of dilation and translation parameters of the wavelets without increasing the number of wavelet bases [38], or it can achieve the same approximation accuracy with a smaller size in comparison with the FNNs/NNs. Second, the appearance of feedback loops with time delay at the CMACs' associate

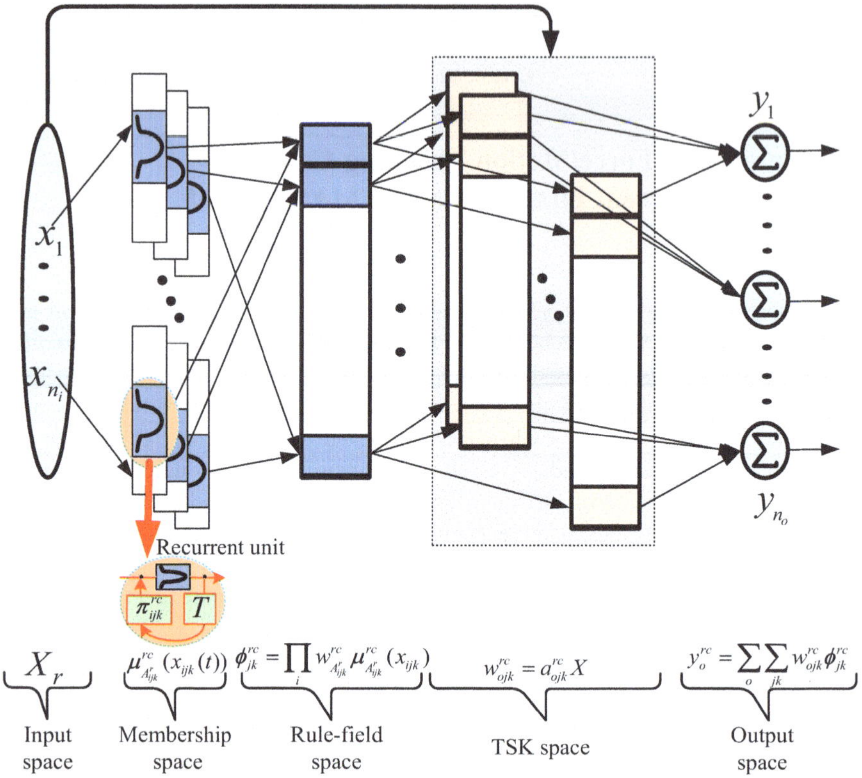

Fig. 3.14 The proposed RFWCMACNN structure

memory will make the RFWCMACNNs become a dynamic structure that can enhance the power to deal with highly coupled dynamics of the CCCMMR/MMR control system in the presence of the time-varying parameter uncertainties.

Figure 3.14 shows the structure of the proposed RFWCMACNNs, which is explained as follows:

Ruler: If x_1 is $\mu^{rc}_{A^r_{ijk}}(x_1)$ and x_2 is $\mu^{rc}_{A^r_{ijk}}(x_2)$, … , x_{n_i} is $\mu^{rc}_{A^r_{ijk}}(x_{n_i})$, then

$$\mathrm{w}^{rc}_{ojk} = a^{rc_o}_{1jk}x_1 + a^{rc_o}_{2jk}x_2 + \ldots + a^{rc_o}_{n_ijk}x_{n_i} \tag{3.58}$$

where $j = 1, \ldots, n_l$, $k = 1, \ldots, n_b$, $i = 1, \ldots, n_i$, $r = 1, \ldots, n_r$, and $o = 1, \ldots, n_o$. $n_i, n_l, n_b, n_o, n_r \in \mathbb{N}$, where n_i is the number of inputs, n_l is the number of layers for each input, n_b is the number of blocks for each layer, n_o is the number of outputs, n_r is the number of fuzzy rules, and $n_r = n_l n_b$. x_i is the i-th fuzzy input. $\mu^{rc}_{A^r_{ijk}}(x_i)$ is the fuzzy membership function for the i-th input, j-th layer, and k-th block. w^{rc}_{ojk} is the TSK-type output weight for the TSK fuzzy output. $a^{rc_o}_{n_ijk}$ is the adjustable factor of the consequence part of fuzzy rules. The proposed RFWCMACNN dynamic model is composed of the input, membership, rule-field, TSK weight memory, and fuzzy output spaces. The detail of the spaces is described as follows:

Input space: For a given input signal $X_r = [x_1, x_2, \ldots, x_{n_i}]^\mathrm{T} \in \mathbb{R}^{n_i \times 1}$, where n_i is the number of the input space (input layer), X_r can be quantized into discrete regions and moved to the next space by the neurons. The number of regions (elements) is represented by n_E (resolution).

Membership space: In this space, some elements can be accumulated as a block that can also be represented as a receptive field basic function. Each block is the input linguistic term, which is performed by its corresponding fuzzy membership wavelet function $\mu^{rc}_{A^r_{ijk}}(x_i)$. Similarly, this section selects the Mexican hat wavelet function to apply in the receptive field basic functions, and the $\mu^{rc}_{A^r_{ijk}}(x_i)$ is calculated by the following formula:

$$\mu^{rc}_{A^r_{ijk}}(x_i) = \left[1 - \left(d^{rc}_{ijk}\right)^2 \left(x_i - c^{rc}_{\mathrm{ijk}}\right)^2\right] e^{-\left(d^{rc}_{\mathrm{ijk}}\right)^2 \left(x_i - c^{rc}_{\mathrm{ijk}}\right)^2} \tag{3.59}$$

where d^{rc}_{ijk} is the dilation parameter and c^{rc}_{ijk} is the translation parameter. Moreover, a local feedback unit with the real time delay method will be added in this layer. Therefore, the input of this space (layer) will be represented as

$$x_{\mathrm{ij}k}(t) = x_i(t) + \pi^{rc}_{\mathrm{ij}k} \mu^{rc}_{A^r_{ijk}}(x_i(t - T)) \tag{3.60}$$

where the membership function $\mu^{rc}_{A^r_{ijk}}(x_i(t-T))$ represents the time delay value of $\mu^{rc}_{A^r_{ijk}}(x_i(t))$ via an interval T. The dynamic of the CMAC system is existent because the input of receptive field contains the past information of the network. The $\pi^{rc}_{\mathrm{ij}k}$ is the link weight (recurrent weight) of the feedback unit.

Rule-field space (receptive field): Areas formed by blocks. Each block is expressed as a rule field. The AND operator is expressed to calculate the outputs of this space:

$$\phi^{rc}_{jk} = \prod_i \mathrm{w}^{rc}_{A^r_{ijk}} \mu^{rc}_{A^r_{ijk}}\left(x_{\mathrm{ij}k}(t)\right) \tag{3.61}$$

The operation of the product $\prod$ can be referred to a fuzzy inference mechanism, and it is used to determine the firing strength. The $\mathrm{w}^{rc}_{A^r_{ijk}}$ presents a weight memory between the membership space and the rule-field space, and it is assumed to be unity. From (3.59), it can show that $\mu^{rc}_{A^r_{ijk}}\left(x_{\mathrm{ij}k}(t)\right)$ is the wavelet basis function. The ϕ^{rc}_{jk} can be represented as the multidimensional wavelet rule-receptive-field function. Multidimensional vector form is described as the following form:

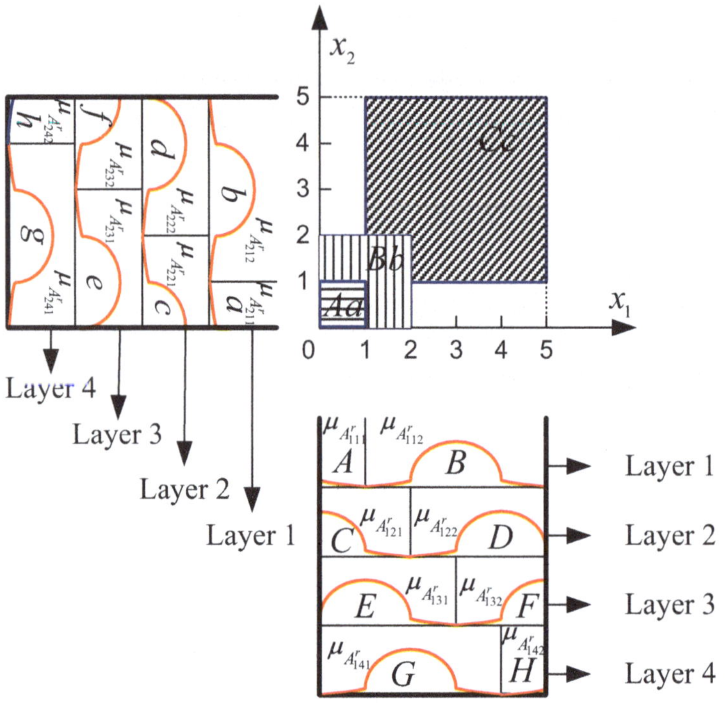

Fig. 3.15 Two-dimensional RFWCMACNNs

$$\phi_{rc} = \left[\phi_{11}^{rc}, \ldots, \phi_{1n_b}^{rc}, \phi_{21}^{rc}, \ldots, \phi_{2n_b}^{rc}, \ldots, \phi_{n_l 1}^{rc}, \ldots, \phi_{n_l n_b}^{rc}\right]^{\mathrm{T}} \in \mathbb{R}^{n_l n_b \times 1} \tag{3.62}$$

In this layer, $\phi_{jk}^{rc}\left(x_{ijk}(t)\right)$ is associated with the j layer and the k block. Figure 3.15 depicts the schematic diagram of two-dimensional RFWCMACNNs' operation with four layers $n_l = 4$ and two blocks $n_b = 2$ in each layer. Areas formed by block are called rule fields. A number of rule fields are equal to $n_r = n_l n_b$.

TSK space (TSK weight memory): The TSK space describes the linear combination functions in the consequent part of the fuzzy system. The fuzzy weight memory can be presented as follows:

$$\mathrm{W}_{rc} = \left[o_{1jk}^{rc}, o_{2jk}^{rc}, \ldots, o_{ojk}^{rc}, \ldots, o_{n_o jk}^{rc}\right]^{\mathrm{T}} \in \mathbb{R}^{n_l n_b \times n_o} \tag{3.63}$$

where $\mathrm{o}_{ojk}^{rc} = \left[\mathrm{w}_{o11}^{rc}, \ldots, \mathrm{w}_{o1n_b}^{rc}, \mathrm{w}_{o21}^{rc}, \ldots, \mathrm{w}_{o2n_b}^{rc}, \ldots, \mathrm{w}_{on_l 1}^{rc}, \ldots, \mathrm{w}_{on_l n_b}^{rc}\right]^{\mathrm{T}} \in \mathbb{R}^{n_l n_b \times 1}$, $o = 1 \ldots n_o$. Following the definition of TSK fuzzy rules (3.58), $\mathrm{w}_{ojk}^{rc} = a_{1jk}^{rc_o} x_1 + a_{2jk}^{rc_o} x_2 + \ldots + a_{n_i jk}^{rc_o} x_{n_i}$. For the purpose of simplicity in the calculating process, assume that $a_{1jk}^{rc_o} = a_{2jk}^{rc_o} = \ldots = a_{n_i jk}^{rc_o} = a_{ojk}^{rc}$. So, w_{ojk}^{rc} can be represented as

$$\mathrm{w}_{ojk}^{rc} = a_{ojk}^{rc}(x_1 + x_2 + \ldots + x_{n_i}) = a_{ojk}^{rc} X \tag{3.64}$$

where $X = x_1 + x_2 + \ldots + x_{n_i}$.

Output space (fuzzy output layer): Each node in this layer expresses the output linguistic variable, and it is computed as the summation of the activated weight with the rule-field space

$$y_o^{rc} = \sum_o \sum_{jk} \mathrm{w}_{ojk}^{rc} \phi_{jk}^{rc} \tag{3.65}$$

In addition, it can be denoted as the following vector form:

$$\boldsymbol{y}_{rc}(x, \boldsymbol{d}_{rc}, \boldsymbol{c}_{rc}, \boldsymbol{\pi}_{rc}, \mathrm{W}_{rc}) = \mathbf{W}_{rc}^{\mathrm{T}} \phi_{rc}(x, \boldsymbol{d}_{rc}, \boldsymbol{c}_{rc}, \boldsymbol{\pi}_{rc}) = \left(\boldsymbol{A}_w^{rc} X\right)^{\mathrm{T}} \phi_{rc}(x, \boldsymbol{d}_{rc}, \boldsymbol{c}_{rc}, \boldsymbol{\pi}_{rc}) \tag{3.66}$$

where $\boldsymbol{A}_w^{rc}$ is the TSK weight matrix and $\boldsymbol{d}_{rc}$, $\boldsymbol{c}_{rc}$, $\boldsymbol{\pi}_{rc}$ are the translation parameter vector, dilation parameter vector, and recurrent parameter vector, respectively. Vector forms are denoted as

$$\begin{cases} \boldsymbol{d}_{rc} = \left[\left(d_{1jk_k}^{rc}\right)^{\mathrm{T}}, \left(d_{2jk}^{rc}\right)^{\mathrm{T}}, \ldots, \left(d_{ijk}^{rc}\right)^{\mathrm{T}}, \ldots, \left(d_{n_ijk}^{rc}\right)^{\mathrm{T}}\right]^{\mathrm{T}} \in \mathbb{R}^{n_i n_l n_b \times 1} \\ \boldsymbol{d}_{ijk}^{rc} = \left[d_{i11}^{rc}, \ldots, d_{i1n_b}^{rc}, d_{i21}^{rc}, \ldots, d_{i2n_b}^{rc}, \ldots, d_{in_l1}^{rc}, \ldots, d_{in_ln_b}^{rc}\right]^{\mathrm{T}} \in \mathbb{R}^{n_l n_b \times 1} \end{cases} \tag{3.67}$$

$$\begin{cases} \boldsymbol{c}_{rc} = \left[\left(c_{1jk_k}^{rc}\right)^{\mathrm{T}}, \left(c_{2jk}^{rc}\right)^{\mathrm{T}}, \ldots, \left(c_{ijk}^{rc}\right)^{\mathrm{T}}, \ldots, \left(c_{n_ijk}^{rc}\right)^{\mathrm{T}}\right]^{\mathrm{T}} \in \mathbb{R}^{n_i n_l n_b \times 1} \\ \boldsymbol{c}_{ijk}^{rc} = \left[c_{i11}^{rc}, \ldots, c_{i1n_b}^{rc}, c_{i21}^{rc}, \ldots, c_{i2n_b}^{rc}, \ldots, c_{in_l1}^{rc}, \ldots, c_{in_ln_b}^{rc}\right]^{\mathrm{T}} \in \mathbb{R}^{n_l n_b \times 1} \end{cases} \tag{3.68}$$

$$\begin{cases} \boldsymbol{\pi}_{rc} = \left[\left(\pi_{1jk}^{rc}\right)^{\mathrm{T}}, \left(\pi_{2jk}^{rc}\right)^{\mathrm{T}}, \ldots, \left(\pi_{ijk}^{rc}\right)^{\mathrm{T}}, \ldots, \left(\pi_{n_ijk}^{rc}\right)^{\mathrm{T}}\right]^{\mathrm{T}} \in \mathbb{R}^{n_i n_l n_b \times 1} \\ \boldsymbol{\pi}_{ijk}^{rc} = \left[\pi_{i11}^{rc}, \ldots, \pi_{i1n_b}^{rc}, \pi_{i21}^{rc}, \ldots, \pi_{i2n_b}^{rc}, \ldots, \pi_{in_l1}^{rc}, \ldots, \pi_{in_ln_b}^{rc}\right]^{\mathrm{T}} \in \mathbb{R}^{n_l n_b \times 1} \end{cases} \tag{3.69}$$

$$\begin{cases} \boldsymbol{A}_w^{rc} = \left[a_{1jk}^{rc}, a_{2jk}^{rc}, \ldots, a_{ojk}^{rc}, \ldots, a_{n_ojk}^{rc}\right] \in \mathbb{R}^{n_l n_b \times n_o} \\ \boldsymbol{A}_{ojk}^{rc} = \left[a_{o11}^{rc}, \ldots, a_{o1n_b}^{rc}, a_{o21}^{rc}, \ldots, a_{o2n_b}^{rc}, \ldots, a_{on_l1}^{rc}, \ldots, a_{on_ln_b}^{rc}\right]^{\mathrm{T}} \in \mathbb{R}^{n_l n_b \times 1} \end{cases} \tag{3.70}$$

In this section, we consider the control problem in cases where the dynamics of the robot control system are unknown. To address this challenge, RFWCMACNNs will be applied. Based on the universal approximation error analysis, there exists an optimal RFWCMACNN structure with its optimal parameters expressed as the following form [37]:

$$y_{\xi}(x(t)) = X\left(A_{W}^{rc*}\right)^{\mathsf{T}} \phi_{rc}\left(x(t), d_{rc}^{*}, c_{rc}^{*}, \pi_{rc}^{*}\right) + \Delta_{rc}(x(t)) \tag{3.71}$$

where $A_{W}^{rc*}, d_{rc}^{*}, c_{rc}^{*}, \pi_{rc}^{*}$ are the unknown optimal parameters of $A_{W}^{rc}, d_{rc}, c_{rc}, \pi_{rc}$, respectively, and $\boldsymbol{\Delta}_{rc}(x(t))$ is the approximation error vector.

Assumption 3.9 The norms of the optimal weights are bounded: $\|A_{W}^{rc}\| < b_{A}^{rc}$, $\|d_{rc}\| < b_{d}^{rc}$, $\|c_{rc}\| < b_{c}^{rc}$, and $\|\pi_{rc}\| < b_{\pi}^{rc}$, where $b_{A}^{rc}, b_{d}^{rc}, b_{c}^{rc}, b_{\pi}^{rc}$ are the positive real values.

Assumption 3.10 The error of approximation process and disturbances of the CCMMR/MMR system are bounded by $\|\Delta_{rc}\| < b_{\Delta}^{rc}$ and $\|\tau_{\xi d}\| < b_{\xi d}^{rc}$, where $b_{\Delta}^{rc}, b_{\xi d}^{rc}$ are the positive real values.

The proposed RFWCMACNN structure is the generation of the fuzzy concept, the theory of multiresolution analysis of the WNNs, and the RCMAC. Therefore, besides the advantages of universal approximation ability, high accuracy, and dynamic structure, a few drawbacks still exist. In the approximation process, this section cannot completely eliminate the approximation errors. The RFWCMACNN multiresolution may create the size of a huge network that the selection of the wavelons becomes very important. The fuzzy controller with linguistic rules alone may fail to control a system under some conditions. To overcome these difficulties, the RFWCMACNNs' parameters/rules need to be adjusted adaptively. Therefore, an adaptive approach for the RFWCMACNNs will be developed in the following section.

Figure 3.16 depicts the structure of the proposed RFWCMACNN controller for the nonholonomic CCMMR/MMR. In this section, the structure of the intelligent adaptive controller-based RFWCMACNNs is similar to that of the adaptive controller-based RFWNNs. However, the RFWCMACNNs are used instead of the RFWNNs/FWNNs. Therefore, the learning rules of the controller-based RFWCMACNNs are different from the controller-based RFWNNs/FWNNs. The designs of the proposed RFWCMACNNs are implemented by the analysis of the stability of the controlled system based on the Lyapunov stability theorem. The detailed designs of the proposed intelligent controllers are presented in the following parts:

3.4.1 Tracking Position Controller Design

An actual torque control law for the CCMMR/MMR's position control is proposed as

$$Z_{q}^{\mathsf{T}} u_{\tau} = \widehat{y}_{rc} + K_{s\xi} s_{\xi} + \widehat{\tau}_{rc} \tag{3.72}$$

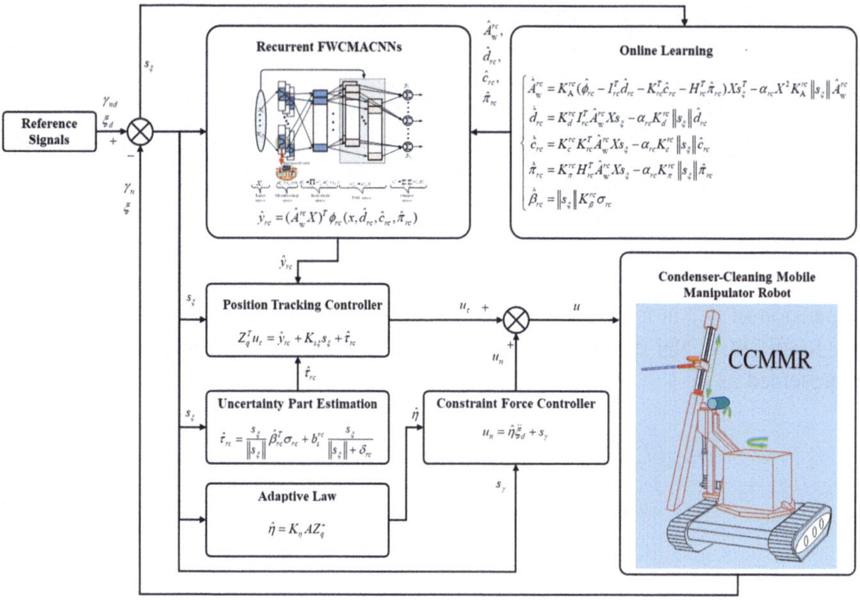

Fig. 3.16 Diagram of the proposed adaptive position/force controller-based RFWCMACNNs

where $\widehat{\tau}_{rc}$ also is the robust term. The part $\widehat{y}_{rc}$ is the RFWCMACNN approximation function of the function y_ξ. Based on (3.66), the $\widehat{y}_{rc}$ can be expressed as the following form:

$$\widehat{y}_{rc} = \left(\widehat{A}_w^{rc} X\right)^{\mathrm{T}} \phi_{rc}\left(x, \widehat{d}_{rc}, \widehat{c}_{rc}, \widehat{\pi}_{rc}\right) \tag{3.73}$$

where $\widehat{A}_w^{rc}, \widehat{d}_{rc}, \widehat{c}_{rc}, \widehat{\pi}_{rc}$ are the approximation values of the RFWCMACNN optimal parameters $A_w^{rc*}, d_{rc}^*, c_{rc}^*, \pi_{rc}^*$, respectively.

Applying the position control law (3.72) into the dynamics of the CCMMR/MMR control system, the closed-loop CCMMR/MMR control system can be expressed as

$$M_\xi \dot{s}_\xi = \widetilde{y}_{rc} - (K_{s\xi} + C_\xi)s_\xi + \tau_{\xi d} - \widehat{\tau}_{rc} \tag{3.74}$$

where the RFWCMACNN approximation function $\widetilde{y}_{rc}$ is defined as

$$\widetilde{y}_{rc} = y_\xi - \widehat{y}_{rc} = X\left(A_w^{rc*}\right)^{\mathrm{T}} \phi_{rc}\left(x, d_{rc}^*, c_{rc}^*, \pi_{rc}^*\right) - X\left(\widehat{A}_w^{rc}\right)^{\mathrm{T}} \phi_{rc}\left(x, \widehat{d}_{rc}, \widehat{c}_{rc}, \widehat{\pi}_{rc}\right) + \Delta_{rc} \tag{3.75}$$

For ease, the notation (3.75) can be rewritten as

$$\widetilde{y}_{rc} = X(A_w^{rc*})^{\mathsf{T}}\widetilde{\phi}_{rc} - X\left(\widetilde{A}_w^{rc}\right)^{\mathsf{T}}\widehat{\phi}_{rc} + \Delta_{rc} \tag{3.76}$$

where $\phi_{rc}^* \equiv \phi_{rc}(x, d_{rc}^*, c_{rc}^*, \pi_{rc}^*)$, $\widehat{\phi}_{rc} \equiv \phi_{rc}\left(x, \widehat{d}_{rc}, \widehat{c}_{rc}, \widehat{\pi}_{rc}\right)$, $\widetilde{A}_w^{rc} = A_w^{rc*} - \widehat{A}_w^{rc}$, and $\widetilde{\phi}_{rc} = \phi_{rc}^* - \widehat{\phi}_{rc}$.

Following (3.74), it can imply that the closed-loop control system from $\widetilde{y}_{rc}$ to s_ξ is a state-strict passive system. The passivity feature exists in this RFWCMACNN control system, which can guarantee the boundedness of the controlled system signals. Similar to the design of RFWNN controller, this section will take the expansion of $\widetilde{\phi}_{rc}$ in a Taylor series to obtain the adaptive online learning rules for the RFWCMACNN control parameters. The Taylor expansion form of $\widetilde{\phi}_{rc}$ can be represented as

$$\begin{aligned}\widetilde{\phi}_{rc} = {} & I_{rc}^{\mathsf{T}}\left(d_{rc}^* - \widehat{d}_{rc}\right) + K_{rc}^{\mathsf{T}}(c_{rc}^* - \widehat{c}_{rc}) + H_{rc}^{\mathsf{T}}(\pi_{rc}^* - \widehat{\pi}_{rc}) \\ & + \boldsymbol{\Theta}_{rc}\left(d_{rc}^* - \widehat{d}_{rc}, c_{rc}^* - \widehat{c}_{rc}, \pi_{rc}^* - \widehat{\pi}_{rc}\right)\end{aligned} \tag{3.77}$$

where $\boldsymbol{\Theta}_{rc}$ is the vector of the higher-order terms in the Taylor series expansion, assuming that I_{rc}, K_{rc}, H_{rc} are bounded by positive constants. The parameters in (3.77) can be denoted as follows:

$$\widetilde{\phi}_{rc} = \left[\widetilde{\phi}_1^{rc}, \ldots, \widetilde{\phi}_r^{rc}, \ldots, \widetilde{\phi}_{n_r}^{rc}\right]^{\mathsf{T}}$$

$$\begin{cases} I_{rc} = \left[\dfrac{\partial \phi_1^{rc}}{\partial d_{rc}}, \ldots, \dfrac{\partial \phi_r^{rc}}{\partial d_{rc}}, \ldots, \dfrac{\partial \phi_{n_r}^{rc}}{\partial d_{rc}}\right]\Bigg|_{d_{rc} = \widehat{d}_{rc}} \\ \left[\dfrac{\partial \phi_r^{rc}}{\partial d_{rc}}\right] = \left[\underbrace{0, \ldots, 0}_{(r-1)(n-m)}, \dfrac{\partial \phi_r^{rc}}{\partial d_{r1}^{rc}}, \ldots, \dfrac{\partial \phi_r^{rc}}{\partial d_{r(n-m)}^{rc}}, \underbrace{0, \ldots, 0}_{(n_r-r)(n-m)}\right]^{\mathsf{T}} \end{cases}$$

$$\begin{cases} K_{rc} = \left[\dfrac{\partial \phi_1^{rc}}{\partial c_{rc}}, \ldots, \dfrac{\partial \phi_r^{rc}}{\partial c_{rc}}, \ldots, \dfrac{\partial \phi_{n_r}^{rc}}{\partial c_{rc}}\right]\Bigg|_{c_{rc} = \widehat{c}_{rc}} \\ \left[\dfrac{\partial \phi_r^{rc}}{\partial c_{rc}}\right] = \left[\underbrace{0, \ldots, 0}_{(r-1)(n-m)}, \dfrac{\partial \phi_r^{rc}}{\partial c_{r1}^{rc}}, \ldots, \dfrac{\partial \phi_r^{rc}}{\partial c_{r(n-m)}^{rc}}, \underbrace{0, \ldots, 0}_{(n_r-r)(n-m)}\right]^{\mathsf{T}} \end{cases}$$

$$\begin{cases} H_{rc} = \left[\dfrac{\partial \phi_1^{rc}}{\partial \pi_{rc}}, \ldots, \dfrac{\partial \phi_r^{rc}}{\partial \pi_{rc}}, \ldots, \dfrac{\partial \phi_{n_r}^{rc}}{\partial \pi_{rc}}\right]\Bigg|_{\pi_{rc} = \widehat{\pi}_{rc}} \\ \left[\dfrac{\partial \phi_r^{rc}}{\partial \pi_{rc}}\right] = \left[\underbrace{0, \ldots, 0}_{(r-1)(n-m)}, \dfrac{\partial \phi_r^{rc}}{\partial \pi_{r1}^{rc}}, \ldots, \dfrac{\partial \phi_r^{rc}}{\partial \pi_{r(n-m)}^{rc}}, \underbrace{0, \ldots, 0}_{(n_r-r)(n-m)}\right]^{\mathsf{T}} \end{cases}$$

Defining $\widetilde{d}_{rc} = d_{rc}^* - \widehat{d}_{rc}$, $\widetilde{c}_{rc} = c_{rc}^* - \widehat{c}_{rc}$, and $\widetilde{\pi}_{rc} = \pi_{rc}^* - \widehat{\pi}_{rc}$, then (3.77) can be rewritten as

$$\widetilde{\phi}_{rc}=I_{rc}^{\mathsf{T}}\widetilde{d}_{rc}+K_{rc}^{\mathsf{T}}\widetilde{c}_{rc}+H_{rc}^{\mathsf{T}}\widetilde{\pi}_{rc}+\boldsymbol{\Theta}_{rc}\left(\widetilde{d}_{rc},\widetilde{c}_{rc},\widetilde{\pi}_{rc}\right) \tag{3.78}$$

From (3.76) and (3.78), it yields

$$\begin{aligned}\widetilde{y}_{rc}+\tau_{\xi d}&=X\left(\widetilde{A}_{w}^{rc}\right)^{\mathsf{T}}\left(\widehat{\phi}_{rc}-I_{rc}^{\mathsf{T}}\widehat{d}_{rc}-K_{rc}^{\mathsf{T}}\widehat{c}_{rc}-H_{rc}^{\mathsf{T}}\widehat{\pi}_{rc}\right)\\&+X\left(\widehat{A}_{w}^{rc}\right)^{\mathsf{T}}\left(I_{rc}^{\mathsf{T}}\widetilde{d}_{rc}+K_{rc}^{\mathsf{T}}\widetilde{c}_{rc}+H_{rc}^{\mathsf{T}}\widetilde{\pi}_{rc}\right)+\Gamma_{rc}\end{aligned} \tag{3.79}$$

where

$$\begin{aligned}\Gamma_{rc}&=X\left(A_{w}^{rc*}\right)^{\mathsf{T}}\left(\widetilde{\phi}_{rc}+I_{rc}^{\mathsf{T}}\widehat{d}_{rc}+K_{rc}^{\mathsf{T}}\widehat{c}_{rc}+H_{rc}^{\mathsf{T}}\widehat{\pi}_{rc}\right)\\&-X\left(\widehat{A}_{w}^{rc}\right)^{\mathsf{T}}\left(I_{rc}^{\mathsf{T}}d_{rc}^{*}+K_{rc}^{\mathsf{T}}c_{rc}^{*}+H_{rc}^{\mathsf{T}}\pi_{rc}^{*}\right)+\Delta_{rc}+\tau_{\xi d}\end{aligned}.$$

By substituting (3.79) into (3.74), (3.74) can be rewritten as

$$\begin{aligned}M\dot{s}_{\xi}&=X\left(\widetilde{A}_{w}^{rc}\right)^{\mathsf{T}}\left(\widehat{\phi}_{rc}-I_{rc}^{\mathsf{T}}\widehat{d}_{rc}-K_{rc}^{\mathsf{T}}\widehat{c}_{rc}-H_{rc}^{\mathsf{T}}\widehat{\pi}_{rc}\right)\\&+X\left(\widehat{A}_{w}^{rc}\right)^{\mathsf{T}}\left(I_{rc}^{\mathsf{T}}\widetilde{d}_{rc}+K_{rc}^{\mathsf{T}}\widetilde{c}_{rc}+H_{rc}^{\mathsf{T}}\widetilde{\pi}_{rc}\right)\\&-\left(K_{s\xi}+C_{\xi}\right)s_{\xi}+\Gamma_{rc}-\widehat{\tau}_{rc}\end{aligned} \tag{3.80}$$

Following Assumptions 3.9 and 3.10, and considering the characteristics of the wavelet basis function as described in [39–41] and (3.79), the following inequality can be obtained:

$$\begin{aligned}\|\Gamma_{rc}\|&=\left\|\begin{matrix}X\left(A_{w}^{rc*}\right)^{\mathsf{T}}\left(\widetilde{\phi}_{rc}+I_{rc}^{\mathsf{T}}\widehat{d}_{rc}+K_{rc}^{\mathsf{T}}\widehat{c}_{rc}+H_{rc}^{\mathsf{T}}\widehat{\pi}_{rc}\right)-X\left(\widehat{A}_{w}^{rc}\right)^{\mathsf{T}}\left(I_{rc}^{\mathsf{T}}d_{rc}^{*}+K_{rc}^{\mathsf{T}}c_{rc}^{*}+H_{rc}^{\mathsf{T}}\pi_{rc}^{*}\right)\\+\Delta_{rc}+\tau_{\xi d}\end{matrix}\right\|\\&\le\left\|X\left(A_{w}^{rc*}\right)^{\mathsf{T}}\widetilde{\phi}_{rc}+\Delta_{rc}+\tau_{\xi d}\right\|+\left\|X\left(A_{w}^{rc*}\right)^{\mathsf{T}}I_{rc}^{\mathsf{T}}\right\|\left\|\widehat{d}_{rc}\right\|+\left\|X\left(A_{w}^{rc*}\right)^{\mathsf{T}}K_{rc}^{\mathsf{T}}\right\|\|\widehat{c}_{rc}\|\\&+\left\|X\left(A_{w}^{rc*}\right)^{\mathsf{T}}H_{rc}^{\mathsf{T}}\right\|\|\widehat{\pi}_{rc}\|+\left\|X\left(I_{rc}^{\mathsf{T}}d_{rc}^{*}+K_{rc}^{\mathsf{T}}c_{rc}^{*}+H_{rc}^{\mathsf{T}}\pi_{rc}^{*}\right)\right\|\left\|\widehat{A}_{w}^{rc}\right\|\end{aligned} \tag{3.81}$$

Adding $\left(\alpha_{rc}\left(Xb_{A}^{rc}\right)^{2}+\alpha_{rc}\left(b_{d}^{rc}\right)^{2}+\alpha_{rc}\left(b_{c}^{rc}\right)^{2}+\alpha_{rc}\left(b_{\pi}^{rc}\right)^{2}\right)/4$ into both sides of (3.81) yields

$$\|\Gamma_{rc}\|+\frac{\alpha_{rc}\left(Xb_{A}^{rc}\right)^{2}}{4}+\frac{\alpha_{rc}\left(b_{d}^{rc}\right)^{2}}{4}+\frac{\alpha_{rc}\left(b_{c}^{rc}\right)^{2}}{4}+\frac{\alpha_{rc}\left(b_{\pi}^{rc}\right)^{2}}{4}\le\beta_{rc}^{*\mathsf{T}}\sigma_{rc} \tag{3.82}$$

where $\beta_{rc}^{*\mathsf{T}}=\left[\beta_{1}^{rc},\beta_{2}^{rc},\beta_{3}^{rc},\beta_{4}^{rc},\beta_{5}^{rc}\right]$ and $\sigma_{rc}^{\mathsf{T}}=\left[1,\left\|\widehat{d}_{rc}\right\|,\|\widehat{c}_{rc}\|,\|\widehat{\pi}_{rc}\|,\left\|\widehat{A}_{w}^{rc}\right\|\right]$, α_{rc} is the positive constant, and $\beta_{1}^{rc},\beta_{2}^{rc},\beta_{3}^{rc},\beta_{4}^{rc},\beta_{5}^{rc}$ are the positive constants and they are uncertain bounds of $\left(\alpha_{rc}(Xb_{A}^{rc})^2+\alpha_{rc}(b_{d}^{rc})^2+\alpha_{rc}(b_{c}^{rc})^2+\alpha_{rc}(b_{\pi}^{rc})^2\right)/4+\left\|X(A_{w}^{rc*})^{\mathrm{T}}\widetilde{\phi}_{rc}+\Delta_{rc}+\tau_{\xi d}\right\|$, $\left\|X(A_{w}^{rc*})^{\mathrm{T}}I_{rc}^{\mathrm{T}}\right\|$, $\left\|X(A_{w}^{rc*})^{\mathrm{T}}K_{rc}^{\mathrm{T}}\right\|$, $\left\|X(A_{w}^{rc*})^{\mathrm{T}}H_{rc}^{\mathrm{T}}\right\|$, and $\left\|X\left(I_{rc}^{\mathrm{T}}d_{rc}^{*}+K_{rc}^{\mathrm{T}}c_{rc}^{*}+H_{rc}^{\mathrm{T}}\pi_{rc}^{*}\right)\right\|$, respectively. Here, the RFWCMACNNs' approximation will be more accurate, and the stability of the RFWCMACNN control system will be guaranteed if the robust term $\widehat{\tau}_{rc}$ can eliminate the uncertainty part Γ_{rc}. Therefore, $\widehat{\tau}_{rc}$ is used to estimate the boundedness of the Γ_{rc}, and it is proposed as

$$\widehat{\tau}_{rc}=\frac{s_{\xi}}{\|s_{\xi}\|}\widehat{\beta}_{rc}^{\mathrm{T}}\sigma_{rc}+b_{s}^{rc}\frac{s_{\xi}}{\|s_{\xi}\|+\delta_{rc}} \tag{3.83}$$

where b_{s}^{rc}, δ_{rc} are the positive constants and $\widehat{\beta}_{rc}$ is an estimated value of β_{rc}^{*}.

With the above analysis, the adaptive online learning algorithms of the controller-based RFWCMACNNs and the robust term are proposed as the following forms:

$$\begin{cases}\dot{\widehat{A}}_{w}^{rc}=K_{A}^{rc}\left(\widehat{\phi}_{rc}-I_{rc}^{\mathrm{T}}\widehat{d}_{rc}-K_{rc}^{\mathrm{T}}\widehat{c}_{rc}-H_{rc}^{\mathrm{T}}\widehat{\pi}_{rc}\right)Xs_{\xi}^{\mathrm{T}}-\alpha_{rc}X^{2}K_{A}^{rc}\|s_{\xi}\|\widehat{A}_{w}^{rc}\\ \dot{\widehat{d}}_{rc}=K_{d}^{rc}I_{rc}^{\mathrm{T}}\widehat{A}_{w}^{rc}Xs_{\xi}-\alpha_{rc}K_{d}^{rc}\|s_{\xi}\|\widehat{d}_{rc}\\ \dot{\widehat{c}}_{rc}=K_{c}^{rc}K_{rc}^{\mathrm{T}}\widehat{A}_{w}^{rc}Xs_{\xi}-\alpha_{rc}K_{c}^{rc}\|s_{\xi}\|\widehat{c}_{rc}\\ \dot{\widehat{\pi}}_{rc}=K_{\pi}^{rc}H_{rc}^{\mathrm{T}}\widehat{A}_{w}^{rc}Xs_{\xi}-\alpha_{rc}K_{\pi}^{rc}\|s_{\xi}\|\widehat{\pi}_{rc}\\ \dot{\widehat{\beta}}_{rc}=\|s_{\xi}\|K_{\beta}^{rc}\sigma_{rc}\end{cases} \tag{3.84}$$

where $K_{A}^{rc}, K_{d}^{rc}, K_{c}^{rc}, K_{\pi}^{rc}, K_{\beta}^{rc}$ are the positive and diagonal constant matrices.

3.4.2 Force Constraint Controller Design

The design of the force constraint controller in the intelligent control system-based RFWCMACNNs is similar to that of the design of the force constraint controller. Therefore, with the assumption about the stability of the signals in the position controller-based RFWCMACNNs, the force constraint controller and the update rule for its parameter are proposed as (3.55) and (3.56), respectively.

3.4.3 Stability Analysis

Theorem 3.3 By considering the CCMMR/MMR model, all the assumptions for the RFWCMACNNs and the CCMMR/MMR hold. If the control law for position tracking RFWCMACNN controller is (3.72), the control law for the constraint force is (3.55), the adaptive online learning algorithms for the controller-based RFWCMACNNs are designed as (3.84), and the adaptive robust update algorithm for the constraint force control is designed as (3.56), then:

1. The parameters of the RFWCMACNNs and the approximation errors are bounded, and the filtered tracking error s_ξ converges to zero.
2. The e_γ and the control inputs of the intelligent controller-based RFWCMACNNs are bounded for $t \geq 0$, and the stability of the proposed control system is guaranteed.

Proof (Theorem 3.3(1)) Define the Lyapunov function candidate as

$$\begin{aligned} V_{rc}\left(s_\xi(t), \widetilde{A}_w^{rc}, \widetilde{d}_{rc}, \widetilde{c}_{rc}, \widetilde{\pi}_{rc}, \widetilde{\beta}_{rc}\right) &= \frac{1}{2} s_\xi^{\mathrm{T}} M_\xi s_\xi + \frac{1}{2}\mathrm{tr}\left(\left(\widetilde{A}_w^{rc}\right)^{\mathrm{T}} K_{\mathrm{A}}^{rc-1} \widetilde{A}_w^{rc}\right) \\ &+ \frac{1}{2}\mathrm{tr}\left(\widetilde{d}_{rc}^{\mathrm{T}} K_d^{rc-1} \widetilde{d}_{rc}\right) + \frac{1}{2}\mathrm{tr}\left(\widetilde{c}_{rc}^{\mathrm{T}} K_c^{rc-1} \widetilde{c}_{rc}\right) \\ &+ \frac{1}{2}\mathrm{tr}\left(\widetilde{\pi}_{rc}^{\mathrm{T}} K_\pi^{rc-1} \widetilde{\pi}_{rc}\right) + \frac{1}{2}\mathrm{tr}\left(\widetilde{\beta}_{rc}^{\mathrm{T}} K_\beta^{rc-1} \widetilde{\beta}_{rc}\right) \end{aligned} \quad (3.85)$$

where $\widetilde{\beta}_{rc} = \beta_{rc}^* - \widehat{\beta}_{rc}$. Differentiating (3.85) with respect to time yields

$$\begin{aligned} \dot{V}_{rc} &= s_\xi^{\mathrm{T}} M_\xi \dot{s}_\xi + \frac{1}{2} s_\xi^{\mathrm{T}} \dot{M}_\xi s_\xi - \mathrm{tr}\left(\left(\widetilde{A}_w^{rc}\right)^{\mathrm{T}} K_{\mathrm{A}}^{rc-1} \dot{\widehat{A}}_w^{rc}\right) - \mathrm{tr}\left(\widetilde{d}_{rc}^{\mathrm{T}} K_d^{rc-1} \dot{\widehat{d}}_{rc}\right) \\ &- \mathrm{tr}\left(\widetilde{c}_{rc}^{\mathrm{T}} K_c^{rc-1} \dot{\widehat{c}}_{rc}\right) - \mathrm{tr}\left(\widetilde{\pi}_{rc}^{\mathrm{T}} K_\pi^{rc-1} \dot{\widehat{\pi}}_{rc}\right) - \mathrm{tr}\left(\widetilde{\beta}_{rc}^{\mathrm{T}} K_\beta^{rc-1} \dot{\widehat{\beta}}_{rc}\right) \end{aligned} \quad (3.86)$$

By substituting (3.80) into (3.86), one can obtain

$$\begin{aligned} \dot{V}_{rc} =& -s_\xi^{\mathrm{T}} K_{s\xi} s_\xi + s_\xi^{\mathrm{T}} X \left(\widetilde{A}_w^{rc}\right)^{\mathrm{T}} \left(\widehat{\phi}_{rc} - I_{rc}^{\mathrm{T}} \widehat{d}_{rc} - K_{rc}^{\mathrm{T}} \widehat{c}_{rc} - H_{rc}^{\mathrm{T}} \widehat{\pi}_{rc}\right) \\ &+ s_\xi^{\mathrm{T}} X \left(\widehat{A}_w^{rc}\right)^{\mathrm{T}} \left(I_{rc}^{\mathrm{T}} \widetilde{d}_{rc} + K_{rc}^{\mathrm{T}} \widetilde{c}_{rc} + H_{rc}^{\mathrm{T}} \widetilde{\pi}_{rc}\right) - \mathrm{tr}\left(\left(\widetilde{A}_w^{rc}\right)^{\mathrm{T}} K_{\mathrm{A}}^{rc-1} \dot{\widehat{A}}_w^{rc}\right) - \mathrm{tr}\left(\widetilde{d}_{rc}^{\mathrm{T}} K_d^{rc-1} \dot{\widehat{d}}_{rc}\right) \\ &- \mathrm{tr}\left(\widetilde{c}_{rc}^{\mathrm{T}} K_c^{rc-1} \dot{\widehat{c}}_{rc}\right) - \mathrm{tr}\left(\widetilde{\pi}_{rc}^{\mathrm{T}} K_\pi^{rc-1} \dot{\widehat{\pi}}_{rc}\right) - \mathrm{tr}\left(\widetilde{\beta}_{rc}^{\mathrm{T}} K_\beta^{rc-1} \dot{\widehat{\beta}}_{rc}\right) + s_\xi^{\mathrm{T}} \Gamma_{rc} - s_\xi^{\mathrm{T}} \widehat{\tau}_{rc} \end{aligned} \quad (3.87)$$

If the online adaptation laws of the controller-based RFWCMACNNs and the robust term are chosen as (3.84), then (3.87) can be written as

$$\begin{aligned} \dot{V}_{rc} =& -s_\xi^{\mathrm{T}} K_{s\xi} s_\xi + X^2 \alpha_{rc} \|s_\xi\| \mathrm{tr}\left(\left(\widetilde{A}_w^{rc}\right)^{\mathrm{T}} \widehat{A}_w^{rc}\right) + \alpha_{rc} \|s_\xi\| \mathrm{tr}\left(\widetilde{d}_{rc}^{\mathrm{T}} \widehat{d}_{rc}\right) + \alpha_{rc} \|s_\xi\| \mathrm{tr}\left(\widetilde{c}_{rc}^{\mathrm{T}} \widehat{c}_{rc}\right) \\ &+ \alpha_{rc} \|s_\xi\| \mathrm{tr}\left(\widetilde{\pi}_{rc}^{\mathrm{T}} \widehat{\pi}_{rc}\right) - \widetilde{\beta}_{rc}^{\mathrm{T}} \|s_\xi\| \sigma_{rc} + s_\xi^{\mathrm{T}} \Gamma_{rc} - s_\xi^{\mathrm{T}} \widehat{\tau}_{rc} \end{aligned} \quad (3.88)$$

By using the results $\mathrm{tr}\left(\left(\widetilde{A}_w^{rc}\right)^{\mathrm{T}}\widehat{A}_w^{rc}\right) \leq b_A^{rc}\left\|\widetilde{A}_w^{rc}\right\| - \left\|\widetilde{A}_w^{rc}\right\|^2$, $\mathrm{tr}\left(\widetilde{d}_{rc}^{\mathrm{T}}\widehat{d}_{rc}\right) \leq b_d^{rc}\left\|\widetilde{d}_{rc}\right\| - \left\|\widetilde{d}_{rc}\right\|^2$, $\mathrm{tr}(\widetilde{c}_{rc}^{\mathrm{T}}\widehat{c}_{rc}) \leq b_c^{rc}\|\widetilde{c}_{rc}\| - \|\widetilde{c}_{rc}\|^2$, $\mathrm{tr}(\widetilde{\pi}_{rc}^{\mathrm{T}}\widehat{\pi}_{rc}) \leq b_\pi^{rc}\|\widetilde{\pi}_{rc}\| - \|\widetilde{\pi}_{rc}\|^2$, and (3.82), then (3.88) can be represented as

$$\begin{aligned}\dot{V}_{rc} \leq & -s_\xi^{\mathrm{T}}K_{s\xi}s_\xi - s_\xi^{\mathrm{T}}\widehat{\tau}_{rc} - \widetilde{\beta}_{rc}^{\mathrm{T}}\|s_\xi\|\sigma_{rc} + \|s_\xi\|\beta_{rc}^{*\mathrm{T}}\sigma_{rc} \\ & -\alpha_{rc}\|s_\xi\|\left[X^2\left(\frac{b_A^{rc}}{2} - \left\|\widetilde{A}_w^{rc}\right\|\right)^2 + \left(\frac{b_d^{rc}}{2} - \left\|\widetilde{d}_{rc}\right\|\right)^2 + \left(\frac{b_c^{rc}}{2} - \|\widetilde{c}_{rc}\|\right)^2 + \left(\frac{b_\pi^{rc}}{2} - \|\widetilde{\pi}_{rc}\|\right)^2\right]\end{aligned} \tag{3.89}$$

By substituting (3.83) into the inequality (3.89), it can be concluded that

$$\begin{aligned}\dot{V}_{rc} \leq & -s_\xi^{\mathrm{T}}K_{s\xi}s_\xi - s_\xi^{\mathrm{T}}\left(\frac{s_\xi}{\|s_\xi\|}\widehat{\beta}_{rc}^{\mathrm{T}}\sigma_{rc} + b_s^{rc}\frac{s_\xi}{\|s_\xi\| + \delta_{rc}}\right) - \widetilde{\beta}_{rc}^{\mathrm{T}}\|s_\xi\|\sigma_{rc} + \|s_\xi\|\beta_{rc}^{*\mathrm{T}}\sigma_{rc} \\ \leq & -s_\xi^{\mathrm{T}}K_{s\xi}s_\xi - \widetilde{\beta}_{rc}^{\mathrm{T}}\|s_\xi\|\sigma_{rc} + \|s_\xi\|\beta_{rc}^{*\mathrm{T}}\sigma_{rc} - \|s_\xi\|\widehat{\beta}_{rc}^{\mathrm{T}}\sigma_{rc} = -s_\xi^{\mathrm{T}}K_{s\xi}s_\xi\end{aligned} \tag{3.90}$$

Based on (3.90), since $\dot{V}_{rc}\left(s_\xi(t), \widetilde{A}_w^{rc}, \widetilde{d}_{rc}, \widetilde{c}_{rc}, \widetilde{\pi}_{rc}, \widetilde{\beta}_{rc}\right)$ is a negative semi-definite function, it follows that, $V_{rc}\left(s_\xi(t), \widetilde{A}_w^{rc}, \widetilde{d}_{rc}, \widetilde{c}_{rc}, \widetilde{\pi}_{rc}, \widetilde{\beta}_{rc}\right) \leq V_{rc}\left(s_\xi(0), \widetilde{A}_w^{rc}, \widetilde{d}_{rc}, \widetilde{c}_{rc}, \widetilde{\pi}_{rc}, \widetilde{\beta}_{rc}\right)$. If $s_\xi(t), \widetilde{A}_w^{rc}, \widetilde{d}_{rc}, \widetilde{c}_{rc}, \widetilde{\pi}_{rc}, \widetilde{\beta}_{rc}$ are bounded at the initial $t = 0$, then they will be bounded for $t \geq 0$. It implies that $s_\xi(t), \widetilde{A}_w^{rc}, \widetilde{d}_{rc}, \widetilde{c}_{rc}, \widetilde{\pi}_{rc}, \widetilde{\beta}_{rc}$ are bounded for $t \geq 0$. Let function $\Omega_{rc}(t) = s_\xi^{\mathrm{T}}K_{s\xi}s_\xi$, then $\Omega_{rc}(t) \leq -\dot{V}_{rc}$. Integrating $\Omega_{rc}(t)$ with respect to time, it can be shown that

$$\int_0^t \Omega_{rc}(v)\mathrm{d}v \leq V_{rc}\left(s_\xi(0), \widetilde{A}_w^{rc}, \widetilde{d}_{rc}, \widetilde{c}_{rc}, \widetilde{\pi}_{rc}, \widetilde{\beta}_{rc}\right) - V_{rc}\left(s_\xi(t), \widetilde{A}_w^{rc}, \widetilde{d}_{rc}, \widetilde{c}_{rc}, \widetilde{\pi}_{rc}, \widetilde{\beta}_{rc}\right) \tag{3.91}$$

Since $V_{rc}\left(s_\xi(0), \widetilde{A}_w^{rc}, \widetilde{d}_{rc}, \widetilde{c}_{rc}, \widetilde{\pi}_{rc}, \widetilde{\beta}_{rc}\right)$ is a bounded function and $V_{rc}\left(s_\xi(t), \widetilde{A}_w^{rc}, \widetilde{d}_{rc}, \widetilde{c}_{rc}, \widetilde{\pi}_{rc}, \widetilde{\beta}_{rc}\right)$ is a non-increasing and bounded function, the following result can be obtained:

$$\lim_{t\to\infty}\int_0^t \Omega_{rc}(v)\mathrm{d}v < \infty \tag{3.92}$$

Thus, since $\dot{\Omega}_{rc}(t)$ is bounded by Barbalat's lemma [42], it can imply that $\lim_{t \to \infty} \Omega_{rc}(t) = 0$. That is, the filtered tracking error $s_\xi(t)$ will converge to zero as time tends to infinity.

Proof (Theorem 3.3(2)) From (3.57) and the results of the proof, $s_\xi(t), \widetilde{A}_w^{rc}, \widetilde{d}_{rc}, \widetilde{c}_{rc}, \widetilde{\pi}_{rc}, \widetilde{\beta}_{rc}$ are bounded. That is, $\Upsilon, Z_q, M_\xi, \ddot{\xi}_d$ and $-\Upsilon Z_q^{+T} M_\xi \ddot{\xi} + \widehat{\eta}\ddot{\xi}_d$ are also bounded. Therefore, the size of e_γ can be adjusted by selecting the proper gain K_γ, K_η. Here, it is easy to conclude the control inputs of the intelligent RFWCMACNN controllers are bounded from (3.72) and (3.55). Based on proof (Theorem 3.3), this section can conclude that the stability of the proposed RFWCMACNN control system is guaranteed.

According to the aforementioned analysis, the design procedure of the RFWCMACNN control system is summarized in the following steps:

Step 1: Specify the structure of the RFWCMACNNs: n_i, n_l, n_b, n_o, n_r. Choose the control parameters: $\lambda_\xi, K_{s\xi}, \alpha_{rc}, K_{\mathrm{A}}^{rc}, K_d^{rc}, K_c^{rc}, K_\pi^{rc}, K_\beta^{rc}, b_s^{rc}, \delta_{rc}, K_\gamma, K_\eta$. Initialize (by random values) the approximating parameters of the RFWCMACNNs $(\widehat{A}_w^{rc}, \widehat{d}_{rc}, \widehat{c}_{rc}, \widehat{\pi}_{rc})$, the robust term $(\widehat{\beta}_{rc})$, and the constraint force controller $(\widehat{\eta})$ from the memory system.

Step 2: Update the inputs of the RFWCMACNNs, the approximating parameters of the RFWCMACNNs, the robust term, and the constraint force controller from memory system.

Step 3: Compute the output of the RFWCMACNNs via (3.73) and the robust term via (3.83). Then, construct the control inputs of the proposed controller via (3.72) and (3.55).

Step 4: Adjust the parameters of the RFWCMACNNs, the robust term, and the constraint force controller with the learning algorithms (3.84) and (3.56). These parameters are saved in the memory system. Then, return to Step 2.

This section notes that the selection of the control parameters of the proposed controller should be chosen through trials in considering the requirement of convergence of the tracking errors, the stability of RFWCMACNN outputs, and the limitation of control inputs.

In accordance with the above analysis, the requirement for bounded uncertainties in the proposed RFWCMACNN control robot system is relaxed. Similar to the proposed RFWNNs for the CCMMR/MMR control, the drawbacks of the previous controller-based NNs for the MMR, the problems of bounded parameters, have been eliminated. The assumptions for the bounds of parameters are used to analyze the stability of the control system. In fact, these assumptions disappear in this designed RFWCMACNN controllers. In addition, according to [37], it is easy to show that the proposed RFWCMACNN controlled system, with the adaptive online update laws, is also strictly state passive such that some internal boundedness properties of the controlled system, without the assumption of observability and stability, are guaranteed. The effectiveness and robustness of the proposed method can be verified by the following numerical simulation and experimental results.

3.4.4 Brief Structure of Adaptive FWCMACNN Controllers

In this section, the structure of the intelligent FWCMACNN controllers for CCMMR/MMR is also presented based on the design of the proposed RFWCMACNN controllers. The purpose of this work is to evaluate the effectiveness of the proposed controllers via comparison with the FWCMACNN controllers. Therefore, the structure of the intelligent controller-based FWCMACNNs is briefed as follows:

1. The proposed FWCMACNNs' structure is similar to the RFWCMACNNs without the recurrent units in the membership space.
2. The proposed controller-based FWCMACNNs are presented as follows:
   The position controller:

$$Z_q^{\mathrm{T}} u_\tau = \widehat{y}_c + K_{s\xi} s_\xi + \widehat{\tau}_c \tag{3.93}$$

   where $\widehat{y}_c$ is the FWCMACNN approximation function of the function y_ξ. $\widehat{\tau}_c$ is the robust term as

$$\widehat{\tau}_c = \frac{s_\xi}{\|s_\xi\|} \widehat{\beta}_c^{\mathrm{T}} \sigma_c + b_s^c \frac{s_\xi}{\|s_\xi\| + \delta_c} \tag{3.94}$$

   where b_s^c, δ_c are the positive constants and $\widehat{\beta}_c$ is an estimated value of β_c^*. $\beta_c^{*\mathrm{T}} \sigma_c$ is the bounds that are similar to $\beta_{rc}^{*\mathrm{T}} \sigma_{rc}$. The adaptive online learning algorithms of the controller-based FWCMACNNs and the robust term are proposed as the following forms:

$$\begin{cases} \dot{\widehat{A}}_w^c = \boldsymbol{K}_{\mathrm{A}}^c \left(\widehat{\phi}_c - I_c^{\mathrm{T}} \widehat{d}_{rc} - K_c^{\mathrm{T}} \widehat{c}_{rc} \right) X s_\xi^{\mathrm{T}} - \alpha_c X^2 \boldsymbol{K}_{\mathrm{A}}^c \|s_\xi\| \widehat{A}_w^c \\ \dot{\widehat{d}}_c = \boldsymbol{K}_d^c I_c^{\mathrm{T}} \widehat{A}_w^c X s_\xi - \alpha_c \boldsymbol{K}_d^c \|s_\xi\| \widehat{d}_c \\ \dot{\widehat{c}}_c = \boldsymbol{K}_c^c K_c^{\mathrm{T}} \widehat{A}_w^c X s_\xi - \alpha_c \boldsymbol{K}_c^c \|s_\xi\| \widehat{c}_c \\ \dot{\widehat{\beta}}_c = \|s_\xi\| \boldsymbol{K}_\beta^c \sigma_c \end{cases} \tag{3.95}$$

   where $\boldsymbol{K}_{\mathrm{A}}^c, \boldsymbol{K}_d^c, \boldsymbol{K}_c^c, \boldsymbol{K}_\beta^c$ are the positive and diagonal constant matrices. $\widehat{A}_w^c$, $\widehat{d}_c$, $\widehat{c}_c$, $\widehat{\phi}_c$, $I_c^{\mathrm{T}}, K_c^{\mathrm{T}}$ are constituted similar to $\widehat{A}_w^{rc}$, $\widehat{d}_{rc}$, $\widehat{c}_{rc}$, $\widehat{\phi}_{rc}$, $I_{rc}^{\mathrm{T}}, K_{rc}^{\mathrm{T}}$.

The constraint force controller is proposed similarly to the constraint force controller of the proposed RFWCMACNN controllers.

The structure of the intelligent controller-based FWCMACNNs will be applied in the following simulation and experimental works.

3.4.5 Simulation Results

This section will consider the 3-DOF CCMMR model as shown in Fig. 2.6, to verify the effectiveness of the proposed intelligent adaptive control method using the RFWCMACNNs. The dynamics of the CCMMR system are not easy to constitute exactly. Thus, for the sake of simplicity in the simulation process, it does not lose the generality of the controlled system, using a simple dynamic model of the 3-DOF MMR. To guarantee this simulation is consistent with experimental works, the detailed mechanical parameters of this MMR are similar to those of the CCMMR. Then, the dynamic parameters of this 3-DOF MMR are given as follows: $q = [q_n, q_h]^{\mathrm{T}}$, $q_n = [X_C, Y_C, \varphi]^{\mathrm{T}}$, $q_h = [\theta_1, \theta_2, \theta_3]^{\mathrm{T}}$, $\theta_2 = \pi/2$, $M_n = [M_{n11}, M_{n12}; M_{n21}, M_{n22}]$, $M_h =$ diag (I_{123}, I_{23}, m_3), $M_{nh} = [M_{nh1}, M_{nh2}, M_{nh3}]$, $M_{hn} = M_{nh}^{\mathrm{T}}$, $C_n = [C_{n11}, C_{n12}; C_{n21}, C_{n22}]$, $C_{nh} = [C_{nh1}, C_{nh2}, C_{nh3}]$, $C_h =$ diag (C_{h1}, C_{h2}, C_{h3}), $C_{hn} = [C_{hn1}, C_{hn2}, C_{hn3}]^{\mathrm{T}}$, $G_n = [0, 0, 0]^{\mathrm{T}}$, $G_h = [0, G_{h1}, G_{h2}]^{\mathrm{T}}$, $B_h =$ diag $(1.0, 1.0, 1.0)$, and $B_n = [\sin\varphi/r_{\mathrm{w}}, \cos\varphi/r_{\mathrm{w}}, -R/r_{\mathrm{w}}; -\sin\varphi/r_{\mathrm{w}}, \cos\varphi/r_{\mathrm{w}}, R/r_{\mathrm{w}}]^{\mathrm{T}}$, where $M_{n11} =$ diag (m_{p123}, m_{p123}), $M_{n12} = [m_{123}d\cos\varphi + m_{23}\cos(\varphi + \theta_1), m_{123}d\sin\varphi + m_{23}\sin(\varphi + \theta_1)]^{\mathrm{T}}$, $M_{n21} = M_{n21}^{\mathrm{T}}$, $M_{n22} = I_p + I_{123} + m_{123}d^2 + m_2\left(l_2^2/4 + l_2 d\cos\theta_1\right) + m_3\left(L_3^2 + 2dL_3\cos\theta_1\right)$, $M_{nh1} = \left[m_{23}\cos(\varphi + \theta_1), m_{23}\sin(\varphi + \theta_1), I_{123} + m_2\left(l_2^2/4 + l_2 d\cos\theta_1\right) + m_3\left(L_3^2 + 2dL_3\cos\theta_1\right)\right]^{\mathrm{T}}$, $M_{nh2} = [0, 0, 0]^{\mathrm{T}}$, $M_{nh3} = [\sin(\varphi + \theta_1), -\cos(\varphi + \theta_1), 0]^{\mathrm{T}}$, $C_{n11} = [0, 0; 0, 0]$, $C_{n12} = \left[-m_{123}d\dot{\varphi}\sin\varphi - m_{23}\sin(\varphi + \theta_1)(\dot{\varphi} + \dot{\theta}_1), -m_{123}d\dot{\theta}\cos\varphi + m_{23}\cos(\varphi + \theta_1)(\dot{\varphi} + \dot{\theta}_1)\right]^{\mathrm{T}}$, $C_{n21} = C_{n12}^{\mathrm{T}}$, $C_{n22} = -2m_{23}\sin(\theta_1)(\dot{\theta}_1)$, $C_{h1} = [-m_{23}\sin(\theta_1)(\dot{\varphi}); 0; 0]$, $C_{h2} = [0, -m_{23}\sin(\theta_1)(\dot{\varphi}), 0]^{\mathrm{T}}$, $C_{h3} = [0, 0, 0]^{\mathrm{T}}$, $C_{nh1} = C_{nh2} = \left[-m_{23}\sin(\varphi + \theta_1)(\dot{\varphi} + \dot{\theta}_1), -m_{23}\cos(\varphi + \theta_1)(\dot{\varphi} + \dot{\theta}_1), 0\right]^{\mathrm{T}}$, $C_{nh3} = \left[-m_3\cos(\varphi + \theta_1)(\dot{\varphi} + \dot{\theta}_1), -m_3\sin(\varphi + \theta_1)(\dot{\varphi} + \dot{\theta}_1), 0\right]^{\mathrm{T}}$, $C_{hn1} = C_{nh1}^{\mathrm{T}}$, $C_{hn2} = C_{nh2}^{\mathrm{T}}$, $C_{hn3} = \left[m_3\cos(\varphi + \theta_1)(\dot{\varphi} + \dot{\theta}_1), m_3\sin(\varphi + \theta_1)(\dot{\varphi} + \dot{\theta}_1), m_3 d\sin(\varphi)(\dot{\theta}_1)\right]$, $m_{23} = l_2 m_2/2 + m_3 L_3$, $L_3 = l_2 + l_3/2 + \theta_3$, $m_{p123} = m_p + m_{123}$, $m_{123} = m_1 + m_2 + m_3$, $I_{123} = I_1 + I_2 + I_3 + m_3 L_3^2$, and $I_{23} = I_2 + I_3 + m_3 L_3^2$.

This 3-DOF MMR is subjected to nonholonomic constraints that are given as

$$\dot{X}_C\cos(\varphi) + \dot{Y}_C\sin(\varphi) = 0$$

Therefore, the nonholonomic constraint matrix is described as

$$\boldsymbol{A}_n = [\cos(\varphi), \sin(\varphi), 0]$$

Then, $\boldsymbol{Z}(q_n)$ is determined as

$$\boldsymbol{Z}(q_n) = \begin{bmatrix} -\tan\varphi & 0 \\ 1 & 0 \\ 0 & 1 \end{bmatrix}$$

In order to exhibit the superior control performance and effectiveness of the proposed scheme, the PIDC, the NN controller [16], and the FWCMACNN controller are examined in the meanwhile. The PIDC law can be expressed as

$$u_\tau = \boldsymbol{K}_{p\tau} e_\xi(t) + \boldsymbol{K}_{i\tau} \int_0^t e_\xi(t)\mathrm{d}t + \boldsymbol{K}_{d\tau} \dot{e}_\xi(t)$$

where $\boldsymbol{K}_{p\tau}$, $\boldsymbol{K}_{i\tau}$, $\boldsymbol{K}_{d\tau}$ denote proportional, integral, and differential gain matrices, respectively, and they are designed by a compromise between the superiority of control performance and the magnitude of control effort. $K_{p\tau}$, $K_{i\tau}$, $K_{d\tau}$ are given as

$$\begin{cases} \boldsymbol{K}_{p\tau} = \mathrm{diag}(100, 90, 70, 70, 50) \\ \boldsymbol{K}_{i\tau} = \mathrm{diag}(0.5, 0.4, 0.3, 0.4, 0.5) \\ \boldsymbol{K}_{d\tau} = \mathrm{diag}(1500, 1200, 800, 700, 600) \end{cases}.$$

The structure of the proposed RFWCMACNNs can be characterized by $n_i = 25$, $n_l = 4$, $n_b = 2$, $n_r = 2 \times 4$, and $n_o = 5$. The control parameters of the proposed RFWCMACNNs are given as follows:

$$\begin{cases} K_{s\xi} = \mathrm{diag}(300, 300, 300, 300, 300), \lambda_\xi = \mathrm{diag}(10, 10, 20, 20, 20) \\ K_{\mathrm{A}}^{rc} = \mathrm{diag}(1, \ldots, 1, \ldots, 1) \in \mathbb{R}^{n_l n_p \times n_l n_p} \\ K_{\mathrm{d}}^{rc} = K_{\mathrm{c}}^{rc} = K_{\pi}^{rc} = \mathrm{diag}(20, \ldots, 20, \ldots, 20) \in \mathbb{R}^{n_i n_l n_b \times n_i n_l n_b} \\ K_{\beta}^{rc} = \mathrm{diag}(0.001, 0.001, 0.001, 0.001, 0.001) \\ \alpha_{rc} = 0.001, \sigma_{rc} = 0.001, \delta_{rc} = 0.001, b_{s\xi}^{rc} = 0.01 \\ K_\eta = 0.01, K_\gamma = 0.02 \end{cases}$$

For recording the respective control performance, the mean square error (MSE) of the position tracking response is defined as

$$\mathrm{MSE} = \frac{1}{T} \sum_{k=1}^{T} [\xi(k) - \xi_d(k)]^2$$

where T is the total sampling instant. Based on this definition, the normalized MSE (NMSE) value of the position tracking response using a per-unit value with 1 rad is used to examine the control performance. The desired joint positions and the nonholonomic constraint are considered as

$$\begin{cases} \xi_d = [4t, 4.5t, \sin(2\pi t/5), pi/2, 0.5\sin(2\pi t/4)](\text{rad}) \\ \gamma_{nd} = 10 \end{cases}$$

To investigate the robustness of the proposed scheme, two simulation cases including the parameter variations and disturbances are considered as follows:

Case 1

There are no tip loads on the links, and the disturbances are given as.

$$\tau_{\xi d} = [0.1\sin(2t); 0.2\cos(2t); 0.4\sin(2t); 0.1\sin(2t); 0.2\cos(2t); 0.4\sin(2t)]$$

Case 2

Tip loads (4.5 kg, 2 kg, and 50 kg on link 1, link 3, and the mobile platform, respectively) will occur at 8 s of the simulation process. The disturbance is given as

$$\tau_{\xi d} = 30[\sin(20t); 0.9\cos(30t); 1.2\sin(18t); \sin(20t); 0.9\cos(30t); 1.2\sin(18t)]$$

In addition, the friction term is also considered in this simulation process:

$$\begin{aligned} F_\xi = \big[&11\dot{\xi}_1 + 0.5\text{sign}(\dot{\xi}_1); 9\dot{\xi}_2 + 0.5\text{sign}(\dot{\xi}_2); 5\dot{\xi}_3 + 0.2\text{sign}(\dot{\xi}_3); \\ &6\dot{\xi}_4 + 0.3\text{sign}(\dot{\xi}_4); 5\dot{\xi}_5 + 0.8\text{sign}(\dot{\xi}_5)\big]. \end{aligned}$$

The simulation is carried out by using the MATLAB package. The time sample for the simulation process is 0.001 s.

The simulation results of the PIDC, NN, FWCMACNN, and proposed RFWCMACNN schemes are depicted in Figs. 3.17 and 3.19 (for case 1) and Figs. 3.18 and 3.20 (for case 2). The simulated NMSE values of each method are presented in Table 3.3. In cases 1 and 2, the tracking performances of all methods are good. However, the tracking errors of the proposed RFWCMACNN and FWCMACNN strategies converge faster than those of the NN and PIDC methods. Moreover, based on the NMSE measures in Table 3.3, the proposed RFWCMACNN strategy has position tracking improvements better than that of the FWCMACNN, PIDC, and NN schemes. Figure 3.17i–l (for case 1) and Fig. 3.18i–l (for case 2) represent the torque control inputs of the RFWCMACNN, FWCMACNN, NN, and PIDC methods. In case 1, the performances of control torque inputs of all the methods are good. In case 2, while the proposed RFWCMACNN, the FWCMACNN, and the NN strategies can show good torque input performance at the time-varying uncertainty conditions (higher disturbance frequency, changing load on links), the PIDC torque input performance has occurred chattering phenomena. The performances of the PIDC, shown in these figures, are the best performance in terms of handling the heuristic process, but the chattering phenomena still exist when the frequency of the external disturbances increases or the load on manipulators is changed suddenly. In the NN simulation, some control parameters, such as

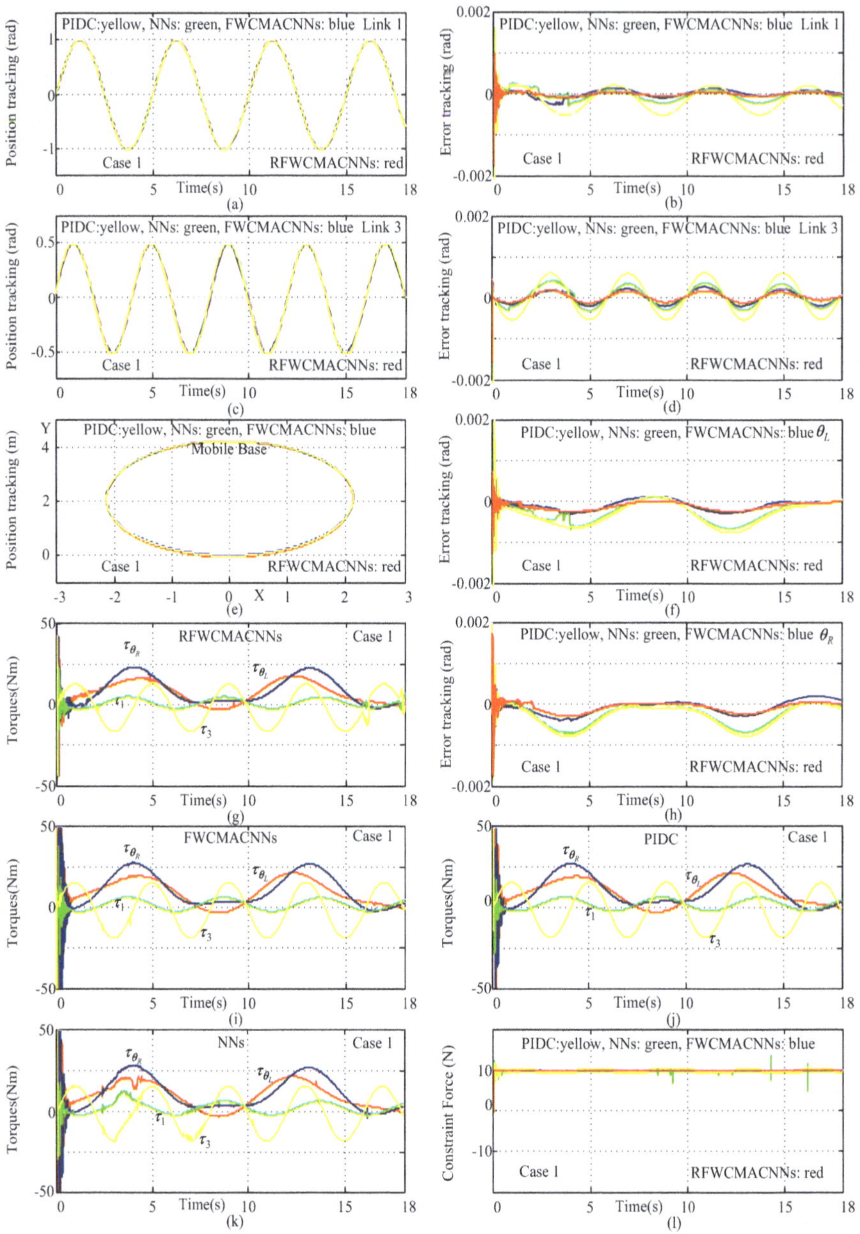

Fig. 3.17 Simulation results with case 1: the tracking positions, position tracking errors, and control inputs. (**a**, **c**, and **e**) Tracking position performances. (**b**, **d**, **f**, and **h**) Tracking error performances. (**g**) Constraint force input. (**i**–**l**) Torque control inputs of RFWCMACNN, FWCMACNN, NN, and PIDC methods

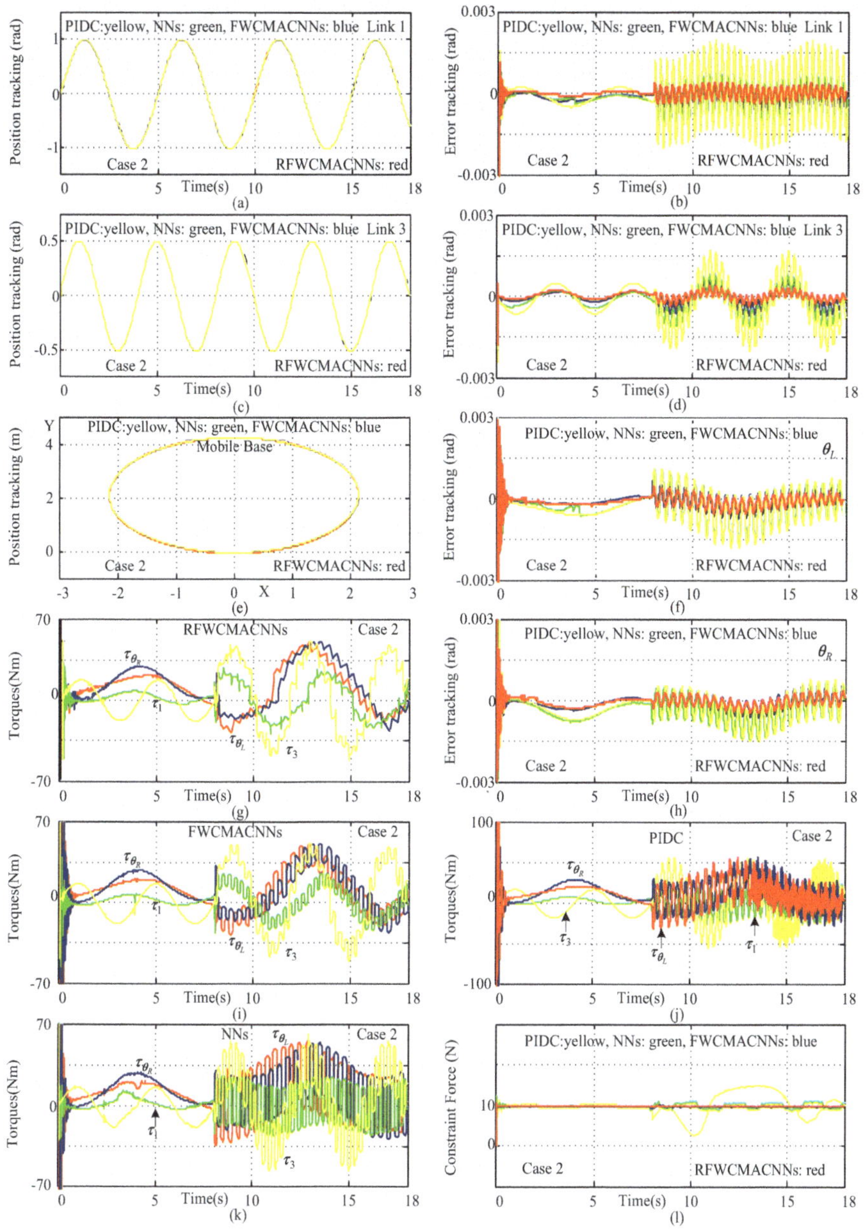

Fig. 3.18 Simulation results with case 2: the tracking positions, position tracking errors, and control inputs. (**a**, **c**, and **e**) Tracking position performances. (**b**, **d**, **f**, and **h**) Tracking error performances. (**g**) Constraint force input. (**i**–**l**) Torque control inputs of RFWCMACNN, FWCMACNN, NN, and PIDC methods

Table 3.3 Simulation performance comparisons of RFWCMACNN, FWCMACNN, NN, and PIDC schemes

Simulation NMSE ($\times 10^{-4}$)	Case 1 (unit: rad)				Case 2 (unit: rad)			
	θ_L	θ_R	Link 1	Link 3	θ_L	θ_R	Link 1	Link 3
RFWCMACNNs	1.543	1.725	0.515	0.358	1.704	1.836	0.819	0.604
FWCMACNNs	1.680	1.887	0.643	0.496	1.823	1.995	0.957	0.944
NNs	2.285	2.613	0.850	0.772	2.703	2.821	1.559	1.475
PIDC	2.402	2.705	1.775	1.657	2.924	2.940	3.256	2.548

RBF function parameters that help achieve high accuracy, are not easy to determine. These drawbacks cause the adaptation of the controller-based NNs or the PIDC to be lower than in the proposed method. In this simulation process, the parameters of the proposed controller are chosen through some trials. The rise time and steady-state error can be reduced by increasing λ_ξ, $K_{s\xi}$, K_η. However, the fast rise time and small steady-state error will increase the control input magnitude. The learning parameters are chosen based on the response of the tuning objects and the accuracy of the approximation process. The selection of these parameters relates to the convergence rate of the state error. High learning rates may cause the RFWCMACNNs to produce unstable output although the convergence speed becomes faster.

The simulation comparisons with the constraint force of all the methods are presented in Figs. 3.17g and 3.18g. The performance of the constraint force control depends on the performance of tracking position control. The design for the constraint force in the NN controller is the same as the RFWCMACNNs. However, the design for the constraint force in the PIDC scheme is different; one more PIDC for the constraint force control is considered. In case 1, the performance of the constraint force of the RFWCMACNNs, FWCMACNNs, and NNs is good, but the simulation result of the PIDC is poor in the steady state. In case 2, while the proposed and FWCMACNN methods can show good constraint force performance at the parameter variation conditions, the simulation results of the NN and PIDC methods are not good, and the undesired response appears when the time increases. The robust characteristic of the proposed RFWCMACNN and FWCMACNN controllers can be set with regard to the parameter variation and external disturbances.

The simulated results of the robust term, the approximating functions, and the estimated parameters of the RFWCMACNN and FWCMACNN methods (for cases 1 and 2) are depicted in Figs. 3.19 and 3.20. The performances of the approximating functions and estimation parameters are good, and they have proved the correctness of the proposed method including the boundedness of the control system parameters. Based on the comparison simulations, the proposed RFWCMACNN controller is suitable to be implemented to control the CCMMR/MMR under the occurrence of parameter variation and external disturbances.

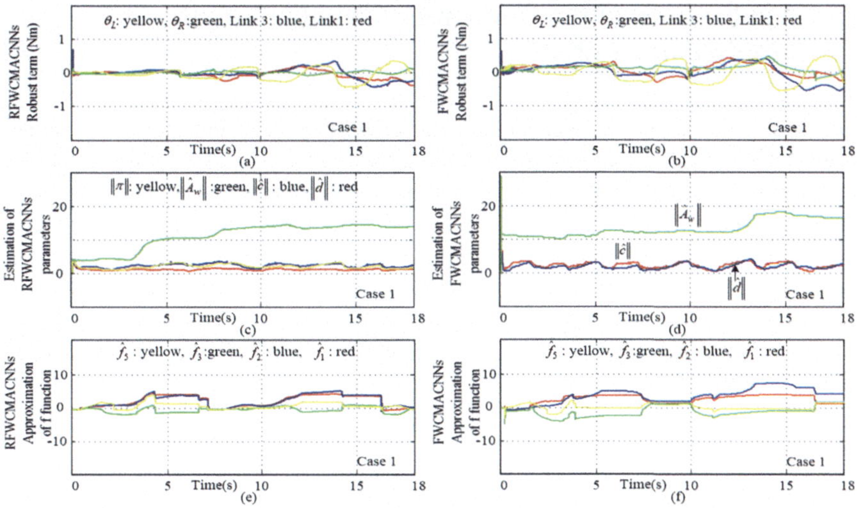

Fig. 3.19 Simulation results with case 1: the estimated parameters, approximation functions, and robust terms. (**a** and **b**) Robust term performances of RFWCMACNN and FWCMACNN methods. (**c** and **d**) Estimated parameter performances of RFWCMACNN and FWCMACNN methods. (**e** and **f**) Approximation functions of RFWCMACNN and FWCMACNN methods

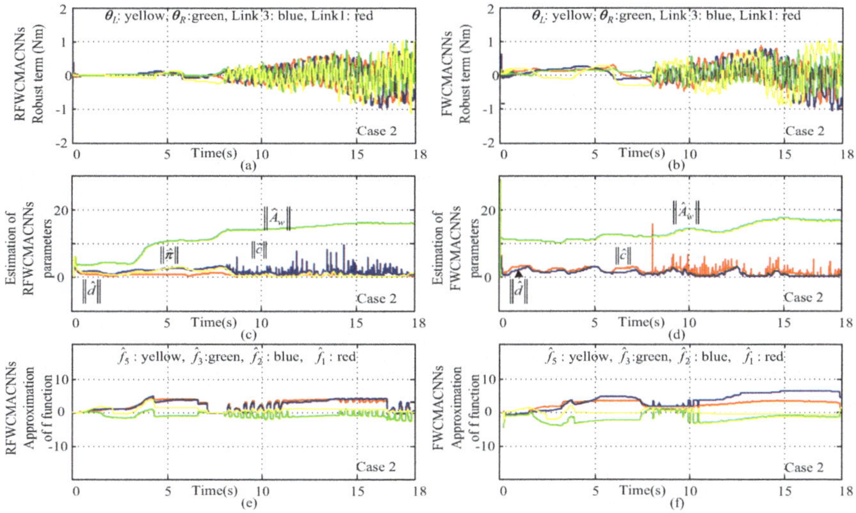

Fig. 3.20 Simulation results with case 2: the estimated parameters, approximation functions, and robust terms. (**a** and **b**) Robust term performances of RFWCMACNN and FWCMACNN methods. (**c** and **d**) Estimated parameter performances of RFWCMACNN and FWCMACNN methods. (**e** and **f**) Approximation functions of RFWCMACNN and FWCMACNN methods

3.5 Applications of the Condenser Cleaning Mobile Manipulator Robot

The above simulations well illustrated the effectiveness of the proposed theory. However, it is well known that good simulation performance of a control law does not guarantee its applicability in real systems due to friction and backlash between joints, unexpected disturbances in signal measurements, difficulty in parameter adjustment, and other limitations. Therefore, in this section, we will introduce the design proposal and its specific implementation of the condenser cleaning mobile manipulator robot. Following that, utilizing the control scheme outlined above, we will perform cleaning tasks on the condenser pipelines and provide specific data to demonstrate the effectiveness and feasibility of this control scheme.

3.5.1 Design and Experimental Setup

In a real industrial power plant, condensers are arranged on an elevated platform, heavily affected by a significant amount of dust, dirt, and deposits. To address this, we employ a condenser cleaning robot equipped with a high-precision end effector. This end effector follows predefined trajectories to carry out the condenser cleaning task.

To carry out the condenser cleaning task described above and validate the effectiveness and feasibility of the three control schemes mentioned in this section, the experimental results are presented in Sect. 3.5.2.

As shown in Fig. 3.21, the CCMMR was used for the experiment. The structure of the CCMMR is described as follows. Using the high-pressure water jet cleaning as the cleaning media, without prejudice to the safety production process of plant, the large-scale CCMMR is arranged according to the condenser tubes (using 30 MW unit as an example, the volume of condenser water room is 5 m × 5 m × 5 m; cold condensate tubes are 21,552; condenser length is 11.46 m; condenser diameter is 25–36 mm; wall thickness is 1.0 mm or 1.5 mm). The condenser cleaning mobile manipulator is movable in the condenser water chamber. It can calculate the exact position of each cleaning condensation tube and automatically determine the optimal pressure washing, cleaning time, or cleaning cycles. Figure 3.23 gives the mechanical structure diagram.

Remark 3.3 The red-end part of the forearm is just used to keep balance. Therefore, we take out such part of Figs. 3.22 and 3.23 for the purpose of simplicity.

Let the desired output trajectory z_d be $z_d = [t \quad 2t \quad \cos(2t)\sin(2t) - \cos(2t)]$. Table 3.4 explains the experimental and CCMMR parameters that were utilized for the experiment and their values.

The electrical control system of the condenser cleaning mobile manipulator is based on the distributed control structure and adopts a multi-parallel processing

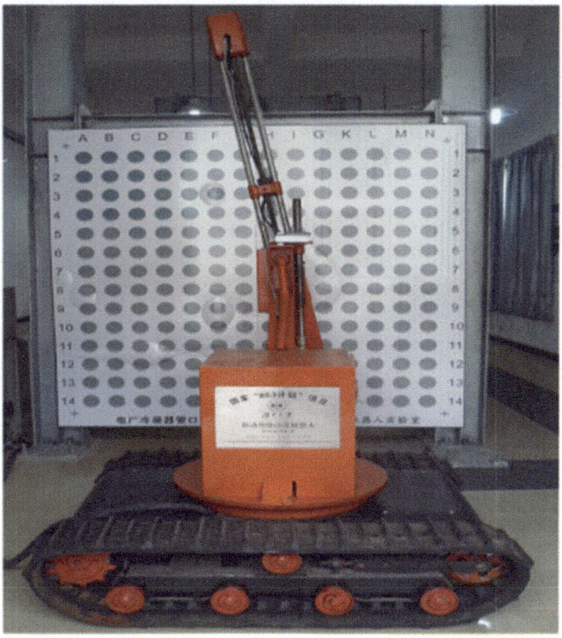

Fig. 3.21 The condenser cleaning mobile manipulator

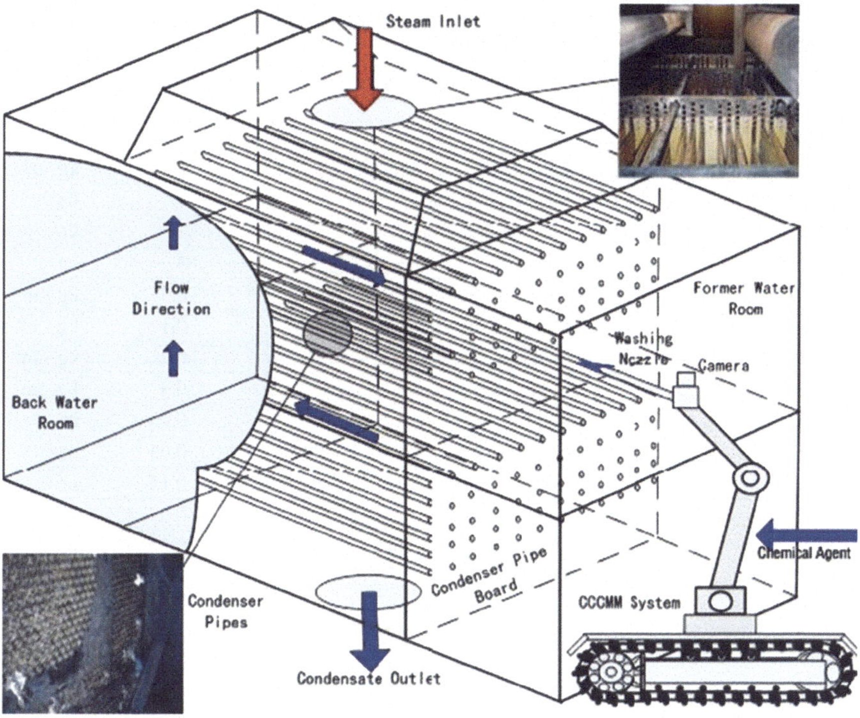

Fig. 3.22 Work principle of the CCMMR

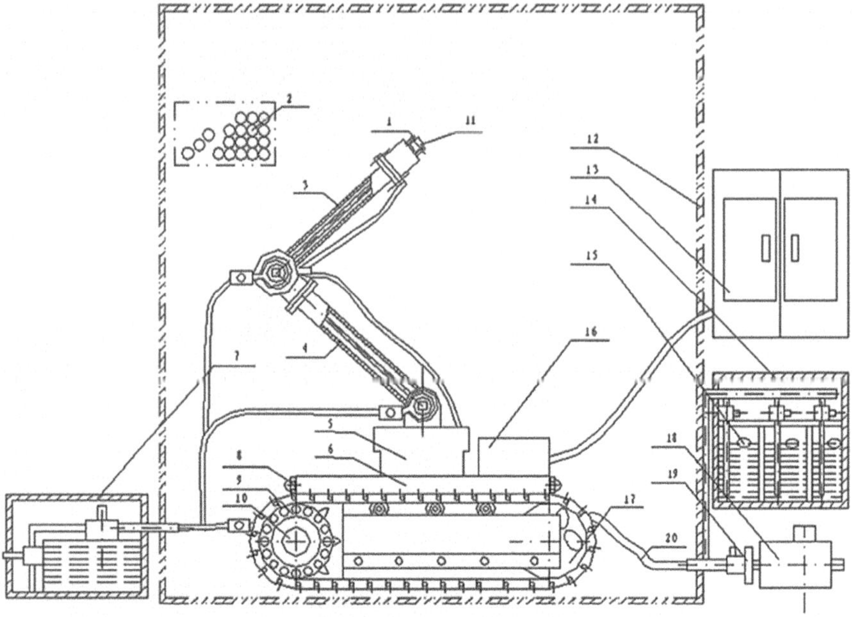

Fig. 3.23 Mechanical structure diagram of CCMMR. (1) underwater camera support; (2) condenser; (3) forearm; (4) big arm; (5) rotating base; (6) chassis; (7) liquid storage tank; (8) sonar sensors; (9) track; (10) wheel; (11) an underwater camera; (12) water room; (13) control cabinet; (14) chemical tank; (15) liquid level sensor; (16) sealed box; (17) the driven wheel; (18) high-pressure pumps; (19) pressure sensor; (20) hose

Table 3.4 Parameters of the mobile manipulator

Parameter	Value	Unit
Mass of the platform m_0	10	kg
Mass of link 1 m_w	1	kg
Mass of link 2 m_w	3	kg
Mass of link 3 m_w	3	kg
Inertia of the platform I_0	0.3	$kg \cdot m^2$
Inertia of the platform I_1	0.1	$kg \cdot m^2$
Inertia of the platform I_2	0.2	$kg \cdot m^2$
Inertia of the platform I_3	0.3	$kg \cdot m^2$
Radius of the wheel r	0.05	m
Length of link 1 l_1	0.05	m
Length of link 2 l_2	0.15	m
Length of link 3 l_3	0.2	m
Distance between the two caterpillars	0.2	m

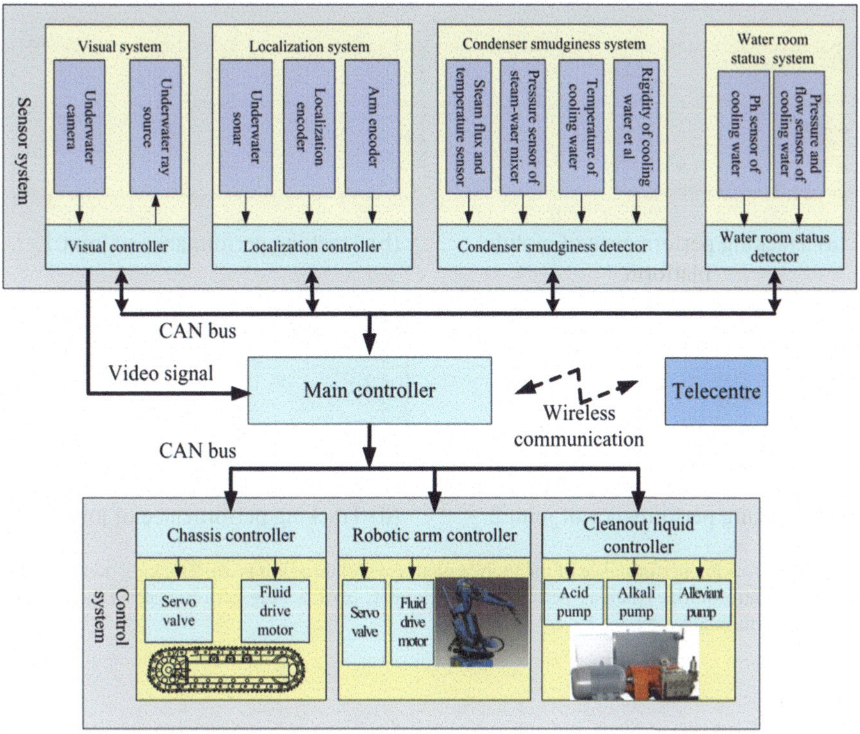

Fig. 3.24 Electrical control structure of CCMMR

system. The internal core of the robot host controller is connected to multiple coprocessor modules through the CAN bus network. The coprocessor modules form the internal network of the robot structure. To meet the remote monitoring and teleoperation needs, the robot body could realize information interaction with the remote monitoring machine through the wireless network. This design enables the robot to have high-speed information processing and real-time communication capabilities. Figure 3.24 describes the electrical control structure of the intelligent cleaning robot.

3.5.2 *Experimental Results of the Control Method for the Condenser Cleaning Mobile Manipulator Robot*

The experiment results are illustrated in Figs. 3.25 and 3.26. The actual computed trajectory of the platform is shown in Fig. 3.25a. The tracking performances of the manipulator are shown in Fig. 3.25b–d. Figure 3.26 shows the control torques of two caterpillars and three joints. It is obvious that the system converges to the desired

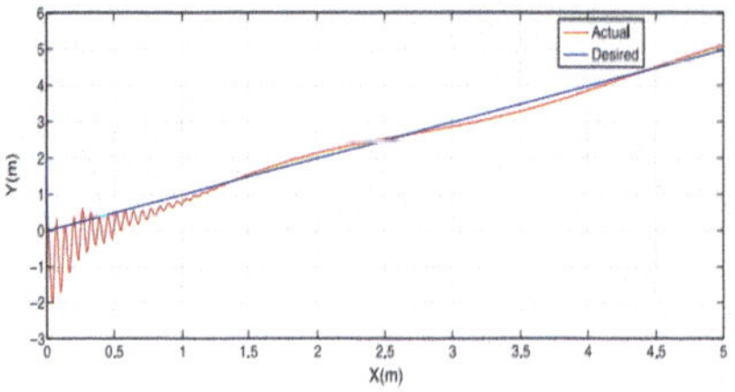

(a) Tracking performance of mobile platform.

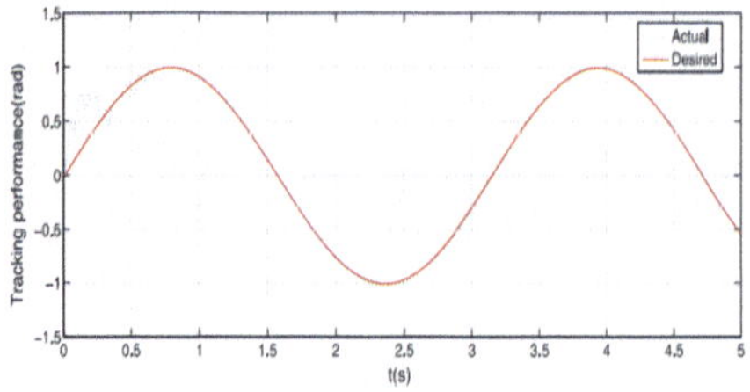

(b) Tracking performance of joint 1.

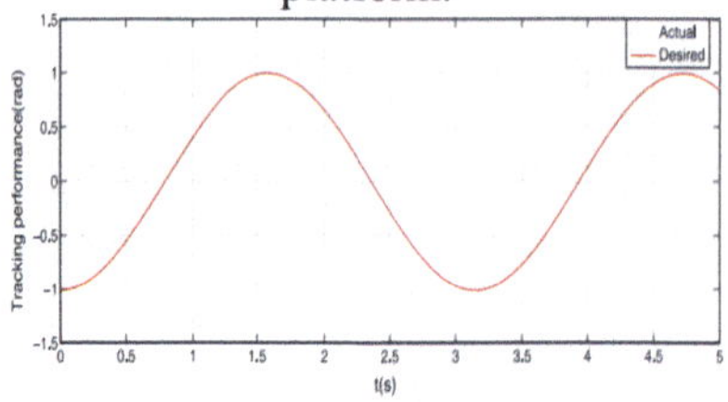

(c) Tracking performance of joint 2.

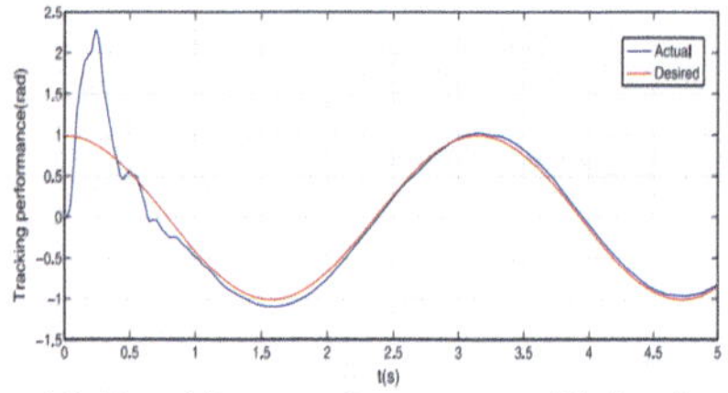

(d) Tracking performance of joint 3.

Fig. 3.25 Tracking performance of the mobile platform and three joints. (**a**) Tracking performance of mobile platform. (**b**) Tracking performance of joint 1. (**c**) Tracking performance of joint 2. (**d**) Tracking performance of joint 3

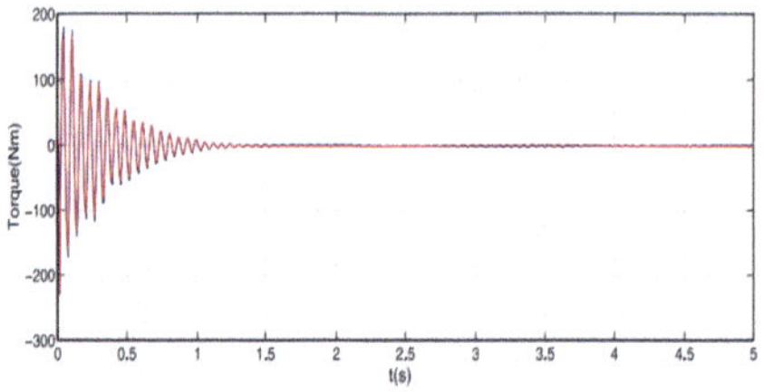

Control torque of two caterpillar.

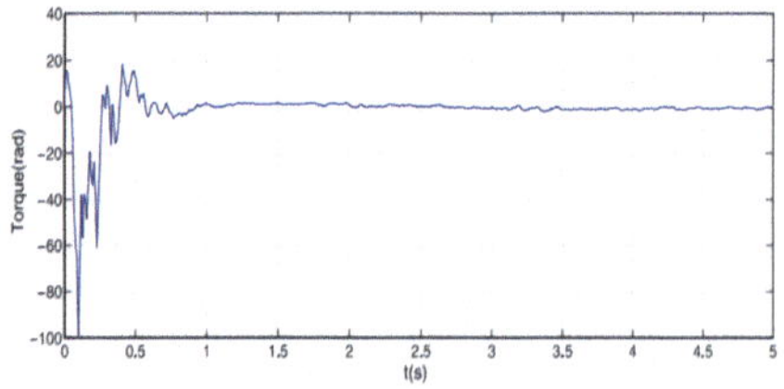

(b) Control torque of Joint 1.

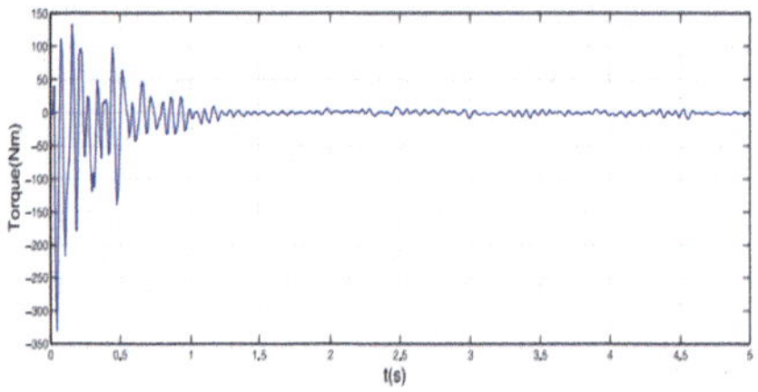

(c) Control torque of Joint 2.

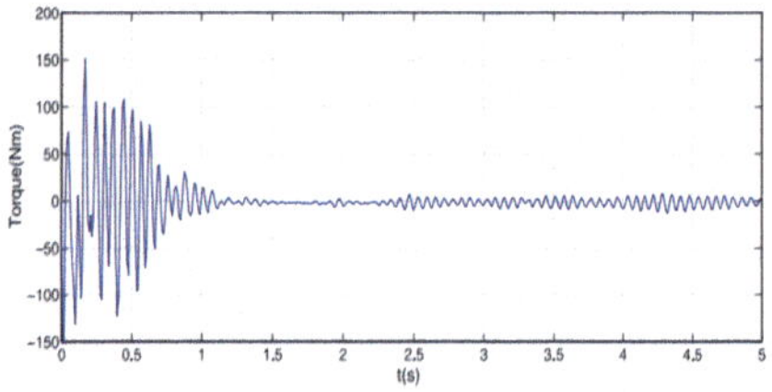

(d) Control torque of Joint 3.

Fig. 3.26 Control torques of two caterpillars and three joints. (**a**) Control torque of two caterpillars. (**b**) Control torque of joint 1. (**c**) Control torque of joint 2. (**d**) Control torque of joint 3 trajectory quickly and achieves good tracking performance. Therefore, the proposed controller is able to reduce chattering on the control input while maintaining a favorable tracking performance of the desired trajectory.

Moreover, the experimental work will also be utilized with the CCMMR control system. To be consistent with the works in the simulation process, link 3 of the CCMMR will not be considered in the experimental process. Some experimental results are provided here to further verify the effectiveness of the proposed method. Two testing cases are implemented to verify the robustness of the proposed control system. The first case is considered with a no-variation condition. The second case is considered with varying conditions at 8 s, i.e., adding more loads on link 2 and mobile base. In addition, this section also adds more external disturbances, i.e., by increasing the friction forces between the joint connections (for the second case). The desired positions and the parameters of the controlled system are chosen similarly to the simulation part.

By analyzing the experimental process of the PIDC system when applied in the CCMMR control system, it is obvious that if a tip load or external disturbance (joint and wheel friction) is inserted into the CCMMR system, then the tracking performance drops significantly. When the PIDC parameters are selected properly, this phenomenon can be improved. However, the control efforts become larger, so the stability of the controlled system cannot be guaranteed. In this experimental work, the PIDC parameters are chosen by the prior knowledge of the CCMMR control system, so the performances of the tracking positions and control efforts are acceptable, and the stability of the CCMMR control system is still guaranteed. The experimental process of the NN, the FWNN, and the proposed RFWNN control systems are implemented more favorably than the PIDC system. However, the determination of the RBF and initialized parameters for the NN control system still need more attention.

In the first case, Fig. 3.27a–h represents the control voltages and tracking errors of the PIDC, NN, FWNN, and RFWNN strategies, respectively. In the second case, Fig. 3.28a–h represents the control voltages and tracking errors of the PIDC, NN, FWNN, and the proposed RFWNN strategies, respectively. By comparison, this section can clearly observe in these figures that good tracking error responses can be obtained with the RFWNN, FWNN, PIDC, and NN methods, but the tracking errors of the proposed RFWNN strategy converge faster than that of the FWNN, NN, and PIDC methods. These improvements are also described by the comparison of the NMSE results in Table 3.5.

The robust terms and estimated parameters (of the wavelet functions) of the proposed RFWNN and FWNN methods, in the first and second cases of the experimentation process, are depicted in Figs. 3.29 and 3.30. Based on these results, they can show that the boundedness of the RFWNN/FWNN parameters is guaranteed. Moreover, the robustness of the proposed RFWNN control system under the occurrence of the parameter variation can be clearly observed by comparing the performance of the tracking errors and control voltages presented in Figs. 3.27 and 3.28. Following the comparative simulation and experimental results, this section can conclude that the control performances and the robustness of the

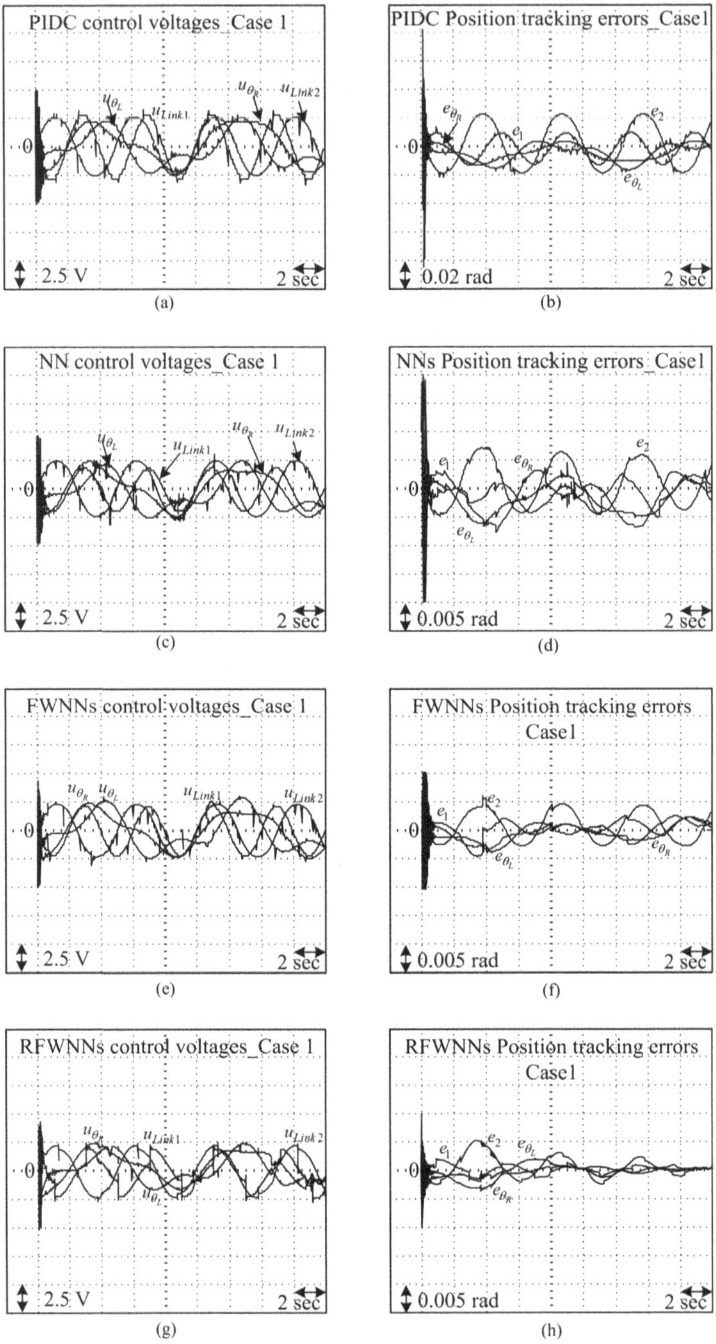

Fig. 3.27 Experimental results with the first case: the position tracking errors and control input voltages. (**a** and **b**) Performances of PIDC method. (**c**) and (**d**) Performances of NN method. (**e** and **f**) Performances of FWNN method. (**g** and **h**) Performances of RFWNN method

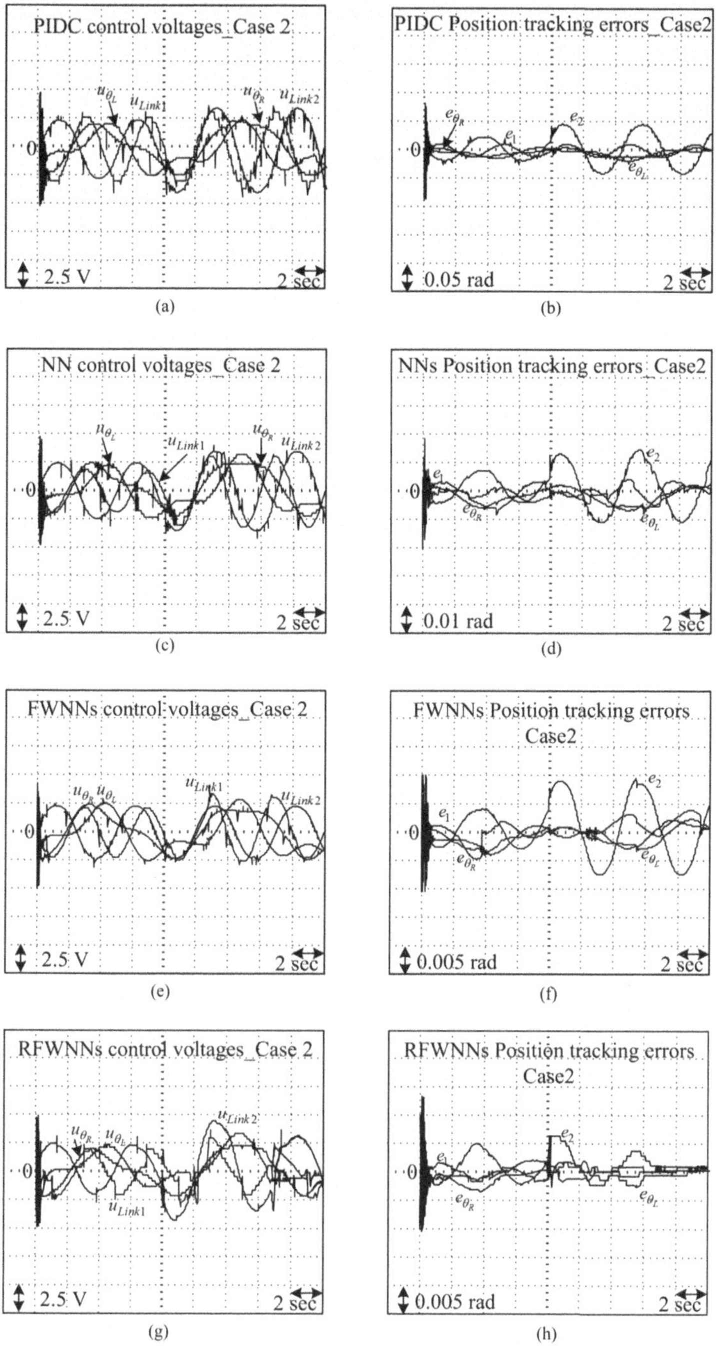

Fig. 3.28 Experimental results with the second case: the position tracking errors and control input voltages. (**a** and **b**) Performances of PIDC method. (**c** and **d**) Performances of NN method. (**e** and **f**) Performances of FWNN method. (**g** and **h**) Performances of RFWNN method

Table 3.5 Experimentation performance comparisons of RFWNN, FWNN, NN, and PIDC schemes

Experimentation NMSE ($\times 10^{-3}$)	Case 1 (unit: rad)				Case 2 (unit: rad)			
	θ_L	θ_R	Link 1	Link 2	θ_L	θ_R	Link 1	Link 2
RFWNNs	1.625	1.850	0.641	0.630	1.792	1.976	0.828	1.314
FWNNs	1.657	1.963	0.889	0.892	1.875	2.273	1.023	1.724
NNs	2.439	2.631	1.245	1.312	2.753	3.180	1.812	2.446
PIDC	4.734	5.601	3.912	5.004	5.534	6.238	4.689	8.783

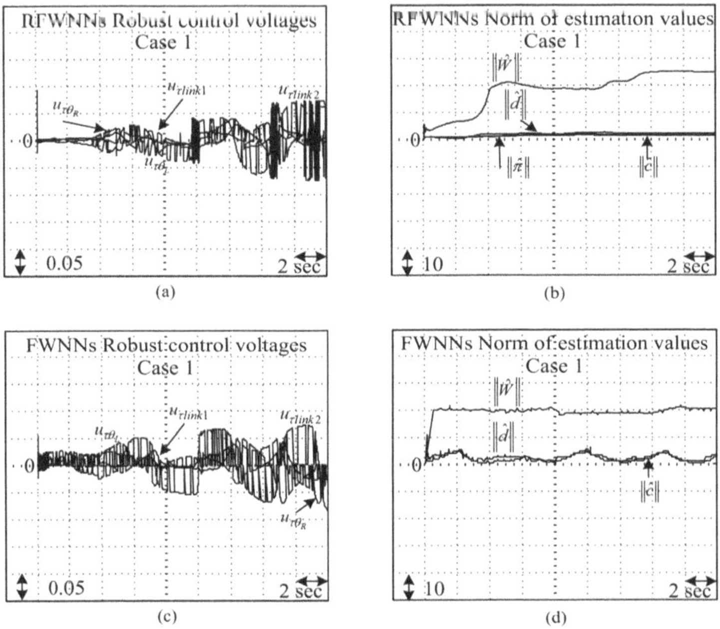

Fig. 3.29 Experimental results with the first case: the robust term voltages and estimated parameters. (**a** and **b**) Performances of RFWNN method. (**c** and **d**) Performances of FWNN method

proposed controller-based RFWNNs of the control of the CCMMR system under the time-varying conditions are better than that of the FWNN, NN, and PIDC strategies.

Furthermore, experimental results are provided to verify the effectiveness of the proposed method. Two testing cases are implemented to verify the robustness of the proposed control system.

The first case is considered with a no-variation condition. The second case is considered with varying conditions at 8 s, i.e., adding more loads on links 1 and 2, and a mobile base. In addition, this section also adds more external disturbances, i.e., by increasing the friction forces between the joint connections (with the second

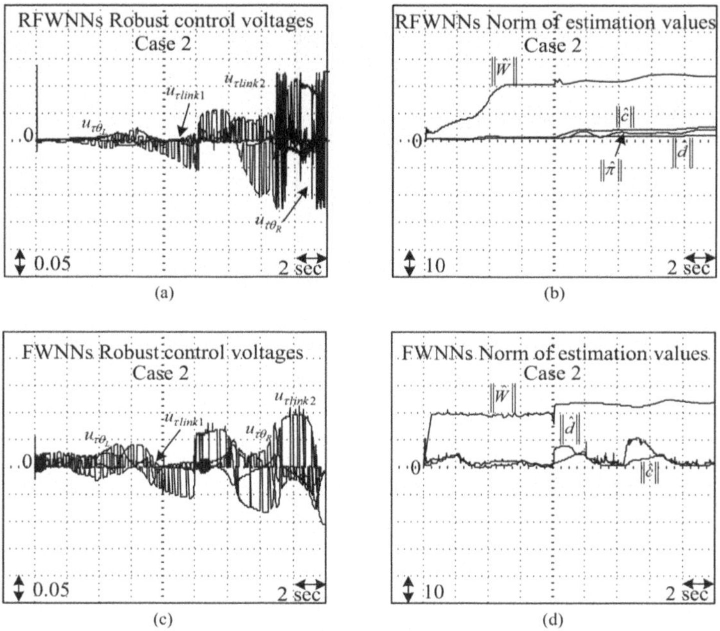

Fig. 3.30 Experimental results with the second case: the robust term voltages and estimated parameters. (**a** and **b**) Performances of RFWNN method. (**c** and **d**) Performances of FWNN method

case). The desired positions and the parameters of the controlled system are similar to the simulation part.

Similarly, the addressed control methods in the simulation part are also examined in this part to exhibit the superior control performance and effectiveness of the proposed RFWCMACNN method. Note that in the experimentation process, the control parameters of all methods are chosen to achieve superior transient control performance by considering the limitation of the control effort, the requirement of stability, and the possible operating conditions.

In the first case, Fig. 3.31a–h represents the control voltages and tracking errors of the PIDC, NN, FWCMACNN, and RFWCMACNN strategies, respectively. In the second case, Fig. 3.32a–h represents the control voltages and tracking errors of the PIDC, NN, FWCMACNN, and RFWCMACNN strategies, respectively. By comparison, this section can clearly conclude that in these figures, good tracking error responses can be obtained with the RFWCMACNN, FWCMACNN, PIDC, and NN methods, but the tracking errors of the proposed RFWCMACNN strategy converge faster than that of the FWCMACNN, NN, and PIDC methods. These improvements are also described by the comparison of the NMSE results in Table 3.6. The robust terms and estimated parameters of the wavelet function of the proposed RFWCMACNN and FWCMACNN methods in the first and second cases are depicted in Figs. 3.33 and 3.34. Based on these results, it is difficult to make a

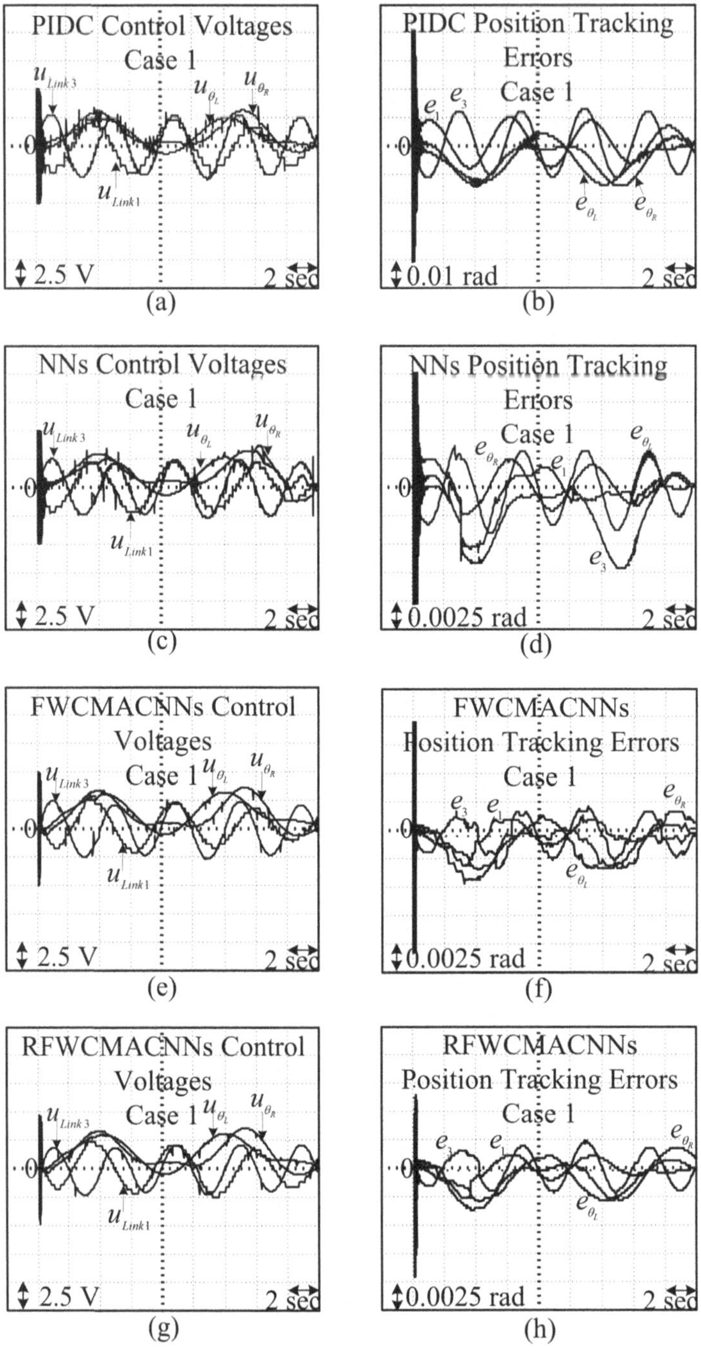

Fig. 3.31 Experimental results with the first case: the position tracking errors and control input voltages. (**a** and **b**) Performances of PIDC method. (**c**) and (**d**) Performances of NN method. (**e** and **f**) Performances of FWCMACNN method. (**g** and **h**) Performances of RFWCMACNN method

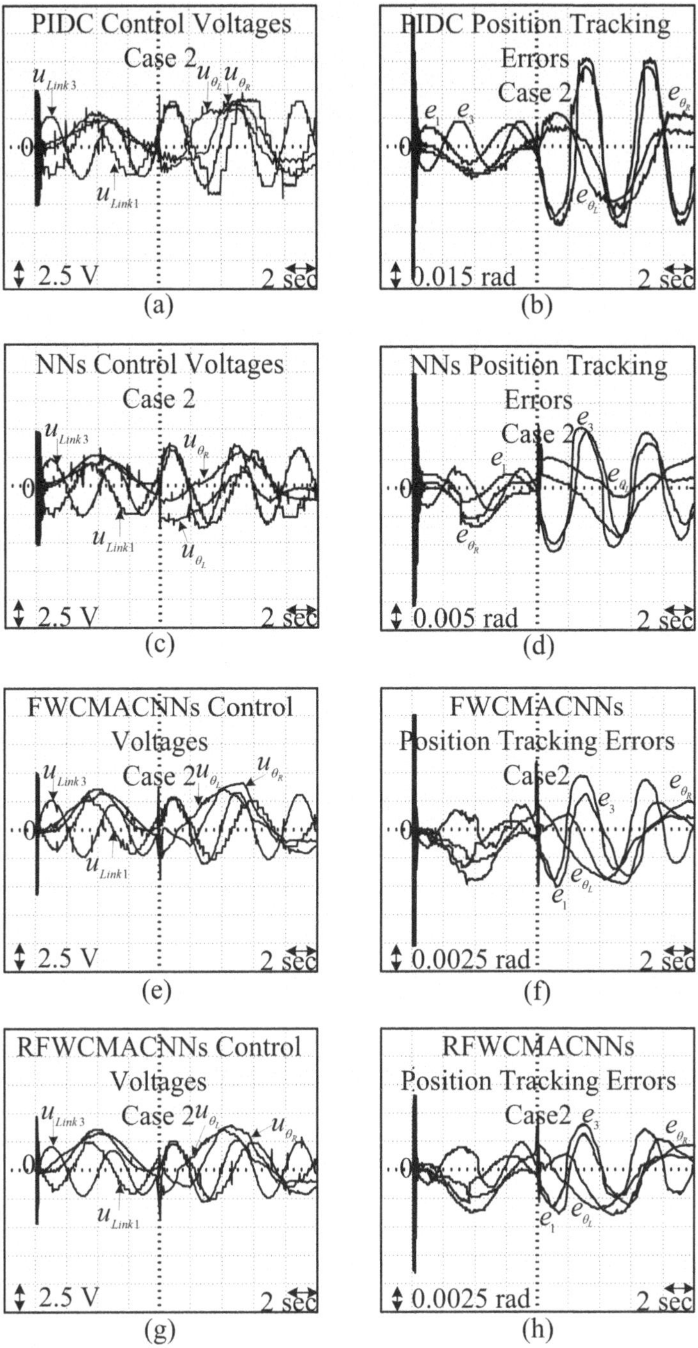

Fig. 3.32 Experimental results with the second case: the position tracking errors and control input voltages. (**a** and **b**) Performances of PIDC method. (**c** and **d**) Performances of NN method. (**e** and **f**) Performances of FWCMACNN method. (**g** and **h**) Performances of RFWCMACNN method

Table 3.6 Experimentation comparisons of RFWCMACNN, FWCMACNN, NN, and PIDC schemes

Experimentation NMSE ($\times 10^{-3}$)	Case 1 (unit: rad)				Case 2 (unit: rad)			
	θ_L	θ_R	Link 1	Link 3	θ_L	θ_R	Link 1	Link 3
RFWCMACNNs	1.324	1.565	0.353	0.456	1.623	1.895	0.675	0.698
FWCMACNNs	1.541	1.832	0.592	0.487	1.862	2.021	0.964	0.989
NNs	2.048	2.632	0.961	0.922	2.381	2.884	1.571	1.472
PIDC	4.765	5.410	3.065	2.821	5.973	5.952	6.042	4.585

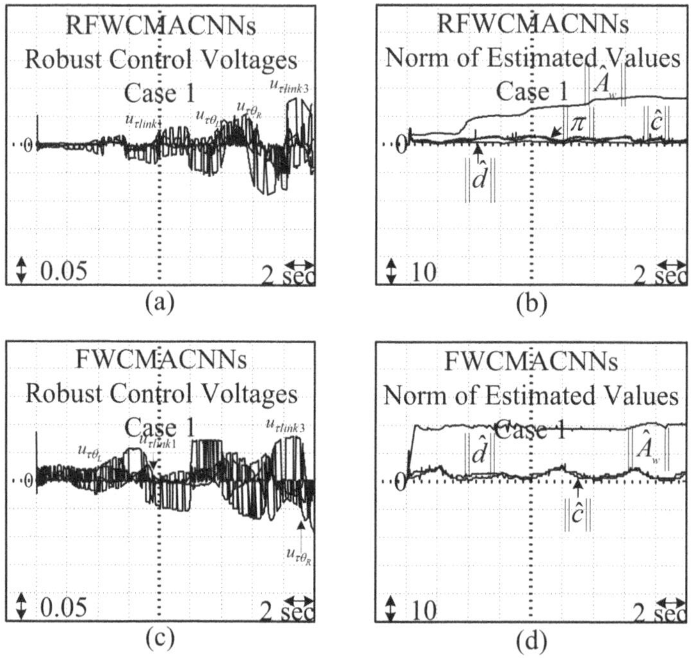

Fig. 3.33 Experimental results with the first case: the robust term voltages and estimated parameters. (**a** and **b**) Performances of RFWCMACNN method. (**c** and **d**) Performances of FWCMACNN method

comparison between the proposed method and the FWCMACNN method, but this section can realize that the boundedness of the RFWCMACNN parameters is guaranteed. Moreover, the robustness of the proposed RFWCMACNN control system under the occurrence of the parameter variation can be clearly observed by comparing the performances of the tracking errors and control voltages presented in Figs. 3.31 and 3.32. Following the comparative simulation and experimental results, this section can conclude that the control performance and robustness of the

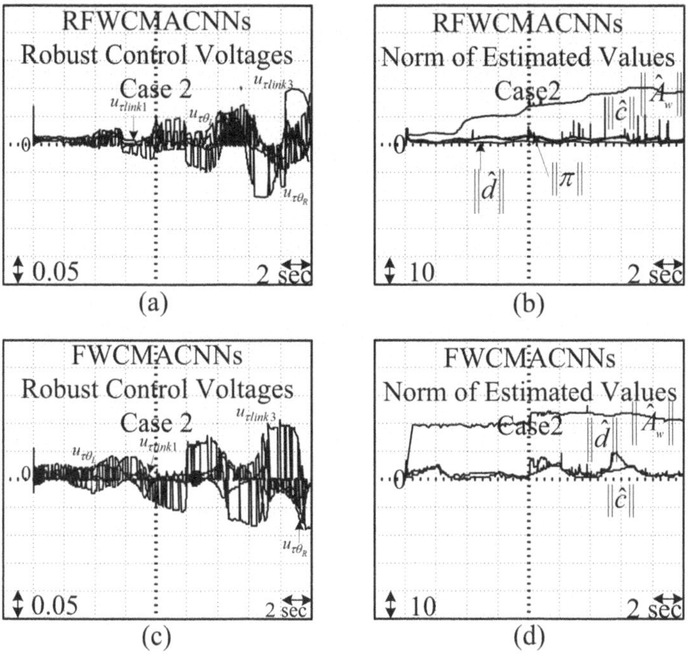

Fig. 3.34 Experimental results with the second case: the robust term voltages and estimated parameters. (**a** and **b**) Performances of RFWCMACNN method. (**c** and **d**) Performances of FWCMACNN method

proposed controller-based RFWCMACNNs for the control of the CCMMR under time-varying conditions are better than that of the FWCMACNN, NN, and PIDC strategies.

References

1. Park BS, Park JB, Choi YH. Robust adaptive formation control and collision avoidance for electrically driven non-holonomic mobile robots. IET Control Theory & Applications, 2011,5 (3):514–522.
2. Li Z, Ge SS, Adams M, Wijesoma WS. Robust adaptive control of uncertain force/motion constrained nonholonomic mobile manipulators. Automatica, 2008,44(1):776–784.
3. Lin S, Goldenberg A. Neural-network control of mobile manipulators. Neural Networks, 2001,12 (5):1121–1133.
4. Li Z, Ge SS, Ming A. Adaptive robust motion/force control of holonomic-constrained nonholonomic mobile manipulators. IEEE Transactions on Systems, Man and Cybernetics. Part B. Cybernetics, 2007,37(3): 607–616.
5. Oya M, Su C, Katoh R. Robust adaptive motion/force tracking control of uncertain nonholonomic mechanical systems. IEEE Transactions on Robotics and Automation. 2003,19 (1):175–181.

6. Zhang H, Wang YN, Peng JZ, Sun W. The design and control on intelligent underwater cleaning robot for power plant condenser. International Journal of Modelling, Identification and Control. 2009;7(4):392–397.
7. Yamamoto Y, Yun X. Coordinating locomotion and manipulation of a mobile manipulator. IEEE Transactions on Automatic Control. 1994, 39(6):1326–1332.
8. Li Z, Chen W, Luo J. Adaptive compliant force-motion control of coordinated non-holonomic mobile manipulators interacting with unknown non-rigid environments. Neurocomputing. 2008,71(10):1330–1344
9. Tsai C, Cheng M, Lin S. Dynamic modeling and tracking control of a nonholonomic wheeled mobile manipulator with dual arms. Journal of Intelligent & Robotic Systems. 2006,47(2):317–340.
10. Becerra HM, Lopez-Nicolas G, Sagues C. A sliding-mode-control law for mobile robots based on epipolar visual servoing from three views. IEEE Transactions on Robotics. 2011,27(1):175–183.
11. Mousavi SH, Noroozi N, Safavi AA, Ebadat A. Modeling and control of nonlinear systems using novel fuzzy wavelet networks: the output adaptive control approach. Communications in Nonlinear Science and Numerical Simulation. 2011,16(9):3798–3814.
12. Lin CK. Nonsingular terminal sliding mode control of robot manipulators using fuzzy wavelet networks. IEEE Transactions on Fuzzy Systems. 2006,14(6):849–859.
13. Lin CK. H∞ reinforcement learning control of robot manipulators using fuzzy wavelet networks. Fuzzy Sets and Systems. 2009,160(12):1765–1786.
14. Lewis FL, Jagannathan S, Yesildirek A. Neural network control of robot manipulators and nonlinear systems. New York: Taylor and Francis, 1998.
15. Omidvar O, Elliott DL. Neural systems for control. New York: Academic, 1997.
16. Lin S, Goldenberg AA. Neural-network control of mobile manipulators. IEEE Transactions on Neural Networks. 2001,12(5):1121–1133.
17. Li Z, Yang C, Gu J. Neuro-adaptive compliant force/motion control for uncertain constrained wheeled mobile manipulator. International Journal of Robotics and Automation. 2007,22 (3):206–214.
18. Xu D, Zhao D, Yi J, et al. Trajectory tracking control of omnidirectional wheeled mobile manipulators: robust neural network-based sliding mode approach. IEEE Transactions on System, Man, and Cybernetics, Part B: Cybernetics. 2009,39(3):788–799.
19. Mbede JB, Ele P, Abia CMM, et al. Intelligent mobile manipulator navigation using adaptive neuro-fuzzy systems. Information Sciences. 2005,171(4):447–474.
20. Albus JS. A new approach to manipulator control: the cerebellar model articulation controller (CMAC). Journal of Dynamic Systems, Measurement, and Control. 1975,97(3):220–227.
21. Jagannathan S. Discrete-time CMAC NN control of feedback linearizable nonlinear systems under a persistence of excitation. IEEE Transactions on Neural Networks. 1999,10(1):128–137.
22. Kim YH, Lewis FL. Optimal design of CMAC neural-network controller for robot manipulators. IEEE Transactions on System, Man, and Cybernetics, Part C: Applications and Reviews. 2000,30(1):22–31.
23. Chen JY, Tsai PS, Wong CC. Adaptive design of a fuzzy cerebellar model arithmetic controller neural network. IEE Proceedings-Control Theory and Applications. 2005,152(2):133–137.
24. Su SF, Lee ZJ, Wang YP. Robust and fast learning for fuzzy cerebellar model articulation controllers. IEEE Transactions on System, Man, and Cybernetics, Part B: Cybernetics. 2006,36 (1):203–208.
25. Kruschke JK, Movellan JR. Benefits of gain: speeded learning and minimal hidden layers in back-propagation networks. IEEE Transactions on Systems, Man, and Cybernetics. 1991,21 (1):273–280.
26. Narendra KS, Annaswamy AM. Stable adaptive systems. Courier Corporation, 2012.
27. Wang S, Lin C. Adaptive tuning of the fuzzy controller for robots. Fuzzy Sets and Systems. 2000,110 (3):351–363.

28. Lewis FL, Liu K, Yesildirek A. Neural net robot controller with guaranteed tracking performance. IEEE Transactions on Neural Networks. 1995,6(3):703–715.
29. Chen S, Ho W, Chou J. Robust controllability of TS fuzzy-model-based control systems using sliding-mode technology. IEEE Transactions on Fuzzy Systems. 2009,17(6):1324–1335.
30. Xi Z, Feng G, Hesketh T. Piecewise integral sliding-mode control for T–S fuzzy systems. IEEE Transactions on Fuzzy Systems. 2010,19(1):65–74.
31. Jin N, Liu D. Wavelet basis function neural networks for sequential learning. IEEE Transactions on Neural Networks. 2008,19(3):523–528.
32. Delyon B, Juditsky A, Benveniste A. Accuracy analysis for wavelet approximations. IEEE Transactions on Neural Networks. 1995,6(2):332–348.
33. Abiyev RH, Kaynak O. Fuzzy wavelet neural networks for identification and control of dynamic plants—a novel structure and a comparative study. IEEE Transactions on Industrial Electronics. 2008,55(8):3133–3340.
34. Yilmaz S, Oysal Y. Fuzzy wavelet neural network models for prediction and identification of dynamical systems. IEEE Transactions on Neural Networks. 2010,21(10):1599–1609.
35. Lu C. Wavelet fuzzy neural networks for identification and predictive control of dynamic systems. IEEE Transactions on Industrial Electronics. 2010,58(7):3046–3058.
36. Hsu C. Adaptive fuzzy wavelet neural controller design for chaos synchronization. Expert Systems with Applications. 2011,38(8):10475–10483.
37. Lewis F, Yesildirek A, Liu K. Multilayer neural-net robot controller with guaranteed tracking performance. IEEE Transactions on Neural Networks. 1996,7(2):388–399.
38. Ho DWC, Zhang P, Xu J. Fuzzy wavelet networks for function learning. IEEE Transactions on Fuzzy Systems. 2001,9(1):200–211.
39. Zhang Q, Benveniste A. Wavelet networks. IEEE Transactions on Neural Networks. 1992,3(6):889–898.
40. Zhang J, Walter GG, Miao Y, et al. Wavelet neural networks for function learning. IEEE Transactions on Signal Processing. 1995,43(6):1485–1497.
41. Zhang Q. Using wavelet network in nonparametric estimation. IEEE Transactions on Neural Networks. 1997,8(2):227–236.
42. Slotine JJE. Applied nonlinear control. Englewood Cliffs: Prentice Hall, 1991.

Chapter 4
Transmission Line De-icing Robot Manipulator

4.1 Background and Motivation

Ice disasters pose a major threat to transmission lines and their associated power equipment [1–5]. Among them, the repair and inspection of transmission lines are the most challenging and have the most significant impact, as shown in Fig. 4.1. Generally, ice accumulation can cause three forms of damage to power lines. The first form occurs when there is a small amount of ice accumulation, and the ice layer wraps around the conductor in an irregular shape. Under the action of external forces such as wind, the conductor will undergo low-frequency and large-amplitude self-excitation vibration, which can lead to line breakage and tower collapse. The second situation is flashover; although the structural damage might not be caused, it results in the loss of insulation and renders the entire power grid unable to supply power normally. The third and most serious situation is that excessive vertical load due to ice accumulation on the line may cause structural damage to the entire power grid. Therefore, strengthening and improving the anti-icing ability of transmission lines, accelerating research on effective anti-icing and de-icing technologies for transmission lines, and minimizing and preventing the harm caused by ice disasters to transmission lines are issues that must be addressed by power companies, cable enterprises, and others in various countries.

The de-icing and anti-icing technologies and methods can be summarized into over 30 types, which can generally fall into 4 major categories [6, 7]: thermal melting method, mechanical ice breaking method, passive de-icing method, and other de-icing methods. Passive de-icing and anti-icing mainly rely on natural methods such as wind, gravity, random scattering, and environmental temperature to remove ice. Its advantages include not requiring additional energy, having low cost, and helping to prevent ice disasters. However, the passive de-icing method is focused on prevention. In the event of large-scale ice disasters, it can only rely on other melting technologies such as the thermal melting method and mechanical ice breaking

Y. Wang et al., *Electric Power Robots*,
https://doi.org/10.1007/978-981-96-2512-3_4

Fig. 4.1 Power transmission lines coated in ice and collapse caused by ice

Fig. 4.2 Manual de-icing on power transmission lines

method [8–10]. The most widely used and effective de-icing method at the present stage is the thermal melting de-icing method [11–13]. The principle of the thermal melting de-icing method is to increase the current in the power line, which in turn generates heat to melt ice. The thermal melting de-icing method mainly includes several approaches, such as adjusting power flow distribution and using phase shifters for load-carrying de-icing, AC short-circuit de-icing, and DC de-icing. However, the first method may cause scheduling and power flow transfer issues, potentially impacting system stability. While using the FACTS equipment in the transmission network increases power grid flexibility, its effectiveness is limited for large ice-covered areas due to restricted power flow transfer. The second method requires insulating split conductors, which can lead to significant costs for line transformation. The challenge with using AC and DC de-icing is the difficulty in finding a suitable power supply for de-icing. Although domestically developed DC de-icing devices can be applied for 220 kV level transmission lines, technical limitations still pose a problem for de-icing technology in China for the transmission networks of 500 kV and above. Moreover, during severe icing weather, standard de-icing methods fail, leading to the inevitable need for manual online de-icing. However, as shown in Fig. 4.2, manual de-icing methods are dangerous and

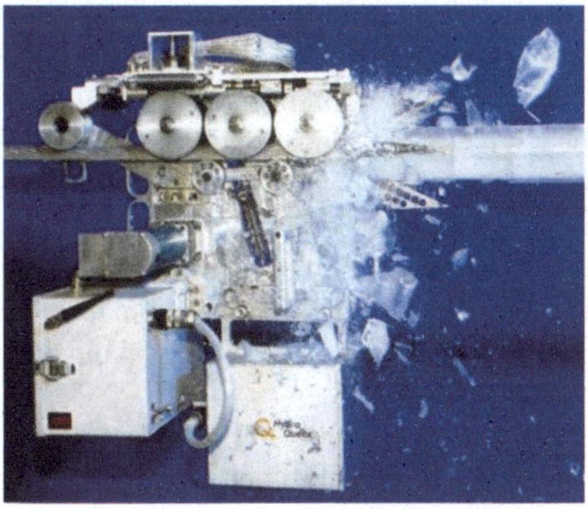

(a) Deicing robot with only line patrol operation between two poles (LineROVer)[14].

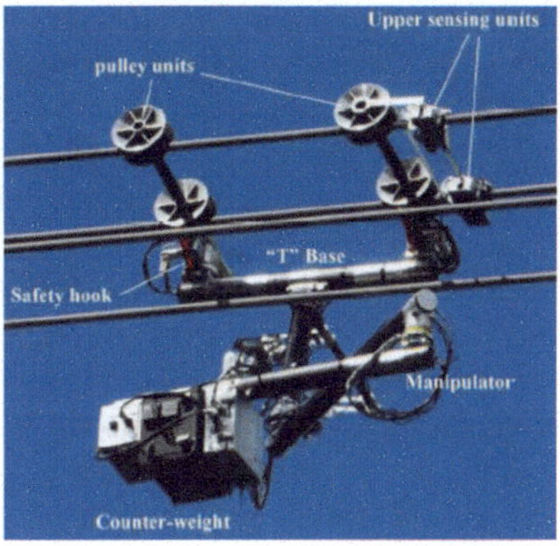

(b) Robot for Inspection of High-Voltage Lines (Expliner)[15].

(c) Inspection robot with the capable of clearing obstacles (LineScout)[16].

Fig. 4.3 Transmission line robot manipulators. (**a**) De-icing robot with only line patrol operation between two poles (LineROVer) [14]. (**b**) Robot for inspection of high-voltage lines (Expliner) [15]. (**c**) Inspection robot capable of clearing obstacles (LineScout) [16]

labor-intensive and have limited de-icing effects, which hinders the quick and effective control of the damage caused by ice on the power grid.

In order to replace manual de-icing and address the above difficulties and challenges, it is crucial to research and develop safe and effective de-icing methods. With the rapid developments of robotics technology, it has become possible to use robots instead of manual labor to complete various tasks in hazardous environments [14–16], as shown in Fig. 4.3. Recent research has been focusing on utilizing scientific and technological applications in the field of robotics to create ice removal robots for power transmission lines [17, 18]. This new technology offers several advantages, including low energy consumption, safety, efficiency, minimal risk and cost, continuous operation, and no impact on the power grid. Additionally, it can be used for live ice removal operations for substations and other power equipment.

When not required for ice removal, the robot can also be used for other purposes such as line inspection, making it highly versatile.

Ice removal robots for transmission lines are different from general ground mobile robots because they are suspended on flexible transmission lines. When these robots move along the transmission line, they have to navigate various obstacles such as vibration dampers, suspension insulators, clamps, and towers [19–24]. In order to enhance the intelligence and automation capabilities of the ice removal robot and enable it to walk, avoid obstacles, and remove ice automatically, it is important to address the motion control, navigation, and obstacle avoidance challenges of the ice removal robot.

Since the ice removal robot is a special suspended structure, all its obstacle-crossing and trajectory planning actions are achieved by controlling the joint angles of the robot's manipulator arm. Therefore, it is necessary to study one of the fundamental problems related to this—the inverse kinematics problem of the robot. The inverse kinematics problem of the robot involves finding the corresponding joint quantities when the desired position and posture of the end effector are given. When the robot follows a continuous trajectory, it is also achieved by calculating the joint angles corresponding to a set of continuous poses of the end effector. Essentially, a continuous trajectory is represented by a series of pose samples that closely resemble the continuous trajectory. The process of completing obstacle-crossing operations along a preplanned path essentially involves calculating the joint positions for each pose sample.

When an ice removal robot needs to work on transmission lines, it not only needs to meet tracking control performance but also needs to consider the actual operating environment and weather conditions of the ice removal robot. In order to navigate obstacles, ice removal robots are typically designed with multi-arm structures that incorporate sensors and mechanical wheel drive mechanisms. It is essential to maintain the weight balance of the robot during complex movements, and factors such as ice breaking resistance and friction need to be carefully considered. As a result, ice removal robots for transmission lines have complex electromechanical structures, making them highly nonlinear and time-varying dynamic systems for motion control.

In general, for systems operating in air, ground, and space environments, especially for specific anti-icing robots, the main limiting factor is the mechanical complexity of their structures and mechanical elements. Therefore, to have an effective control system for flexible systems, compensating through flexibility and vibration suppression is an important part of the development direction of compact and high-performance systems. However, achieving this is a difficult and challenging task. The difficulties associated with designing flexible systems that include uncertainties include changes in structural parameter uncertainty over time, low damping in multimodal frequency ranges, unmodeled closely spaced modes, structural nonlinearity, and operation under a wide range of loads and external disturbances. Based on the above analysis, this chapter focuses on the inverse kinematics research of the manipulator to achieve the tracking control performance of the ice removal robot for transmission lines.

4.2 Modeling of De-icing Robot Manipulator System

4.2.1 De-icing Robot

4.2.1.1 Forward Kinematics

The forward kinematics problem for a serial-chain manipulator is to find the position and orientation of the end effector relative to the base using the joint positions and the geometric link parameters. The frame fixed in the end effector is referred to as the tool frame. This frame is also fixed in the final link *N* and generally has a constant offset in both position and orientation relative to frame *N*. Similarly, the station frame is located in the base to establish the location of the task to be performed. This frame generally has a constant offset in its pose relative to frame 0, which is also fixed in the base.

A more general expression of the forward kinematics problem is to find the relative position and orientation of any two designated members given the geometric structure of the manipulator and the values of several joint positions equal to the number of degrees of freedom of the mechanism. The forward kinematics problem is critical for developing manipulator coordination algorithms because joint positions are typically measured by sensors mounted on the joints and it is necessary to calculate the positions of the joint axes relative to the fixed reference frame.

In practice, the forward kinematics problem is solved by calculating the transformation between a reference frame fixed in the end effector and another reference frame fixed in the base, i.e., between the tool and station frames. This is straightforward for a serial chain since the transformation describing the position of the end effector relative to the base is obtained by simply concatenating transformations between frames fixed in adjacent links of the chain. The convention for the geometric representation of a manipulator presented in [25] reduces this to finding an equivalent 4 × 4 homogeneous transformation matrix that relates the spatial displacement of the end-effector reference frame to the base reference frame. According to [25], the 4 × 4 homogeneous transformation matrix is represented as the following form:

$$
{}^{i-1}\boldsymbol{T}_i = \begin{pmatrix} \cos(\theta_i) & -\sin(\theta_i) & 0 & a_{i-1} \\ \sin(\theta_i)\cos(\alpha_{i-1}) & \cos(\theta_i)\cos(\alpha_{i-1}) & -\sin\alpha_{i-1} & -\sin(\alpha_{i-1})d_i \\ \sin(\theta_i)\sin(\alpha_{i-1}) & \cos(\theta_i)\sin(\alpha_{i-1}) & \cos(\alpha_{i-1}) & \cos(\alpha_{i-1})d_i \\ 0 & 0 & 0 & 1 \end{pmatrix} \tag{4.1}
$$

(a) Single de-icing robot.

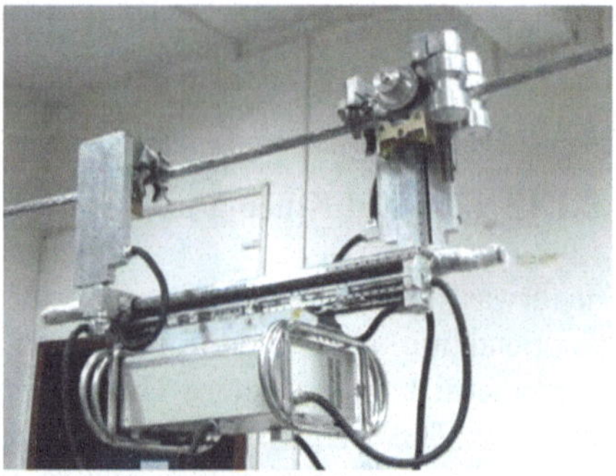

(b) Two-arm deicing robot.

(c) Three-arm deicing robot.

Fig. 4.4 Articulation of 3-DOF de-icing robot manipulator. (**a**) Single de-icing robot. (**b**) Two-arm de-icing robot. (**c**) Three-arm de-icing robot

Table 4.1 D-H parameters of the de-icing robot manipulator in Fig. 4.4b

i	α_i	a_i	d_i	θ_i
1	0	0	0	θ_1
2	$\frac{\pi}{2}$	l_1	0	θ_2
3	0	l_2	0	$-\theta_2$
4	$-\frac{\pi}{2}$	l_3	d_3	0

In this section, the geometric parameters for the de-icing robot manipulator are shown in Fig. 4.4b and listed in Table 4.1. The joints of this manipulator are two revolute and a prismatic. Joint 1 has a vertical orientation. Joint 2 is perpendicular to joint 1 and intersects it. Joint 3 is parallel to joint 1, and the lengths of links 1 and 2 are l_1 and l_2.

Using Table 4.1 and (4.1), the forward kinematics of de-icing robot manipulator is derived, and the elements of the homogeneous transformation matrix are calculated as follows:

$$
{}^{0}\boldsymbol{T}_1 = \begin{pmatrix} \cos(\theta_1) & -\sin(\theta_1) & 0 & 0 \\ \sin(\theta_1) & \cos(\theta_1) & 0 & 0 \\ 0 & 0 & 1 & 0 \\ 0 & 0 & 0 & 1 \end{pmatrix}, \quad {}^{1}\boldsymbol{T}_2 = \begin{pmatrix} \cos(\theta_2) & -\sin(\theta_2) & 0 & l_1 \\ 0 & 0 & -1 & 0 \\ \sin(\theta_2) & \cos(\theta_2) & 0 & 0 \\ 0 & 0 & 0 & 1 \end{pmatrix},
$$
$$
{}^{2}\boldsymbol{T}_3 = \begin{pmatrix} \cos(\theta_2) & \sin(\theta_2) & 0 & l_2 \\ -\sin(\theta_2) & \cos(\theta_2) & 0 & 0 \\ 0 & 0 & 1 & 0 \\ 0 & 0 & 0 & 1 \end{pmatrix}, \quad {}^{3}\boldsymbol{T}_4 = \begin{pmatrix} 1 & 0 & 0 & l_2 \\ 0 & 0 & 1 & d_3 \\ 0 & -1 & 0 & 0 \\ 0 & 0 & 0 & 1 \end{pmatrix}. \tag{4.2}
$$

The serial-chain manipulator is shown in Fig. 4.4b, and neglecting the addition of tool and station frames, the transformation is

$$
\boldsymbol{T}_1 = {}^{0}\boldsymbol{T}_1 = \begin{pmatrix} c_1 & -s_1 & 0 & 0 \\ s_1 & c_1 & 0 & 0 \\ 0 & 0 & 1 & 0 \\ 0 & 0 & 0 & 1 \end{pmatrix},
$$

$$
\boldsymbol{T}_2 = {}^{0}\boldsymbol{T}_1{}^{1}\boldsymbol{T}_2 = \begin{pmatrix} c_1c_2 & -c_1s_2 & s_1 & c_1l_1 \\ s_1c_2 & -s_1s_2 & -c_1 & s_1l_1 \\ s_2 & c_2 & 0 & 0 \\ 0 & 0 & 0 & 1 \end{pmatrix},
$$

$$
\boldsymbol{T}_3 = {}^{0}\boldsymbol{T}_1{}^{1}\boldsymbol{T}_2{}^{2}\boldsymbol{T}_3 = \begin{pmatrix} c_1c_2^2 + c_1s_2^2 & 0 & s_1 & c_1c_2l_2 + c_1l_1 \\ s_1c_2^2 + s_1s_2^2 & 0 & -c_1 & s_1c_2l_2 + s_1l_1 \\ 0 & s_2^2 + c_2^2 & 0 & s_2l_2 \\ 0 & 0 & 0 & 1 \end{pmatrix},
$$

$$
\begin{aligned}
\boldsymbol{T}_4 &= {}^0\boldsymbol{T}_1{}^1\boldsymbol{T}_2{}^2\boldsymbol{T}_3{}^3\boldsymbol{T}_4 \\
&= \begin{pmatrix} \overbrace{\begin{pmatrix} r_{11} \\ r_{21} \\ r_{31} \end{pmatrix}}^{\boldsymbol{r}_1} & \overbrace{\begin{pmatrix} r_{12} \\ r_{22} \\ r_{32} \end{pmatrix}}^{\boldsymbol{r}_2} & \overbrace{\begin{pmatrix} r_{13} \\ r_{23} \\ r_{33} \end{pmatrix}}^{\boldsymbol{r}_3} & \overbrace{\begin{pmatrix} {}^0p_4^x \\ {}^0p_4^y \\ {}^0p_4^z \end{pmatrix}}^{\boldsymbol{x}_p} \\ 0 & 0 & 0 & 1 \end{pmatrix} \\
&= \begin{pmatrix} \overbrace{\begin{pmatrix} c_1c_2^2+c_1s_2^2 \\ s_1c_2^2+s_1s_2^2 \\ 0 \end{pmatrix}}^{\boldsymbol{r}_1} & \overbrace{\begin{pmatrix} -s_1 \\ c_1 \\ 0 \end{pmatrix}}^{\boldsymbol{r}_2} & \overbrace{\begin{pmatrix} 0 \\ 0 \\ s_2^2+c_2^2 \end{pmatrix}}^{\boldsymbol{r}_3} & \overbrace{\begin{pmatrix} (c_1c_2^2+c_1s_2^2)l_3+c_1c_2l_2+c_1l_1 \\ (s_1c_2^2+s_1s_2^2)l_3+s_1c_2l_2+s_1l_1 \\ (s_2^2+c_2^2)d_3+s_2l_2 \end{pmatrix}}^{\boldsymbol{x}_p} \\ 0 & 0 & 0 & 1 \end{pmatrix} \in \mathbb{R}^{4\times 4}.
\end{aligned}
\tag{4.3}
$$

with abbreviations $c_i = \cos(\theta_i)$ and $s_i = \sin(\theta_i)$.

From (4.3), the position and orientation of end effector can be represented in vector form as the following:

$$
\boldsymbol{x} = \begin{pmatrix} \boldsymbol{x}_p \\ \boldsymbol{r}_1 \\ \boldsymbol{r}_2 \\ \boldsymbol{r}_3 \end{pmatrix} = \begin{pmatrix} \overbrace{\begin{pmatrix} {}^0p_4^x \\ {}^0p_4^y \\ {}^0p_4^z \end{pmatrix}}^{\boldsymbol{x}_p} \\ \overbrace{\begin{pmatrix} r_{11} \\ r_{21} \\ r_{31} \end{pmatrix}}^{\boldsymbol{r}_1} \\ \overbrace{\begin{pmatrix} r_{12} \\ r_{22} \\ r_{32} \end{pmatrix}}^{\boldsymbol{r}_2} \\ \overbrace{\begin{pmatrix} r_{13} \\ r_{23} \\ r_{33} \end{pmatrix}}^{\boldsymbol{r}_3} \end{pmatrix} = \begin{pmatrix} \overbrace{\begin{pmatrix} (c_1c_2^2 + c_1s_2^2)l_3 + c_1c_2l_2 + c_1l_1 \\ (s_1c_2^2 + s_1s_2^2)l_3 + s_1c_2l_2 + s_1l_1 \\ (s_2^2 + c_2^2)d_3 + s_2l_2 \end{pmatrix}}^{\boldsymbol{x}_p} \\ \overbrace{\begin{pmatrix} c_1c_2^2 + c_1s_2^2 \\ s_1c_2^2 + s_1s_2^2 \\ 0 \end{pmatrix}}^{\boldsymbol{r}_1} \\ \overbrace{\begin{pmatrix} -s_1 \\ c_1 \\ 0 \end{pmatrix}}^{\boldsymbol{r}_2} \\ \overbrace{\begin{pmatrix} 0 \\ 0 \\ s_2^2 + c_2^2 \end{pmatrix}}^{\boldsymbol{r}_3} \end{pmatrix} \in \mathbb{R}^{12\times 1}.
\tag{4.4}
$$

Forward Instantaneous Kinematics

Differentiation eq. (4.4) with respect to time of the forward position kinematics equations yields a set of equations of the form

$$\begin{pmatrix} \dot{\boldsymbol{x}}_{\boldsymbol{P}} \\ \dot{\boldsymbol{x}}_{\boldsymbol{R}} \end{pmatrix} = \begin{pmatrix} \boldsymbol{J}_{X_p}(\boldsymbol{q}) \\ \boldsymbol{J}_{X_R}(\boldsymbol{q}) \end{pmatrix} \dot{\boldsymbol{q}} \tag{4.5}$$

where $\dot{\boldsymbol{x}}_{\boldsymbol{P}}$ and $\dot{\boldsymbol{x}}_{\boldsymbol{R}}$ are the linear and angular velocity of the end effector, respectively. $\dot{\boldsymbol{q}}$ is the n-dimensional vector composed of the joint rates, and $\boldsymbol{J}_{X_P}(\boldsymbol{q})$is a Jacobian matrix whose elements are represent in (4.6) and (4.7) as

$$\begin{aligned} \boldsymbol{J}_{X_P} &= \begin{pmatrix} \frac{\partial^0 p_4^x}{\partial q_1} & \frac{\partial^0 p_4^x}{\partial q_2} & \frac{\partial^0 p_4^x}{\partial d_3} \\ \frac{\partial^0 p_4^y}{\partial q_1} & \frac{\partial^0 p_4^y}{\partial q_2} & \frac{\partial^0 p_4^y}{\partial d_3} \\ \frac{\partial^0 p_4^z}{\partial q_1} & \frac{\partial^0 p_4^z}{\partial q_2} & \frac{\partial^0 p_4^z}{\partial d_3} \end{pmatrix} \\ &= \begin{pmatrix} -\left(s_1 c_2^2 + s_1 s_2^2\right) l_3 - s_1 c_2 l_2 - s_1 l_1 & -c_2 s_2 l_2 & 0 \\ \left(c_1 c_2^2 + c_1 s_2^2\right) l_3 + c_1 c_2 l_2 + c_1 l_1 & -s_1 s_2 l_2 & 0 \\ 0 & c_2 l_2 & s_2^2 + c_2^2 \end{pmatrix} \in R^{3\times 3} \end{aligned} \tag{4.6}$$

$$\boldsymbol{J}_{X_R} = \begin{pmatrix} \frac{\partial r_{11}}{\partial q_1} & \frac{\partial r_{11}}{\partial q_2} & \frac{\partial r_{11}}{\partial d_3} \\ \frac{\partial r_{21}}{\partial q_1} & \frac{\partial r_{21}}{\partial q_2} & \frac{\partial r_{21}}{\partial d_3} \\ \frac{\partial r_{31}}{\partial q_1} & \frac{\partial r_{31}}{\partial q_2} & \frac{\partial r_{31}}{\partial d_3} \\ \frac{\partial r_{12}}{\partial q_1} & \frac{\partial r_{12}}{\partial q_2} & \frac{\partial r_{12}}{\partial d_3} \\ \frac{\partial r_{22}}{\partial q_1} & \frac{\partial r_{22}}{\partial q_2} & \frac{\partial r_{22}}{\partial d_3} \\ \frac{\partial r_{32}}{\partial q_1} & \frac{\partial r_{32}}{\partial q_2} & \frac{\partial r_{32}}{\partial d_3} \\ \frac{\partial r_{13}}{\partial q_1} & \frac{\partial r_{13}}{\partial q_2} & \frac{\partial r_{13}}{\partial d_3} \\ \frac{\partial r_{23}}{\partial q_1} & \frac{\partial r_{23}}{\partial q_2} & \frac{\partial r_{23}}{\partial d_3} \\ \frac{\partial r_{33}}{\partial q_1} & \frac{\partial r_{33}}{\partial q_2} & \frac{\partial r_{33}}{\partial d_3} \end{pmatrix} = \begin{pmatrix} -s_1 c_2^2 - s_1 s_2^2 & 0 & 0 \\ cc + cs & 0 & 0 \\ 0 & 0 & 0 \\ -c_1 & 0 & 0 \\ -s_1 & 0 & 0 \\ 0 & 0 & 0 \\ 0 & 0 & 0 \\ 0 & 0 & 0 \\ 0 & 0 & 0 \end{pmatrix} \in R^{3\times 3} \tag{4.7}$$

4.2.2 Dynamic Models of De-icing Robot Manipulator

4.2.2.1 Canonical Equations

The motion equations of a robot mechanism are usually presented in canonical forms (joint-space formulation) as follows:

$$M(q)\ddot{q} + C(q,\ \dot{q})\dot{q} + G(q) = \tau \tag{4.8}$$

The symbols q, $\dot{q}$, $\ddot{q}$, and τ denote the n-dimensional vectors of joint position, velocity, acceleration, and force variables, respectively, where n is the number of degrees of motion freedom of the robot mechanism. The inertia matrix $M(q)$ is an $n \times n$ symmetric and positive definite matrix, $C(q,\ \dot{q})$ is an $n \times n$ matrix and represents the Coriolis and centrifugal terms, and $G(q)$ is the vector of the gravity term.

4.2.2.2 Lagrange Formulation

Various methods exist for deriving the terms in (4.8). The two that are commonly used in robotics are the Newton-Euler formulation and the Lagrange formulation. In this section, the dynamic model of de-icing robot manipulator is shown in Fig. 4.4b, which is derived based on the Lagrange formulation in [25].

The Lagrange formulation proceeds via the Lagrangian of the robot mechanism:

$$\mathrm{L} = \mathrm{T} - \mathrm{U} \tag{4.9}$$

where T and U are the total kinetic and potential energy of the mechanism, respectively. Then, the dynamic equations can be developed using the Lagrange equations for each generalized coordinate:

$$\frac{\mathrm{d}}{\mathrm{d}t}\frac{\partial \mathrm{L}}{\partial \dot{q}_i} - \frac{\partial \mathrm{L}}{\partial q_i} = \tau_i, \quad i = 1, 2, \cdots, n_i. \tag{4.10}$$

where n_i is the number of links. The general robot manipulator dynamic equation is obtained by determining the manipulator kinetic and potential energies and then substituting into the Lagrange equation (4.10) to obtain the final results.

4.2.2.3 Kinetic Energy

Consider a point on the i-th link. The coordinates of this point with respect to the i-th attached frame are denoted as ${}^{i}r$, as shown in Fig. 4.4b. Thus, the coordinates can be presented in the base coordinates as form

$$r = \boldsymbol{T}_i {}^i\boldsymbol{r} \tag{4.11}$$

where $\boldsymbol{T}_i$ are the 4×4 homogeneous transformation matrix, which is defined in (4.3). The velocity of the point in base coordinate is described as follows:

$$\boldsymbol{v} = \frac{\mathrm{d}\boldsymbol{r}}{\mathrm{d}t} = \sum_{j=1}^{i} \left[\frac{\partial \boldsymbol{T}_i}{\partial \boldsymbol{q}_j} \dot{\boldsymbol{q}}_j \right] {}^i\boldsymbol{r} \tag{4.12}$$

Since $\partial \boldsymbol{T}_i / \partial \boldsymbol{q}_j = 0$ with $j > i$, the upper summation limit can be replaced by the number of links n. The matrix $\partial \boldsymbol{T}_i / \partial \boldsymbol{q}_j = 0$ can be computed when the matrix $\boldsymbol{T}_i$ are known.

The kinetic energy of an infinitesimal mass $\mathrm{d}m$ at coordinates ${}^i\boldsymbol{r}$ that has a velocity $\boldsymbol{v} = [v_x, v_y, v_z]^\top$ is

$$\begin{aligned} \mathrm{d}\boldsymbol{K}_i &= \frac{1}{2}\left(v_x^2 + v_y^2 + v_z^2\right)\mathrm{d}m = \frac{1}{2}\mathrm{tr}(\boldsymbol{v}^\top \boldsymbol{v})\mathrm{d}m \\ &= \frac{1}{2}\mathrm{tr}\left[\sum_{j=1}^{n} \sum_{k=1}^{n} \frac{\partial \boldsymbol{T}_i}{\partial \boldsymbol{q}_j} \left({}^i\boldsymbol{r}{}^i\boldsymbol{r}^\top \mathrm{d}m\right) \frac{\partial \boldsymbol{T}_i^\top}{\partial \boldsymbol{q}_k} \dot{\boldsymbol{q}}_j \dot{\boldsymbol{q}}_k \right] \mathrm{d}m \end{aligned} \tag{4.13}$$

The total kinetic energy for each i-th link is computed as follows:

$$\boldsymbol{K}_i = \int_{\mathrm{link}\,i} \mathrm{d}\boldsymbol{K}_i \tag{4.14}$$

By substituting (4.13) into (4.14), the integral can be moved inside the summations. Finally, the kinetic energy of the link can be rewritten as follows:

$$\boldsymbol{K}_i = \frac{1}{2}\mathrm{tr}\left[\sum_{j=1}^{n} \sum_{k=1}^{n} \frac{\partial \boldsymbol{T}_i}{\partial \boldsymbol{q}_j} \boldsymbol{I}_i \frac{\partial \boldsymbol{T}_i^\top}{\partial \boldsymbol{q}_k} \dot{\boldsymbol{q}}_j \dot{\boldsymbol{q}}_k \right] \tag{4.15}$$

where I_i is a pseudo-inertia matrix for each joint, which is described as follows:

$$\boldsymbol{I}_i = \int_{\mathrm{link}\,i} {}^i\boldsymbol{r}{}^i\boldsymbol{r}^\top \mathrm{d}m \tag{4.16}$$

Now, the pseudo-inertia matrix will be introduced before finding the manipulator kinetic energy. Let ${}^i\boldsymbol{r} = [x, y, z, 1]^\top$ be the coordinates in i-th frame of the infinitesimal mass $\mathrm{d}m$. Then, expanding (4.16) yields

$$
\boldsymbol{I}_i = \begin{pmatrix} \int x^2 \mathrm{d}m & \int yx\mathrm{d}m & \int zx\mathrm{d}m & \int x\mathrm{d}m \\ \int xy\mathrm{d}m & \int y^2\mathrm{d}m & \int zy\mathrm{d}m & \int y\mathrm{d}m \\ \int xz\mathrm{d}m & \int yz\mathrm{d}m & \int z^2\mathrm{d}m & \int z\mathrm{d}m \\ \int x\mathrm{d}m & \int y\mathrm{d}m & \int x\mathrm{d}m & \int \mathrm{d}m \end{pmatrix} \tag{4.17}
$$

The integrals are taken over the volume of the i-th link. This is a constant matrix that depends on the geometry and mass distribution that is computed once for each i-th link. In fact, for each term of the i-th link, the moments of inertia are described as follows:

$$
\begin{aligned} I_{xx} &= \int (y^2 + z^2)\mathrm{d}m \\ I_{yy} &= \int (x^2 + z^2)\mathrm{d}m \\ I_{zz} &= \int (x^2 + y^2)\mathrm{d}m \end{aligned} \tag{4.18}
$$

Cross-products of inertia

$$
\begin{aligned} I_{xy} &= \int xy\mathrm{d}m \\ I_{xz} &= \int xz\mathrm{d}m \\ I_{yz} &= \int yz\mathrm{d}m \end{aligned} \tag{4.19}
$$

And first moments

$$
\begin{aligned} m\overline{x} &= \int x\mathrm{d}m \\ m\overline{y} &= \int y\mathrm{d}m \\ m\overline{z} &= \int z\mathrm{d}m \end{aligned} \tag{4.20}
$$

where m is the total mass of the i-th link and ${}^i\overline{\boldsymbol{r}} = [\overline{x}, \overline{y}, \overline{z}, 1]^\top$.

The coordinates in the i-th frame of the center of gravity of the i-th link can be rewritten as

$$I_i = \begin{pmatrix} \frac{-I_{xx}+I_{yy}+I_{zz}}{2} & I_{xy} & I_{xz} & m\bar{x} \\ I_{xy} & \frac{I_{xx}-I_{yy}+I_{zz}}{2} & I_{yz} & m\bar{y} \\ I_{xz} & I_{yz} & \frac{I_{xx}+I_{yy}-I_{zz}}{2} & m\bar{z} \\ m\bar{x} & m\bar{y} & m\bar{z} & m \end{pmatrix} \tag{4.21}$$

Now, the computation of the manipulator kinetic energy will be continued, which can be rewritten as

$$\boldsymbol{K} = \sum_{i=1}^{n} \boldsymbol{K}_i = \frac{1}{2}\sum_{i=1}^{n} \operatorname{tr}\left[\sum_{j=1}^{n}\sum_{k=1}^{n} \frac{\partial \boldsymbol{T}_i}{\partial \boldsymbol{q}_j} \boldsymbol{I}_i \frac{\partial \boldsymbol{T}_i^{\top}}{\partial \boldsymbol{q}_k} \dot{\boldsymbol{q}}_j \dot{\boldsymbol{q}}_k\right] \tag{4.22}$$

Since the trace of a sum of matrices is the sum of the individual traces, summations and the trace operator can be interchanged to obtain

$$\begin{aligned} \boldsymbol{K} &= \frac{1}{2}\sum_{j=1}^{n}\sum_{k=1}^{n} m_{jk}(\boldsymbol{q})\dot{\boldsymbol{q}}_j\dot{\boldsymbol{q}}_k \\ \boldsymbol{K} &= \frac{1}{2}\dot{\boldsymbol{q}}^{\top}\boldsymbol{M}(\boldsymbol{q})\dot{\boldsymbol{q}} \end{aligned} \tag{4.23}$$

where the $n \times n$ manipulator inertia matrix $\boldsymbol{M}(\boldsymbol{q})$ has elements defined as

$$m_{jk}(\boldsymbol{q}) = \sum_{i=1}^{n} \operatorname{tr}\left[\frac{\partial \boldsymbol{T}_i}{\partial \boldsymbol{q}_j} \boldsymbol{I}_i \frac{\partial \boldsymbol{T}_i^{\top}}{\partial \boldsymbol{q}_k}\right] \tag{4.24}$$

Since $\partial \boldsymbol{T}_i / \partial \boldsymbol{q}_j = 0$ with $j > i$, it can be written more efficiently as

$$m_{jk}(\boldsymbol{q}) = \sum_{i=\max(j,k)}^{n} \operatorname{tr}\left[\frac{\partial \boldsymbol{T}_i}{\partial \boldsymbol{q}_j} \boldsymbol{I}_i \frac{\partial \boldsymbol{T}_i^{\top}}{\partial \boldsymbol{q}_k}\right] \tag{4.25}$$

From Eq. (4.22), the manipulator kinetic energy can be calculated in terms of known quantities and the joint variables $\boldsymbol{q}$. Since $m_{jk} = m_{kj}$, the inertia matrix $\boldsymbol{M}(\boldsymbol{q})$ is symmetric and positive definite.

4.2.2.4 Potential Energy

Assuming the mass of the i-th link and the center of gravity ${}^{i}\bar{\boldsymbol{r}}$ represented in the coordinates of the i-th frame are known, the potential energy of the arm link is given as follows:

$$\boldsymbol{U}_i = -m_i \boldsymbol{g}^\top \boldsymbol{T}_i {}^i\bar{\boldsymbol{r}} \tag{4.26}$$

where the gravity vector is represented in the base coordinate as

$$\boldsymbol{g} = \begin{bmatrix} g_x & g_y & g_z & 0 \end{bmatrix}^\top \tag{4.27}$$

Finally, the total potential energy of robot manipulator is given as

$$\boldsymbol{U}(\boldsymbol{q}) = -\sum_{i=1}^{n} m_i \boldsymbol{g}^\top \boldsymbol{T}_i {}^i\bar{\boldsymbol{r}} \tag{4.28}$$

Note that the potential energy depends only on the joint variable q, not on the joint velocity variable $\dot{\boldsymbol{q}}$.

4.2.2.5 Equation

By substituting (4.23) and (4.28) into (4.9), the manipulator Lagrangian is rewritten as

$$\mathrm{L}(\boldsymbol{q}, \dot{\boldsymbol{q}}) = \mathrm{K}(\boldsymbol{q}, \dot{\boldsymbol{q}}) - \mathrm{U}(\boldsymbol{q}) = \frac{1}{1} \dot{\boldsymbol{q}}^\top \boldsymbol{M}(\boldsymbol{q}) \dot{\boldsymbol{q}} - \boldsymbol{U}(\boldsymbol{q}) \tag{4.29}$$

From (4.10), the Lagrange motion equations can be given by

$$\begin{aligned} \frac{\partial \mathrm{L}}{\partial \dot{\boldsymbol{q}}} &= \frac{\partial \mathrm{K}}{\partial \dot{\boldsymbol{q}}} = \boldsymbol{M}(\boldsymbol{q}) \dot{\boldsymbol{q}} \\ \frac{d}{dt} \frac{\partial \mathrm{L}}{\partial \dot{\boldsymbol{q}}} &= \boldsymbol{M}(\boldsymbol{q}) \ddot{\boldsymbol{q}} + \dot{\boldsymbol{M}}(\boldsymbol{q}) \dot{\boldsymbol{q}} \\ \frac{\partial \mathrm{L}}{\partial \boldsymbol{q}} &= \frac{1}{2} \frac{\partial}{\partial \boldsymbol{q}} \left(\dot{\boldsymbol{q}}^T \boldsymbol{M}(\boldsymbol{q}) \dot{\boldsymbol{q}} \right) - \frac{\partial \boldsymbol{U}(\boldsymbol{q})}{\partial \boldsymbol{q}} \end{aligned} \tag{4.30}$$

Therefore, the manipulator dynamic equation is presented in canonical forms (joint-space formulation) as (4.8)

$$\underbrace{\boldsymbol{M}(\boldsymbol{q}) \ddot{\boldsymbol{q}}}_{\boldsymbol{M}(\boldsymbol{q})\ddot{\boldsymbol{q}}} + \underbrace{\dot{\boldsymbol{M}}(\boldsymbol{q}) \dot{\boldsymbol{q}} - \frac{1}{2} \frac{\partial}{\partial \boldsymbol{q}} \left(\dot{\boldsymbol{q}}^\top \boldsymbol{M}(\boldsymbol{q}) \dot{\boldsymbol{q}} \right)}_{\boldsymbol{C}(\boldsymbol{q}, \dot{\boldsymbol{q}})\dot{\boldsymbol{q}}} + \underbrace{\frac{\partial \boldsymbol{U}(\boldsymbol{q})}{\partial \boldsymbol{q}}}_{\boldsymbol{G}(\boldsymbol{q})} = \boldsymbol{\tau} \tag{4.31}$$

The remainder of this section establishes the dynamic model of the three-link de-icing robot manipulator, as shown in Fig. 4.5b, based on the method introduced above. First, the kinematics for a three-link de-icing robot manipulator must be given in (4.3). Then, to determine the manipulator dynamics, it is necessary to calculate the

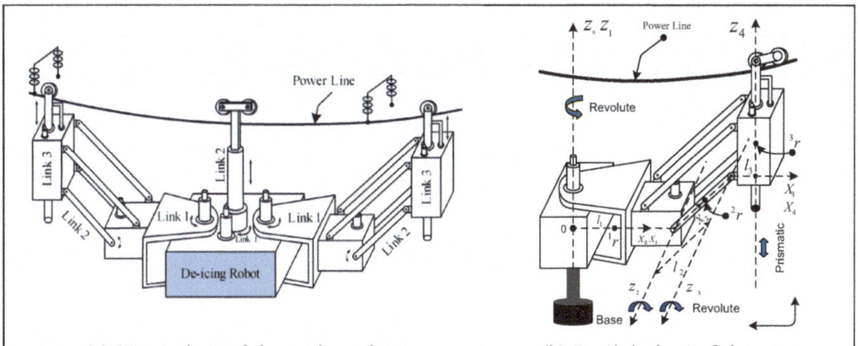

(a) Front view of the entire robot. (b) Partial view of the arm.

Fig. 4.5 De-icing robot for power transmission lines (Hunan University) [26]. (**a**) Single de-icing robot. (**b**) Two-arm de-icing robot. (**c**) Three-arm de-icing robot

kinetic and potential energy, which are required in the Lagrange equation. Figure 4.3b shows that the joint variables and joint velocity vectors are $\boldsymbol{q} = [q_1, q_2, d_3]^\top$ and $\dot{\boldsymbol{q}} = [\dot{q}_1, \dot{q}_2, \dot{d}_3]^\top$. Assuming the moments of inertia equal to zero, the coordinates in each frame of the center of gravity and center of mass of each link are selected in the center of each link. Based on Eqs. (4.22), (4.28), and (4.31), the kinetic energy, potential energy, and gravity matrices can be computed, respectively. Finally, the dynamic model of de-icing robot manipulator can be represented in Table 4.2.

4.3 Inverse Kinematic Solution for Robot Manipulator

The inverse kinematics problem of a robot is to determine the corresponding joint variables based on the known pose (position and orientation) of the end effector. It is one of the key technologies in robot trajectory planning and motion control. There are two main categories of methods for solving the inverse kinematics problem: the closed-form solutions (analytical methods) and numerical solutions [27]. Closed-form solutions can be further divided into algebraic methods and geometric methods. They are characterized by fast computation speed and the ability to find all possible solutions conveniently. However, this method is highly dependent on the robot's structure and has relatively low generality. For example, for the commonly used 6-degrees-of-freedom robot (such as the PUMA robot), closed-form solutions can only be obtained when its geometric structure satisfies the Pieper criterion. Currently, the most effective method to find closed-form solutions for robot inverse kinematics is to utilize the special characteristics of the robot's structure and employ an "ad hoc" approach.

Table 4.2 The parameters of the dynamic model in (4.8)

Inertia matrix $\boldsymbol{M}(\boldsymbol{q}) \in \mathbb{R}^{3 \times 3}$	
$\boldsymbol{M}(\boldsymbol{q}) = \begin{pmatrix} \boldsymbol{M}_{11} & \boldsymbol{M}_{12} & \boldsymbol{M}_{13} \\ \boldsymbol{M}_{21} & \boldsymbol{M}_{22} & \boldsymbol{M}_{23} \\ \boldsymbol{M}_{31} & \boldsymbol{M}_{32} & \boldsymbol{M}_{33} \end{pmatrix}$	$\boldsymbol{M}_{11} = 9/4m_1l_1 + m_2\left(1/4c_2l_2 + l_1^2 + l_2l_1\left(c_1^2 - s_1^2\right)\right) + m_3\left(c_2l_2^2 + l_2^2 + 2c_2l_1l_2\right)$
	$\boldsymbol{M}_{22} = 1/4m_2l_2^2 + m_3l_2^2 + 4/3m_1l_1^2$
	$\boldsymbol{M}_{23} = \boldsymbol{M}_{32} = m_3c_2l_2$
	$\boldsymbol{M}_{33} = m_3$
	$\boldsymbol{M}_{12} = \boldsymbol{M}_{13} = \boldsymbol{M}_{21} = \boldsymbol{M}_{31} = 0$
Matrix of centripetal and Coriolis forces $\boldsymbol{C}(\boldsymbol{q}, \dot{\boldsymbol{q}}) \in \mathbb{R}^{3 \times 3}$	
$\boldsymbol{C}(\boldsymbol{q}, \dot{\boldsymbol{q}}) = \begin{pmatrix} \boldsymbol{C}_{11} & \boldsymbol{C}_{12} & \boldsymbol{C}_{13} \\ \boldsymbol{C}_{21} & \boldsymbol{C}_{22} & \boldsymbol{C}_{23} \\ \boldsymbol{C}_{31} & \boldsymbol{C}_{32} & \boldsymbol{C}_{33} \end{pmatrix}$	$\boldsymbol{C}_{11} = -8m_2l_1l_2c_1s_1\dot{q}_1 + \left(-1/2m_2s_2c_2l_2^2 + m_3\left(-2s_2c_2l_2^2 - 2s_2l_1l_2\right)\dot{q}_2\right.$
	$\boldsymbol{C}_{21} = \left(-1/2m_2s_2c_2l_2^2\right) + m_3\left(-2s_2c_2l_2^2 - 2s_2l_1l_2\right)\dot{q}_1$
	$\boldsymbol{C}_{22} = -m_3s_2l_2\dot{d}_3$
	$\boldsymbol{C}_{23} = -2m_3s_2l_2\dot{q}_2$
	$\boldsymbol{C}_{32} = -m_3s_2l_2\dot{q}_2$
	$\boldsymbol{C}_{12} = \boldsymbol{C}_{13} = \boldsymbol{C}_{31} = \boldsymbol{C}_{33} = 0$
Gravity vector $\boldsymbol{G}(\boldsymbol{q}) \in \mathbb{R}^{3 \times 1}$	
$\boldsymbol{G}(\boldsymbol{q}) = \begin{pmatrix} \boldsymbol{G}_1 \\ \boldsymbol{G}_2 \\ \boldsymbol{G}_3 \end{pmatrix}$	$\boldsymbol{G}_1 = (1/2c_1c_2l_2 + c_1l_1)m_2g$
	$\boldsymbol{G}_2 = (-1/2s_1s_2l_2m_2 + c_2l_2m_3)g$
	$\boldsymbol{G}_3 = m_3g$

4.3.1 Innovative Hybrid Electromagnetism-Like Algorithm for Robot Inverse Kinematics

4.3.1.1 Problem Formulation

As shown in Fig. 4.6, the desired position vector and orientation matrix of a manipulator end effector are denoted by $\boldsymbol{P}_d$ and $[\boldsymbol{R}_d] = [d_1, d_2, d_3]$, where $\boldsymbol{d}_j(j = 1, 2, 3)$ is the unit vector along the $x_d - y_d - z_d$ axes. $\boldsymbol{P}_h$ is the current position vector of the end effector. The current orientation matrix is defined by $[\boldsymbol{R}_h] = [h_1, h_2, h_3]$, where $\boldsymbol{h}_j(j = 1, 2, 3)$ is the unit vector along the $x_h - y_h - z_h$ axes and the joint variables $\boldsymbol{\theta} = [\theta_1, \theta_2, \ldots, \theta_n]^{\mathrm{T}}$ are denoted by the $n \times 1$ vector.

Therefore, the error between the current and the desired locations of the end effector can be described by the following functions:

Position error:

$$\Delta p(\boldsymbol{\theta}) = (\boldsymbol{P}_d - \boldsymbol{P}_h(\boldsymbol{\theta}))^2 \tag{4.32}$$

Orientation error:

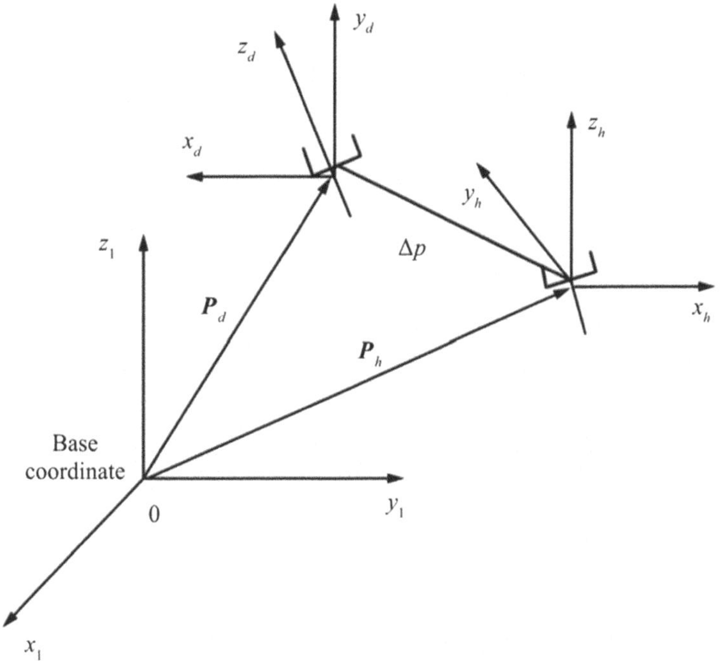

Fig. 4.6 The current and the desired end-effector configurations

$$\Delta o(\boldsymbol{\theta}) = \sum_{j=1}^{3} \left(\boldsymbol{d}_j \cdot \boldsymbol{h}_j(\boldsymbol{\theta}) - 1\right)^2 \tag{4.33}$$

The total error:

$$e(\boldsymbol{\theta}) = \Delta p(\boldsymbol{\theta}) + \Delta o(\boldsymbol{\theta}) \tag{4.34}$$

where $(\cdot)$ denotes the vector dot product. The goal is reachable if and only if both the position error and the orientation error are equal to zero. In practice, to enforce one length unit compatible with d degrees in angle, Eq. (4.33) needs to be multiplied by a factor c_d such that [28]

$$\frac{1}{c_d} = \frac{2\pi}{360} d$$

With minor modifications, the form of the objective function (4.34) can be rewritten as

$$e(\boldsymbol{\theta}) = w_p(\boldsymbol{p_d} - \boldsymbol{p_h})^2 + w_o \sum\nolimits_{j=1}^{3} c_{dj}^2 \left(\boldsymbol{d}_j \cdot \boldsymbol{h}_j - 1\right)^2 \tag{4.35}$$

where w_p and w_o are weighting factors assigned to position and orientation, respectively, such that $w_p + w_o = 1$.

Then, the inverse kinematics problem is to find a solution $\boldsymbol{\theta}$, such that $e(\boldsymbol{\theta}^*) \leq \varepsilon$(-$\varepsilon \to 0$). Obviously, this problem can be transformed into the following minimization problem:

$$\min e(\boldsymbol{\theta}) s.t. \boldsymbol{\theta} \in \mathbb{R}^n | l_k \leq \theta_k \leq u_k, \quad i = 1, \ldots, n \tag{4.36}$$

where l_k and u_k are the lower and upper bound of the i-th joint variable, respectively.

4.3.1.2 Principles of Basic Electromagnetism-Like

The electromagnetism-like method (EM) is a new stochastic search method for global optimization [29, 30]. Similar to the genetic algorithm (GA), the population-based algorithm starts with randomly sampling points from the feasible region. According to the objective function values of these sample points, the regions of attraction are determined. Then a mechanism, similar to the attraction-repulsion mechanism of the electromagnetism theory, is invoked for further exploitation of these candidate regions. In EM, the charge of each point relates to the objective function value being optimized. This charge also determines the magnitude of attraction or repulsion of the point over the sample population—the better the objective function value, the higher the magnitude of attraction. The direction for

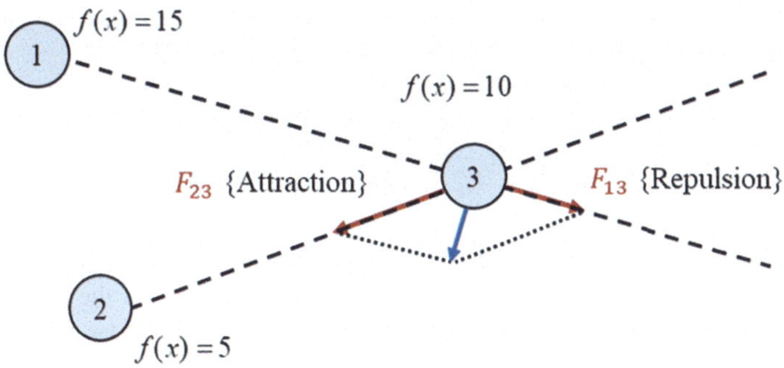

Fig. 4.7 An example of attract-repulse effect on particle number 3

each point to move in subsequent iterations can be determined by evaluating a combination force exerted on the point via other points. Like the electromagnetic forces, this force is calculated by adding vectorially the forces from each of the other points calculated separately. Then this attraction-repulsion mechanism encourages the points to converge to the optimal regions rapidly.

Taking the following figure for an example, there are three particles and their own objective values are 15, 10, and 5, respectively. Since particle 1 is worse than particle 3, while particle 2 is better than particle 3, particle 1 represents a repulsion force, which is the F_{13}, and particle 2 encourages particle 3 that moves to the neighborhood region of particle 2. Consequently, particle 3 moves along with the total force F (Fig. 4.7).

Using the EM to solve problem (4.36), it consists of four steps: initialization, local search, calculation of charge and total force vector, and movement according to the total force.

Step 1. Initialization

The procedure initialization is used to sample m points, $\{\theta^1, \ldots, \theta^m\}$, randomly from the feasible domain of the joint variables, where $\theta^i = \left[\theta^i_1, \ldots, \theta^i_n\right] (i = 1, \ldots, m)$. The procedure uniform sampling can be determined by the following:

$$\theta^i_k = l_k + \text{rand}\cdot(u_k - l_k), \quad k = 1, \ldots, n$$

The procedure ends with m points identified, and the point that has the best function value is the current best point and denoted by $\boldsymbol{\theta}^*$.

Step 2. Local search

The local search procedure is used to gather the local information and improve the current solutions. The iteration procedure is as follows: a random number is selected as a step length and the current point i is moved along this direction.

If point i observes a better point within a given number of iterations, point i is replaced by this better point. Finally, some points are updated.

Step 3. Calculation of charge and total force vector

In each iteration, the charges of the points are calculated according to their objective function values, and the charge of each point is not constant and changes from iteration to iteration. The charge of each point determines the power of attraction or repulsion of point i. This charge is evaluated as follows:

$$q^i = \exp\left(-n\frac{e(\boldsymbol{\theta}^i) - e(\boldsymbol{\theta}^{\text{best}})}{\Sigma_{k=1}^{m}\left(e(\boldsymbol{\theta}^k) - e(\boldsymbol{\theta}^{\text{best}})\right)}\right), \forall i \tag{4.37}$$

In this way, points that have better objective values possess higher charges. Notice that, unlike electrical charges, no signs are attached to the charge of an individual point in Eq. (4.37). Instead, the direction of a particular force between two points is decided after comparing their objective function values. Hence, the total force $\boldsymbol{F}^i$ exerted on point i is computed by the following equation:

$$\boldsymbol{F}^i = \begin{cases} \sum_{j \neq i}^{m} (\boldsymbol{\theta}^j - \boldsymbol{\theta}^i)\dfrac{q^i q^j}{\|\boldsymbol{\theta}^j - \boldsymbol{\theta}^i\|^2} & \text{if } e(\boldsymbol{\theta}^j) < e(\boldsymbol{\theta}^i) \\ \sum_{j \neq i}^{m} (\boldsymbol{\theta}^i - \boldsymbol{\theta}^j)\dfrac{q^i q^j}{\|\boldsymbol{\theta}^j - \boldsymbol{\theta}^i\|^2} & \text{elseif } e(\boldsymbol{\theta}^j) \geq e(\boldsymbol{\theta}^i) \end{cases}, \forall i \tag{4.38}$$

According to Eq. (4.38), the point that has a better objective function value attracts the other one. Contrarily, the point with a worse objective function value repels the other. Since $\boldsymbol{\theta}^{\text{best}}$ has the minimum objective function value, it acts as an absolute point of attraction. Then, it attracts all other points in the population to better region.

Step 4. Movement according to the total force

After evaluating the total force vector $\boldsymbol{F}^i$, point i is moved in the direction of the force by a random step length in Eq. (4.39). Here, the random step length λ is assumed to be uniformly distributed between 0 and 1:

$$\boldsymbol{\theta}^i = \boldsymbol{\theta}^i + \lambda\frac{\boldsymbol{F}^i}{\|\boldsymbol{F}^i\|}(\boldsymbol{R}_{\text{NG}}) \quad i = 1, 2, \ldots, m \tag{4.39}$$

where $\boldsymbol{R}_{\text{NG}}$ is a vector, whose components denote the allowed feasible movement toward the upper bound or the lower bound of the joint variables.

After completing the above procedures, the positions of points are updated, and one iteration calculation of EM is finished. According to the test results in [29], in general, 25 iterations per dimension are satisfactory for converging to the optimum point for the moderate difficulty functions. Therefore, a solution vector that is near the true solution to the inverse kinematics problem of n-DOF robot manipulator can be obtained in finite iterations ($\leq 25n$).

For convenience's sake, an example for solving the inverse kinematics problem of a general 3R manipulator has been used, as shown in Fig. 4.8. This example demonstrates a typical run of EM by presenting the positions of all particles in each iteration.

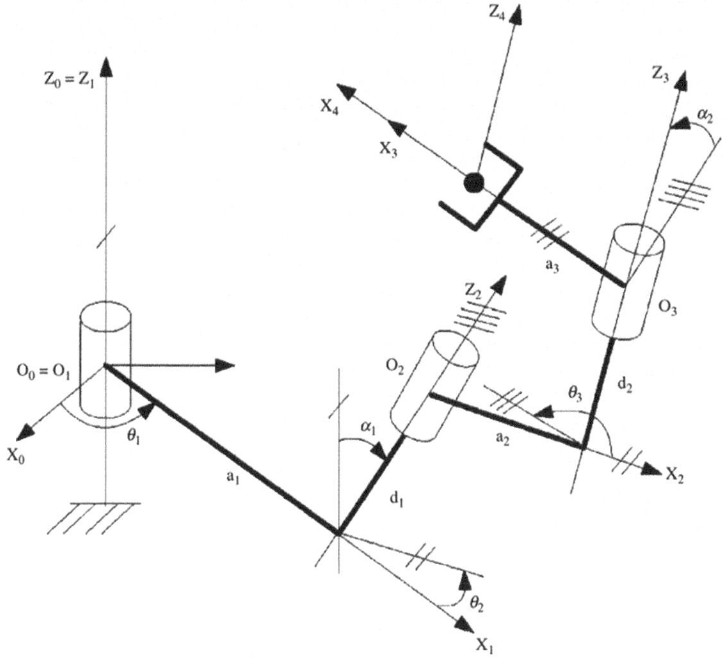

Fig. 4.8 A kinematic scheme for a general 3R manipulator

Table 4.3 The link parameters of the 3R manipulator

Joint	Link length $a(m)$	Twist angle $\alpha(°)$	Offset length d(m)	Rotation angle (°)
1	2.4354	27.0245	0.2513	θ_1
2	1.0826	15.7664	0.1509	θ_2
3	0.9748	0	0	θ_3

For this example, the H-D parameters are shown in Table 4.3. The joint limitations are $\theta_i \in [-\pi/2, \pi]$, $i = 1, 2, 3$. The desired configuration of the end effector is given by $\boldsymbol{P}_{\mathbf{d}} = [0.0084, 2.2992, 1.2962]^{\top}$ and $[\boldsymbol{R}_{\mathbf{d}}] = [\boldsymbol{d}_1, \boldsymbol{d}_2, \boldsymbol{d}_3]$, where $\boldsymbol{d}_1 = [-0.6863, -0.5636, 0.4598]^{\top}$, $\boldsymbol{d}_2 = [0.4415, -0.8251, -0.3525]^{\top}$, and $\boldsymbol{d}_3 = [0.5781, -0.0389, 0.8151]^{\top}$, which corresponds to an exact solution of $\boldsymbol{\theta} = [60°, 70°, 80°]$. In this example, the population size m is set to be 20.

In Figs. 4.8, 4.9, and 4.10, * represents the location of true solution and ⋄ shows the current best point. Figure 4.9 shows the location of particles when the algorithm is started by randomly sampling points from the feasible region. The points in the population move toward the region around the current best point.

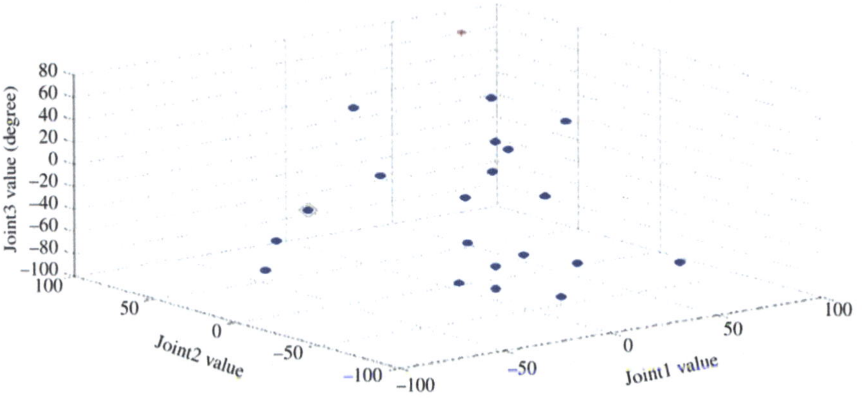

Fig. 4.9 The random initial positions of the particles

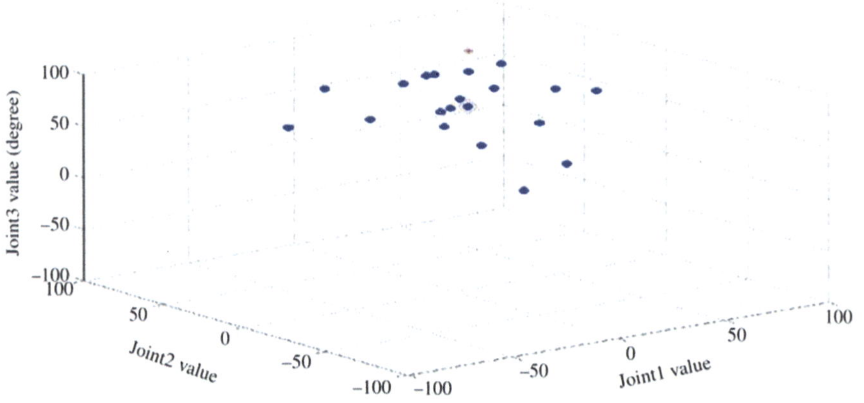

Fig. 4.10 Points start to attract and repel each other

Figure 4.10 shows a new best point $\boldsymbol{\theta}^{\mathrm{best}}$ is observed and the points start to converge toward the new best point. Finally, the optimum solution is located as shown in Fig. 4.11.

4.3.1.3 Hybrid Methods Using EM for Solving the Inverse Kinematics Problem

Our experience shows that applying only a few iterations of EM will bring at least one sample point into the neighborhood of the true solution, but it may become slow thereafter. On the other hand, when the solution vector is near the true solution, the Davidon-Fletcher-Powell (DFP) method is generally expected to give a rapid convergence rate of 154. Combining the above two optimization methods will result in

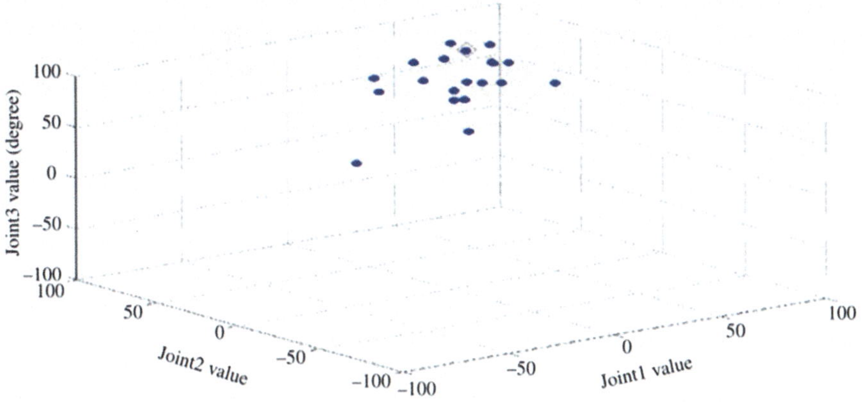

Fig. 4.11 One of the points finds the true solution

higher accuracy and real-time performance. The procedure is given as follows: it uses the EM to rapidly find a feasible point that is near to the true solution and uses the DFP method to obtain a solution at the desired degree of precision. In this section, approximated formulas are developed to calculate the gradient vector and the optimum search step size.

Gradient Computation

For Eq. (4.34), the gradient vector is defined by the following equation:

$$\nabla e(\boldsymbol{\theta}) = \left[\frac{\partial e(\boldsymbol{\theta})}{\partial \theta_1}, \ldots, \frac{\partial e(\boldsymbol{\theta})}{\partial \theta_n}\right]^{\mathrm{T}} \tag{4.40}$$

Substituting (4.32) and (4.33) into (4.34), the elements of the gradient vector are given as

$$\frac{\partial e(\boldsymbol{\theta})}{\partial \theta_i} = 2\left\{[\boldsymbol{P}_h(\boldsymbol{\theta}) - \boldsymbol{P}_d]\cdot\frac{\partial \boldsymbol{P}_h(\boldsymbol{\theta})}{\partial \theta_i} + \sum_{j=1}^{3} \boldsymbol{d}_j\cdot\frac{\partial \boldsymbol{h}_j(\boldsymbol{\theta})}{\partial \theta_i}[\boldsymbol{d}_j\cdot\boldsymbol{h}_j(\boldsymbol{\theta}) - 1]\right\}, \quad i = 1, \ldots, n \tag{4.41}$$

Given $\boldsymbol{\theta}$ and $\boldsymbol{P}_h(\boldsymbol{\theta})$ can be computed by using the forward kinematic equations of robot. The function $h_j(\boldsymbol{\theta})(j = 1, 2, 3)$ is obtained as a result of the n individual rotations $\{Q_i\}_1^n$ about each revolute axis, in a sequential order, from 1 to n [31]. Now the problem can be transformed into how to calculate $((\partial \boldsymbol{P}_h(\boldsymbol{\theta}))/(\partial \theta_i))$ and $((\partial \boldsymbol{h}_j(\boldsymbol{\theta}))/(\partial \theta_i))$.

Figure 4.12 shows the vector relations when rotating the i-th joint about z_i axis from positions 1 to 2 with an angle $\delta\theta_i$, where the vector $\boldsymbol{P}_i$ indicates the current

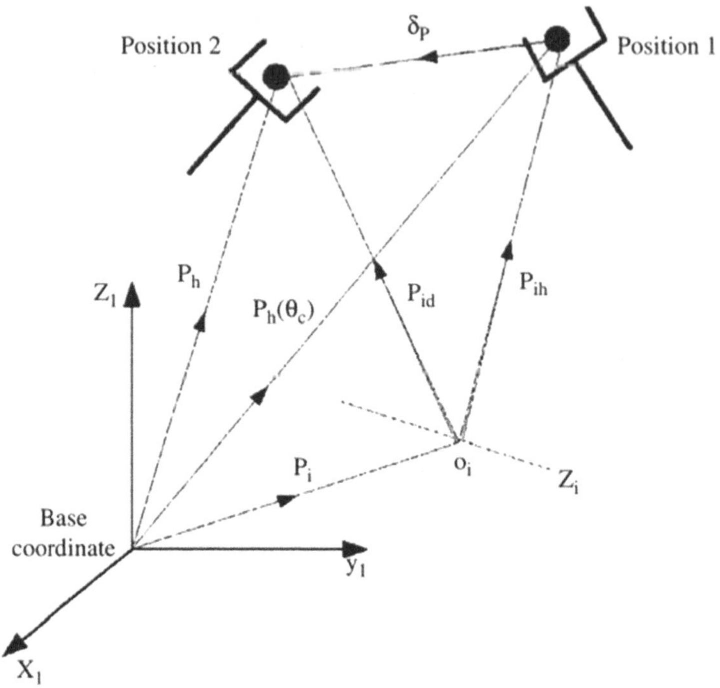

Fig. 4.12 The vector relations about rotation joint i

position of point o_i (the origin of the i-th coordinate frame), $\boldsymbol{P}_{ih}$ is the vector from o_i to the current position of the end effector, $\boldsymbol{P}_{id}$ is a vector from o_i to the new position of the end effector, and $\boldsymbol{P}_h(\theta_c)$ and $\boldsymbol{P}_h$ are the current and the new position vector of the end effector, respectively.

According to the vector relationship as shown in Fig. 4.12, one can obtain

$$\boldsymbol{P}_h = \boldsymbol{P}_i + \boldsymbol{P}_{id} \tag{4.42}$$

and vector $\boldsymbol{P}_{id}$ can be written by

$$\boldsymbol{P}_{id} = \boldsymbol{R}(z_i, \delta\theta_i)\boldsymbol{P}_{ih} \tag{4.43}$$

where $\boldsymbol{R}(z_i, \delta\theta_i)$ is a 3×3 spatial rotation matrix about the z_i axis with rotation angle $\delta\theta_i$ and z_i is a unit vector along the z_i axis. According to the vector form of Rodrigues' equation (4.42) [32] becomes

$$\boldsymbol{P}_h = \boldsymbol{P}_i + \mathbf{z}_i(\boldsymbol{P}_{ih} \times \mathbf{z}_i)(1 - \cos\delta\theta_\mathbf{i}) + \boldsymbol{P}_{ih}\cos\delta\theta_i + (\mathbf{z}_i \times \boldsymbol{P}_{ih})\sin\delta\theta_i \tag{4.44}$$

where $\times$ denotes the vector cross product. Noting that the vector $\boldsymbol{P}_i$ remains constant when rotation of the i-th joint. The partial derivative of (4.44) about variable θ_i is

$$\frac{\partial \boldsymbol{P}_h}{\partial \theta_i} = \mathbf{z}_i(\boldsymbol{P}_{ih} \times \mathbf{z}_i)\sin\delta\theta_i - \boldsymbol{P}_{ih}\sin\delta\theta_i + (\mathbf{z}_i \times \boldsymbol{P}_{ih})\cos\delta\theta_i \tag{4.45}$$

Since the initial guess value of DFP is close to the true solution, the rotation angle of the i-th joint is sufficiently small. The following approximate equation can be obtained:

$$\begin{cases} \sin\delta\theta_i \approx \delta\theta_i \approx 0 \\ \cos\delta\theta_i \approx 1 \end{cases} \quad \text{if} \quad \delta\theta_i \to 0 \tag{4.46}$$

Substituting (4.46) into (4.45), (4.45) can be approximated by

$$\frac{\partial \boldsymbol{P}_h}{\partial \theta_i} = \mathbf{z}_i \times \boldsymbol{P}_{ih} \tag{4.47}$$

Similar to (4.43), the orientation vectors can be written as

$$\boldsymbol{h}_j = R(\mathbf{z}_i, \delta\theta_i)\boldsymbol{h}_j(\theta_c) \quad j = 1, 2, 3 \tag{4.48}$$

where $\boldsymbol{h}_j(\theta_c)$ is the current orientation vector of the end effector; similarly, the following can be obtained:

$$\frac{\partial \boldsymbol{h}_j}{\partial \theta_i} = \mathbf{z}_i \times \boldsymbol{h}_j(\theta_c) \quad j = 1, 2, 3 \tag{4.49}$$

Substituting (4.47) and (4.49) into (4.41), after rearranging and combining terms, one can obtain

$$\frac{\partial e(\theta)}{\partial \theta_i} = 2\mathbf{z}_i \cdot \left\{ (\boldsymbol{P}_d - \boldsymbol{P}_h(\theta)) \times \boldsymbol{P}_{ih} + \sum_{\mathrm{i}=1}^{3} (\boldsymbol{d}_j \times \boldsymbol{h}_j(\theta) - 1)(\boldsymbol{h}_j(\theta) \times \boldsymbol{d}_j) \right\} \tag{4.50}$$

Figure 4.13 shows the vector relationships after translating the translation joint about axis z_i from positions 1 to 2 with an angle $\delta\theta_i$. Then

$$\boldsymbol{P}_{id} = \boldsymbol{P}_{ih} + \delta \boldsymbol{P}_i = \boldsymbol{P}_{ih} + \delta\theta_i z_i \tag{4.51}$$

where the variable $\delta\theta_i$ is a displacement amount along axis z_i. According to vector relations as shown in Fig. 4.13,

$$\boldsymbol{P}_h = \boldsymbol{P}_i + (\boldsymbol{P}_{ih} + \delta\theta_i z_i) \tag{4.52}$$

Notice that $\boldsymbol{P}_i$, $\boldsymbol{P}_{ih}$, and z_i remain unchanged when translation. Then, the partial derivative of (4.52) about variable θ_i is

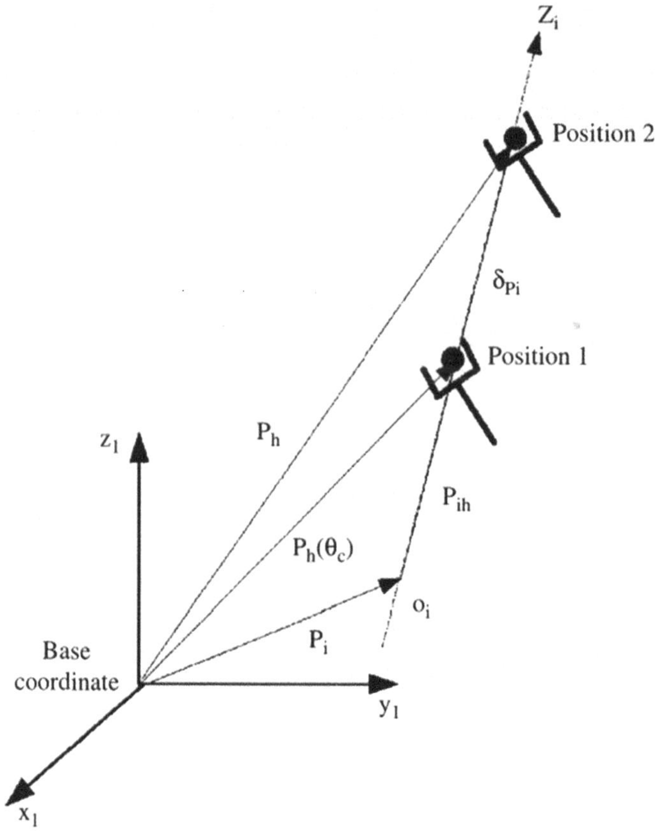

Fig. 4.13 The vector relations about translation joint *i*

$$\frac{\partial \boldsymbol{P}_h}{\partial \theta_i} = z_i \tag{4.53}$$

For the translation joint, the orientation of the end effector remains unchanged when translation; then

$$\frac{\partial \boldsymbol{h}_j}{\partial \theta_i} = 0 \quad j = 1, 2, 3 \tag{4.54}$$

Substituting (4.53) and (4.54) into (4.41),

$$\frac{\partial e(\theta)}{\partial \theta_i} = 2z_i \cdot (\boldsymbol{P}_h(\theta) - \boldsymbol{P}_d) \tag{4.55}$$

According to the type of joint, the gradient vector (4.40) can be efficiently computed by (4.50) or (4.55).

Evaluation of Step Size in Search

The DFP method uses an approximated Hessian matrix and the gradient vector to obtain the search direction [31]. After the search direction λ is obtained, the objective function (4.34) is reduced to a single variable function about λ:

$$e(\lambda) = \Delta p(\lambda) + \Delta o(\lambda) \tag{4.56}$$

where

$$\Delta p(\lambda) = (\boldsymbol{P}_d - \boldsymbol{P}_h(\theta_c + \lambda w))^2 \tag{4.57}$$

and

$$\Delta o(\lambda) = \sum_{j=1}^{3} (\boldsymbol{d}_j \cdot \boldsymbol{h}_j(\theta_c + \lambda \boldsymbol{w}) - 1)^2 \tag{4.58}$$

where $\boldsymbol{w} \in R^n$ is the search direction vector, θ_c is the current joint positions, and λ is the search step size. Now the problem is to find an optimum step size λ^* such that λ^* minimizes $e(\lambda)$.

The differential kinematic equations of the n-DOF manipulator can be written as [16]

$$\begin{bmatrix} \delta \boldsymbol{p} \\ \delta \boldsymbol{o} \end{bmatrix} = \begin{bmatrix} \mathbf{J}_{pc} \\ \mathbf{J}_{oc} \end{bmatrix} \delta \boldsymbol{\theta} = \begin{bmatrix} \mathbf{J}_{pc} \\ \mathbf{J}_{oc} \end{bmatrix} [\lambda \boldsymbol{w}] \tag{4.59}$$

where $\boldsymbol{J}_{pc}$ and $\boldsymbol{J}_{oc}$ are the upper and lower $3 \times n$ submatrices of Jacobian matrix and $\delta \boldsymbol{p}$ and $\delta \boldsymbol{o}$ are, respectively, the corresponding Cartesian space differential displacement and rotation of the end effector.

According to (4.59), one obtains

$$\begin{cases} \delta \boldsymbol{p} = \lambda [\boldsymbol{J}_{pc}] \boldsymbol{w} \\ \delta \boldsymbol{o} = \lambda [\boldsymbol{J}_{oc}] \boldsymbol{w} \end{cases} \tag{4.60}$$

According to the vector relationship (Figs. 4.11 and 4.12), the new position vector of end effector is

$$\boldsymbol{P}_h = \boldsymbol{P}_h(\theta_c) + \delta\boldsymbol{p} = \boldsymbol{P}_h(\theta_c) + \lambda[\boldsymbol{J}_{\mathbf{pc}}]\boldsymbol{w} \tag{4.61}$$

Substituting (4.61) into (4.57),

$$\Delta p(\lambda) = ((\boldsymbol{P}_d - \boldsymbol{P}_h(\theta_c)) - \lambda[\boldsymbol{J}_{pc}]\boldsymbol{w})^2 \tag{4.62}$$

Rotation about the k axis with an angle $\delta\theta$ is denoted by Rot(k, $\delta\theta$). If the rotation amount (radian) is sufficiently small, eq. (4.46) can be rewritten as

$$\begin{cases} \sin\delta\theta \approx \delta\theta \\ \cos\delta\theta \approx 1 \end{cases} \quad \text{if} \quad \delta\theta \to 0 \tag{4.63}$$

where δx, δy, and δz are the corresponding differential rotation of end effector about x, y, and z axes, respectively. Then, the total rotation matrix Rot about x, y, and z axes with the angle δx, δy, and δz can be written as

$$\mathrm{Rot} = \boldsymbol{Rot}(x, \delta x)\boldsymbol{Rot}(y, \delta y)\boldsymbol{Rot}(z, \delta z) \tag{4.64}$$

According to Eq. (4.63), the matrix representation of rotation transformation (4.64) can be approximated by

$$\boldsymbol{R} = \begin{bmatrix} 1 & -\delta z & \delta y \\ \delta z + \delta x\delta y & 1 - \delta x\delta y\delta z & -\delta x] \\ -\delta y + \delta x\delta z & \delta x + \delta y\delta z & 1 \end{bmatrix} \tag{4.65}$$

If the differential value $\delta$is sufficiently small, the higher-order differential $(\delta)^2$ can be neglected. Then, (4.65) can be approximated by the following matrix:

$$\boldsymbol{h}_j(\theta_c) = [x_j, y_j, z_j]^\top \quad (j = 1, 2, 3) \tag{4.66}$$

The orientation vector after rotation becomes

$$\boldsymbol{h}_j(\theta_c + \lambda\mathrm{w}) = \boldsymbol{R} \times \boldsymbol{h}_j(\theta_c) \quad j = 1, 2, 3 \tag{4.67}$$

or in the matrix form

$$\boldsymbol{h}_j(\boldsymbol{\theta}_c + \boldsymbol{\lambda}\mathbf{w}) = \begin{bmatrix} x_j + (z_j\delta y - y_j\delta z) \\ y_j + (x_j\delta z - z_j\delta x) \\ z_j + (y_j\delta x - x_j\delta y) \end{bmatrix}, \quad j = 1, 2, 3 \tag{4.68}$$

Notice that

$$\delta \boldsymbol{o} \times \boldsymbol{h}_j(\theta_c) = \begin{bmatrix} z_j\delta y - y_j\delta z \\ x_j\delta z - z_j\delta x \\ y_j\delta x - x_j\delta y \end{bmatrix}, \quad j=1,2,3 \tag{4.69}$$

According to (4.60), (4.68), and (4.69),

$$\begin{aligned} \boldsymbol{h}_j(\theta_c + \lambda \boldsymbol{w}) &= \boldsymbol{h}_j(\theta_c) + \delta \boldsymbol{o} \times \boldsymbol{h}_j(\theta_c) \\ &= \boldsymbol{h}_j(\theta_c) + \lambda\big(([\boldsymbol{J}_{oc}]\boldsymbol{d}) \times \boldsymbol{h}_j(\theta_c)\big) \end{aligned} \tag{4.70}$$

Substituting (4.70) into (4.58), after rearranging and combining terms,

$$\Delta o(\lambda) = \sum_{i=1}^{3} \big(\boldsymbol{d}_j \cdot \boldsymbol{h}_j(\theta_c) + \boldsymbol{d}_j \cdot \lambda\big(([\boldsymbol{J}_{oc}]\boldsymbol{d}) \times \boldsymbol{h}_j(\theta_c)\big) - 1\big)^2 \tag{4.71}$$

Substituting (4.62) and (4.71) into (4.56),

$$e(\lambda) = k_1\lambda^2 - 2k_2\lambda + k_3 \tag{4.72}$$

where

$$\begin{cases} k_1 = \big([\boldsymbol{J}_{pc}]\boldsymbol{w}\big)\cdot\big([\boldsymbol{J}_{pc}]\boldsymbol{w}\big) + \sum\limits_{j=1}^{3} \big[\big(\big(([\boldsymbol{J}_{oc}]\boldsymbol{w}\big) \times \boldsymbol{h}_j(\theta_c)\big)\cdot\boldsymbol{d}_j\big]^2 \\ k_2 = \big([\boldsymbol{J}_{pc}]\boldsymbol{w}\big)\cdot(\boldsymbol{P}_d - \boldsymbol{P}_h(\theta_c)) + ([\boldsymbol{J}_{oc}]\boldsymbol{w})\cdot\left(\sum\limits_{j=1}^{3} \big(\boldsymbol{h}_j(\theta_c)\cdot\boldsymbol{d}_j - 1\big)\big(\boldsymbol{d}_j \times \boldsymbol{h}_j(\theta_c)\big)\right) \\ k_3 = (\boldsymbol{P}_d - \boldsymbol{P}_h(\theta_c))\cdot(\boldsymbol{P}_d - \boldsymbol{P}_h(\theta_c)) + \sum\limits_{j=1}^{3} \big(\boldsymbol{h}_j(\theta_c)\cdot\boldsymbol{d}_j - 1\big)^2 \end{cases} \tag{4.73}$$

Obviously, (4.73) is minimized when

$$\begin{cases} \dfrac{\partial e(\lambda)}{\partial \lambda} = 2k_1\lambda - 2k_2 = 0 \\ \dfrac{\partial^2 e(\lambda)}{\partial \lambda^2} = 2k_1 > 0 \end{cases} \tag{4.74}$$

Then the approximate optimum search step size can be obtained:

$$\lambda^* = \frac{k_1}{k_2} \tag{4.75}$$

4.3.1.4 Numerical Examples

This example demonstrates that the proposed method in this section can be used for continuous joint space trajectory planning of redundant manipulators. A planar 2-DOF manipulator with one-redundant degree of freedom is used as an illustrative example, as shown in Fig. 4.14. Joints 1 and 2 are revolute joints, and link 2 has a prismatic joint. The position of the end effector is a function of the fixed link length and three variables: θ_1, θ_2, and l_2. The forward kinematic equations are

$$\begin{cases} x = l_1 \cos(\theta_1) + l_2 \cos(\theta_1 + \theta_2) \\ y = l_1 \sin(\theta_1) + l_2 \sin(\theta_1 + \theta_2) \end{cases} \tag{4.76}$$

As shown in Fig. 4.14, the desired trajectory of the end effector is a circle centered at the point (x_0, y_0) and a radius equal to r. The trajectory is discretized into 72 equally spaced points, and the coordinates of these discrete points are

$$\begin{cases} x_n = x_0 + r\cos\left((n-1)\cdot\frac{\pi}{36}\right) \\ y_n = y_0 + r\sin\left((n-1)\cdot\frac{\pi}{36}\right) \end{cases}, \quad n = 1, 2, \cdots, 72 \tag{4.77}$$

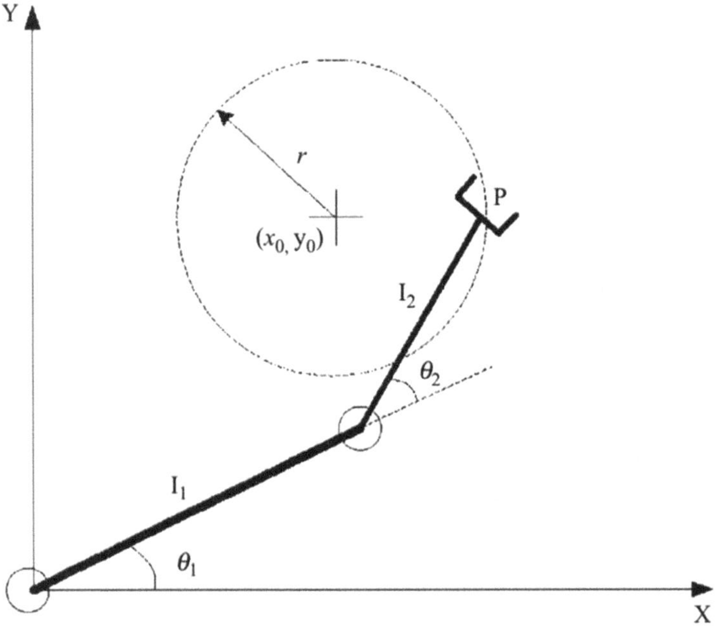

Fig. 4.14 The two-link manipulator and a circular trajectory

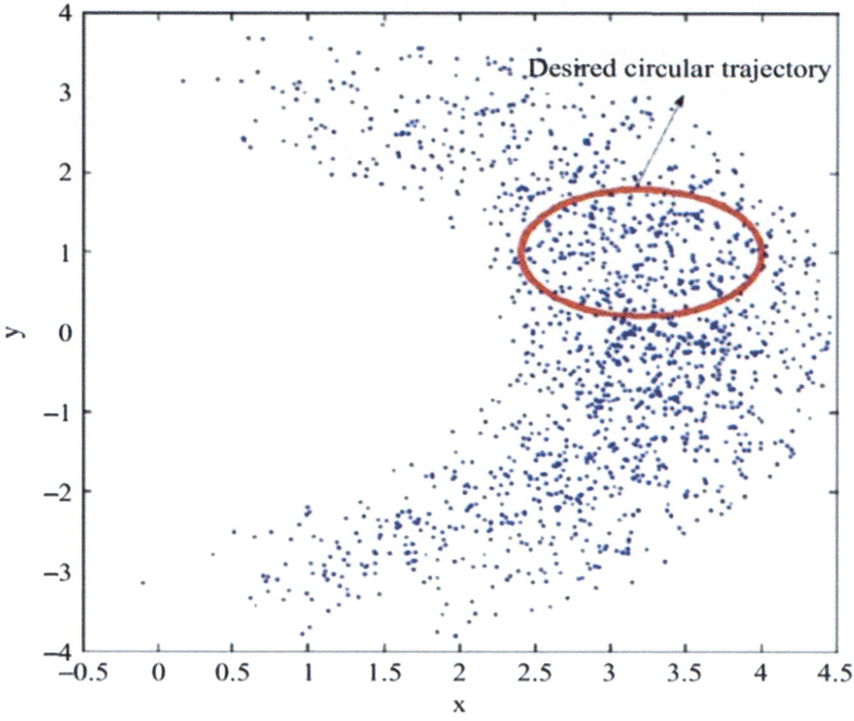

Fig. 4.15 The workspace of manipulator

For this example, the joint limitations are $-\pi/4 \le \theta_1 \le 3\pi/4$, $-\pi/2 \le \theta_2 \le 3\pi/2$, and $1.0 \le l_2 \le 2.5$. The desired trajectory of the end effector is a circle centered at (3.2, 1.0). The radius r equals to 0.8 and $l_1 = 2.0$.

The Monte Carlo method of random sampling is applied to the joint space of the manipulator to approximate the workspace [33]. Figure 4.15 shows the workspace and the desired trajectory of the manipulator. Obviously, all the desired poses are reachable. For this example, EM was tested with two different cases.

First, the procedure of hybrid optimization is excluded, and only the standalone EM is used. Second, the hybrid optimization method developed in this section is used to solve the same problem. For each desired position in (4.77), the algorithm would converge to a solution of $\theta^* = \left[\theta_1^*, \theta_2^*, l_2^*\right]$. It should be noted that this is just one solution among an infinite number of solutions for a redundant manipulator. Submitting into (4.76), the discrete space trajectory of the manipulator end effector will be obtained. In this example, the population size m is set to be 30.

Figure 4.16a–c shows the results when iteration number is 30, 50, and 100, respectively, by EM only. Figure 4.16d shows the result when the hybrid optimization method is used with iteration number 30. The results in Fig. 4.16 show that EM can approximate the optimum if a large number of iterations are to be used. Compared to EM, application of the hybrid method developed in this section can get better precision at less iteration number.

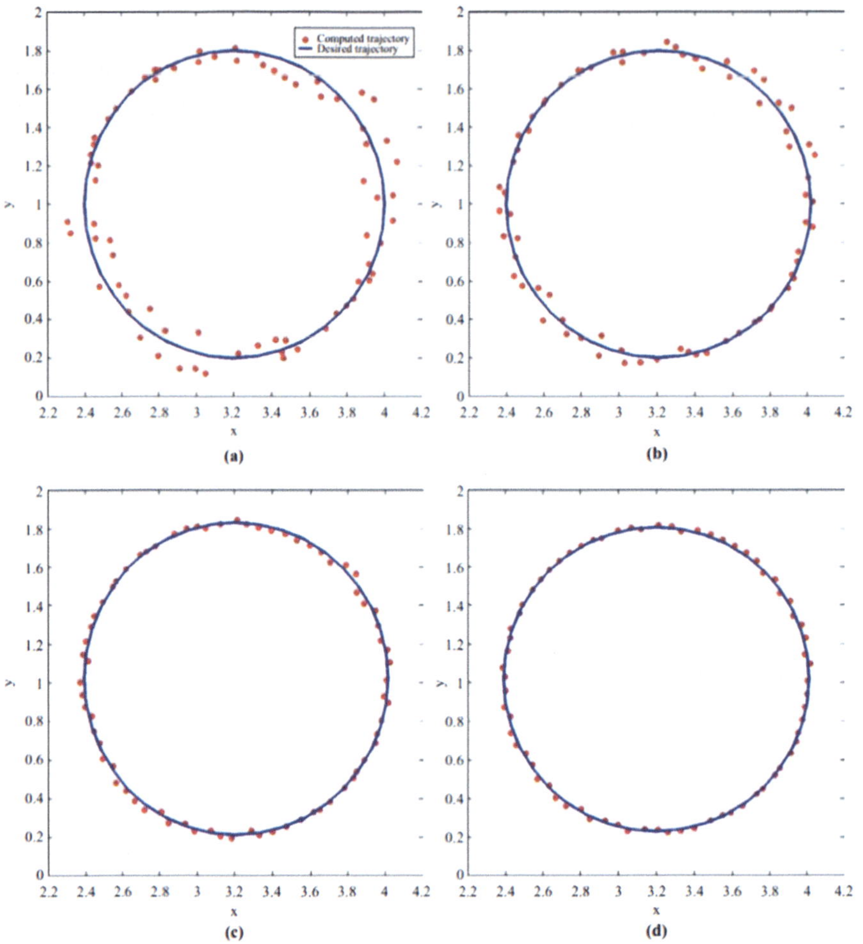

Fig. 4.16 Computed trajectory of the manipulator end effector. (**a**) Iteration numbers are 30. (**b**) Iteration numbers are 50. (**c**) Iteration numbers are 100. (**d**) Iteration numbers are 30 by hybrid method

4.3.2 *Solving Robot Manipulator Inverse Kinematics with Electromagnetism-like and Modified DFP Algorithms*

4.3.2.1 Problem Formulation

As shown in Fig. 4.6, the desired position vector and orientation matrix of a manipulator end effector are denoted by $\boldsymbol{P}_d$ and $[\boldsymbol{R}_d] = [\boldsymbol{d}_1, \boldsymbol{d}_2, \boldsymbol{d}_3]$, where $\boldsymbol{d}_j (j = 1, 2, 3)$ are unit vectors along the x_d, y_d, z_d axes. $\boldsymbol{P}_h$ is the current position vector of the end effector. The current orientation matrix is defined by $[\boldsymbol{R}_h] = [\boldsymbol{h}_1,$

$\boldsymbol{h}_2, \boldsymbol{h}_3]$, where $\boldsymbol{h}_j (j = 1, 2, 3)$ are unit vectors along the x_h, y_h, z_h axes and the joint variables are denoted by the $n \times 1$ vector, $\boldsymbol{\theta} = [\theta_1, \theta_2, \cdots, \theta_n]^\top$.

Therefore, the error between the current and the desired locations of the end effector can be described by the following functions:

Position error:

$$\Delta p(\boldsymbol{\theta}) = \|\boldsymbol{P}_d - \boldsymbol{P}_h(\boldsymbol{\theta})\| \tag{4.78}$$

Orientation error:

$$\Delta o(\boldsymbol{\theta}) = \sum_{j=1}^{3} \left(\boldsymbol{d}_j \cdot \boldsymbol{h}_j(\boldsymbol{\theta}) - 1\right)^2 \tag{4.79}$$

The total error:

$$e(\boldsymbol{\theta}) = \Delta p(\boldsymbol{\theta}) + \Delta o(\boldsymbol{\theta}) \tag{4.80}$$

where $(\cdot)$ denotes the vector dot product. Now, the inverse kinematics problem is to find a solution θ^* such that $e(\boldsymbol{\theta}^*) \le \varepsilon (\varepsilon \to 0)$. This problem can be transformed into the following equivalent minimization problem:

$$\min e(\boldsymbol{\theta}) \text{s.t.} \boldsymbol{\theta} \in \boldsymbol{R}^n | l_k \le \theta_k \le u_k, \quad k = 1, 2, \cdots, n \tag{4.81}$$

where l_k and u_k are the lower and upper bounds of the i-th joint variable, respectively.

4.3.2.2 Review of EM

EM is a new stochastic search method similar to the GA [29]. To solve the problem in (4.81), the general scheme of EM is given by the following procedure:

Algorithm 4.1

Initialize;
  While termination criteria are not satisfied **do**;
    Local search;
    Calculation of charge and total force vector;
    Movement according to the total force;
  End while

The procedure initialization is used to sample m points, $\theta^1, \cdots, \theta^m$, randomly from the feasible domain of the joint variables, where $\theta^i = \left[\theta_1^i, \cdots, \theta_n^i\right] (i = 1, \cdots, m)$. The procedure of uniform sampling can be determined by

$$\theta_k^i = l_k + \text{rand}\cdot(u_k - l_k), \quad k = 1, 2, \cdots, n \tag{4.82}$$

The procedure ends with m points identified, and the point that has the best function value is stored in θ^{best}.

The local search procedure is used to gather the local information and improve the current solutions. It can be applied to one or many points for local refinement at each iteration. The selection of these two procedures does not affect the convergence result.

The charges of the points are calculated according to their objective function values, and the charge of each point is not constant and changes from iteration to iteration. The charge of the i-th point $q^i$is evaluated as

$$q^i = \exp\left[-n\frac{\left(e(\theta^i) - e(\theta^{\text{best}})\right)}{\sum\limits_{k=1}^{m}\left(e(\theta^k) - e(\theta^{\text{best}})\right)}\right], \quad i = 1, 2, \cdots, m \tag{4.83}$$

In this way, the points that have better objective values possess higher charges. Notice that, unlike electrical charges, no signs are attached to the charge of an individual point in (4.83). Instead, the direction of a particular force between two points is decided after comparing their objective function values. Hence, the total force $\boldsymbol{F}^i$ exerted on point i is computed by the following equation:

$$\boldsymbol{F}^i = \begin{cases} \sum\limits_{j\neq i}^{m}\left(\theta^j - \theta^i\right)\dfrac{q^j q^i}{\left\|\theta^j - \theta^i\right\|^2}, & \text{if} \quad e(\theta^j) < e(\theta^i) \\ \sum\limits_{j\neq i}^{m}\left(\theta^i - \theta^j\right)\dfrac{q^j q^i}{\left\|\theta^j - \theta^i\right\|^2}, & \text{others} \end{cases} \tag{4.84}$$

According to (4.84), the point that has a better objective function value attracts the other one. Contrarily, the point with worse objective function value repels the other. Since θ^{best} has the minimum objective function value, it acts as an absolute point of attraction. Then, it attracts all other points in the population to better region. The procedure of calculation for total force vector is as follows:

Algorithm 4.2

for $i = 1$ to m do

$$q^i = \exp\left[-n\frac{e(\theta^i) - e(\theta^{\text{best}})}{\sum\limits_{k=1}^{m}\left(e(\theta^k) - e(\theta^{\text{best}})\right)}\right], \quad \boldsymbol{F}^i = 0$$

end for

for $i = 1$ to m do

for $j = 1$ to m do
  if $i \neq j$ then
    $\boldsymbol{F}_j^i = (\theta^j - \theta^i) \frac{q^i q^j}{|x^j - x^i|^2}$;
    if $f(x^j) < f(x^i)$ then
    $\boldsymbol{F}^i = \boldsymbol{F}^i + \boldsymbol{F}_j^i$ Attraction;
    else
    $\boldsymbol{F}^i = \boldsymbol{F}^i - \boldsymbol{F}_j^i$ Repulsion;
    end if
  end if
end for
end for

After evaluating the total force vector $\boldsymbol{F}^i$, point i is moved in the direction of the force by a random step length in (4.85). Here, the random step length λ is assumed to be uniformly distributed between 0 and 1.

$$\theta^i = \theta^i + \lambda \frac{\boldsymbol{F}^i}{\|\boldsymbol{F}^i\|} \boldsymbol{R}_{\mathrm{NG}}, \quad i = 1, 2, \cdots, m \tag{4.85}$$

In (4.85), $\boldsymbol{R}_{\mathrm{NG}}$ is a vector whose components denote the allowed feasible movement toward the upper bound u_k or the lower bound l_k of the joint variables. The procedure is as follows:

Algorithm 4.3

for $i = 1$ to m do
  if $i \neq$ best then
    $\lambda = U(0, 1)$;
    $\boldsymbol{F}^i = \frac{\boldsymbol{F}^i}{\|\boldsymbol{F}^i\|}$;
    for $k = 1$ to n do
      if $F_k^i > 0$ then
      $x_k^i = x_k^i + \lambda F_k^i (u_k - x_k^i)$;
      else
      $x_k^i = x_k^i + \lambda F_k^i (x_k^i - l_k)$;
      end if
    end for
  end if
end for

After finishing the above procedures, the positions of points are updated, and we have finished one iteration calculation of EM. Taking Fig. 4.7 for an example, there are three particles, and their own objective values are 15, 10, and 5, respectively. Because particle 1 is worse than particle 3, while particle 2 is better than particle

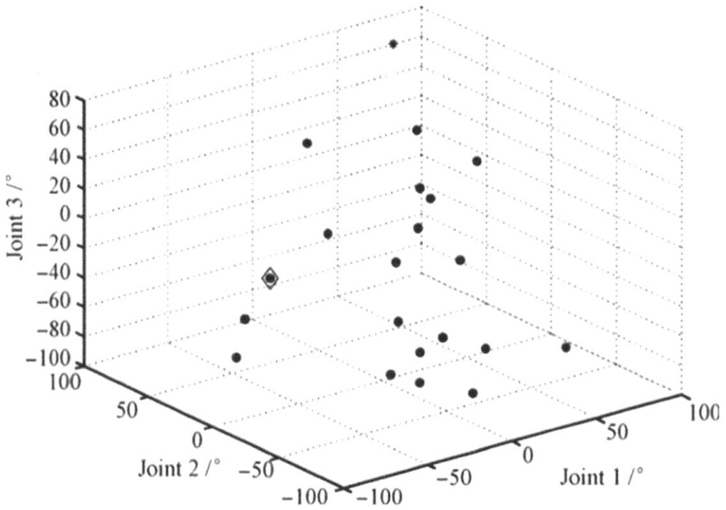

Fig. 4.17 The random initial positions of the particles

3, particle 1 represents a repulsion force, which is $\boldsymbol{F}_{13}$, and particle 2 encourages particle 3 that moves to the neighborhood region of particle 2. Consequently, particle 3 moves along with the total force $\boldsymbol{F}$.

According to test results in [29], in general, 25 iterations per dimension are satisfactory for converging to the optimum point for moderately difficult functions. Therefore, an approximate solution vector, which is near the true solution to the inverse kinematics problem of n-DOF robot manipulator, can be obtained in finite iterations (≤ 25). An example for solving the inverse kinematics problem of a general 3R manipulator has been used, as shown in Fig. 4.8. This example demonstrates a typical run of EM by presenting the positions of all particles in iterations.

For this example, the Denavit-Hartenberg parameters are shown in Table 4.3. The joint limitations are $\theta \in [-\pi/2, \pi]$. The desired configuration is $\boldsymbol{P}_d = [0.0084, 2.2992, 1.2962]^\top$ and $\boldsymbol{R}_d = [\boldsymbol{d}_1, \boldsymbol{d}_2, \boldsymbol{d}_3]$ with $\boldsymbol{d}_1 = [-0.6863, -0.5636, 0.4598]^\top$, $\boldsymbol{d}_2 = [0.4415, -0.8251, -0.3525]^\top$, and $\boldsymbol{d}_3 = [0.578, -0.0389, 0.815]^\top$, which corresponds to an exact solution of $\boldsymbol{\theta} = [\theta_1, \theta_2, \theta_3] = [60°, 70°, 80°]$. In this example, we have selected the sample point m to be 20.

In the following figures, "*" represents the location of true solution and "$\diamond$" shows the current best point. Figure 4.17 shows the location of particles when the algorithm is started by randomly sampling points from the feasible region. The points in the population move toward the region around the current best point. Figure 4.18 shows a new best point θ^{best} is observed and the points start to converge toward the new best point. Finally, the optimum solution is located as shown in Fig. 4.19.

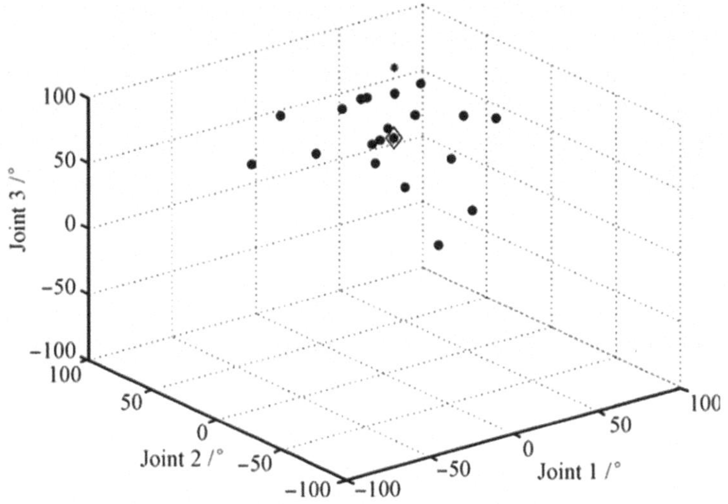

Fig. 4.18 Points start to attract and repel each other

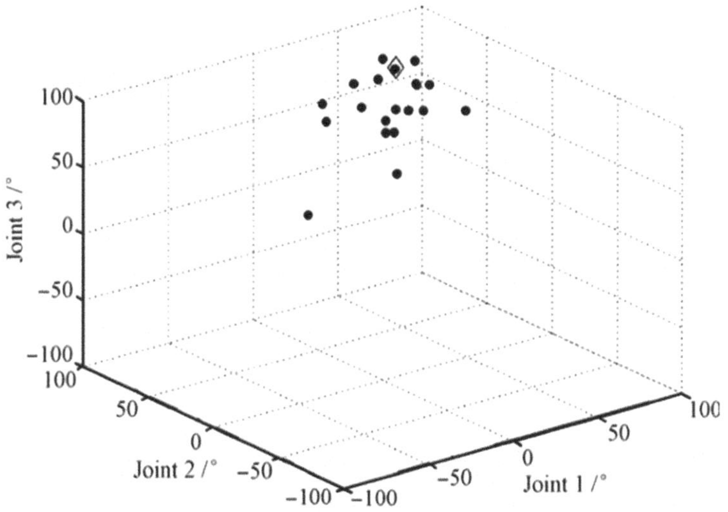

Fig. 4.19 One of the points finds the true solution

4.3.2.3 Hybrid Methods Using EM and Modified DFP for Solving the Inverse Kinematics Problem

Our experience shows that applying only a few iterations of EM will bring at least one sample point into the neighborhood of the true solution. In general, four results are usually obtained by EM:

Case 1. A true solution is obtained within a small number of iterations.
Case 2. A true solution is obtained within a desired number of iterations.
Case 3. A true solution is obtained when the number of iterations is large enough.
Case 4. An approximate solution close to the true solution is obtained within a desired number of iterations.

Without loss of generality, we will consider the most general Case 4. For Case 4, although the result can be improved by increasing the number of iterations, this may increase the running time. Notice that an approximate solution near to the true solution can be obtained rapidly by EM. On the other hand, when the solution vector is near the true solution, the DFP method is generally expected to give a rapid convergence rate [31]. Therefore, if we combine the above two optimization methods, this hybrid approach is more efficient and precise.

The DFP method finds the solution to the secant equation that is closest to the current estimate and satisfies the curvature condition (see below). It was the first quasi-Newton method that generalizes the secant method to a multidimensional problem. This update maintains the symmetry and positive definiteness of the Hessian matrix. Given a function $f(x)$, its gradient ∇f, and positive definite Hessian matrix B, the Taylor series is

$$f(x_k + s_k) = f(x_k) + \nabla f(x_k)^\top s_k + \frac{1}{2} s_k^\top B s_k \tag{4.86}$$

And the Taylor series of the gradient itself is

$$\nabla f(x_k + s_k) = \nabla f(x_k) + B s_k \tag{4.87}$$

which is used to update dB. The DFP method finds a solution that is symmetric, positive, and closest to the current approximation of $\boldsymbol{B}_k$:

$$\boldsymbol{B}_{k+1} = (I - \gamma_k y_k s_k^\top)\boldsymbol{B}_k(I - \gamma_k s_k y_k^\top) + \gamma_k y_k y_k^\top \tag{4.88}$$

$$y_k = \nabla f(x_k + s_k) - \nabla f(x_k), \gamma_k = \frac{1}{y_k^\top s_k} \tag{4.89}$$

where $\boldsymbol{B}_k$ is a symmetric and positive definite matrix. The corresponding update to the inverse Hessian approximation $H_k = B_k^{-1}$ is given by

$$H_{k+1} = H_k - \frac{H_k \boldsymbol{y}_k \boldsymbol{y}_k^\top H_k}{\boldsymbol{y}_k^\top H_k y_k} + \frac{s_k \boldsymbol{s}_k^\top}{\boldsymbol{y}_k^\top s_k} \tag{4.90}$$

$\boldsymbol{B}$ is assumed to be positive definite and the vectors $\boldsymbol{s}_k^\top$ and $\boldsymbol{y}$ must satisfy the curvature condition:

$$s_k^\top y_k = s_k^\top B s_k > 0 \tag{4.91}$$

The DFP method is quite effective. However, due to the highly nonlinear nature of the manipulator kinematic equations, the gradient vector of the objective function and the optimum step size in the search direction are difficult to obtain in DFP. To solve these two problems, the analytical form of the gradient vector and an approximation to the optimum search step size had been developed in [34]. However, since the Jacobian matrix is used in [34], the computational process is more complex. To reduce the computational complexity, a modified DFP algorithm is developed in this section. Unlike the traditional DFP algorithm, this modified DFP algorithm randomly chooses the search step size between 0 and 1. Thus, the computational complexity is greatly reduced, and more importantly, this method is not only computationally efficient but also numerically stable. In order to ensure that the modified DFP algorithm converges to the optimum solution, the following assumptions are made:

Assumption 4.1 The pre-specified initial value of the solution is close to the optimum solution and the search procedure is restricted to a small area near to the true solution in modified algorithm.

Assumption 4.2 The optimum search direction is known per iteration.

From the foregoing discussions, the initial guess value can be given by EM. The search direction in Assumption 4.2 can be obtained using an approximated Hessian matrix and the gradient vector of the objective function [31]. Based on the above assumptions, the true solution can be approached gradually using a small random search step size between 0 and 1. Since the search domain is small, the optimum solution can be efficiently calculated. The complete solution procedure of the modified DFP (MDFP) algorithm is summarized below:

Algorithm 4.4

Step 1. Given the initial value $x^{(1)} \in \boldsymbol{R}^n$ and termination criteria $\varepsilon > 0$

Step 2. $\boldsymbol{H}_1 = \boldsymbol{I}_n$ (unit matrix);

Step 3. Given a random search step size: $\lambda = \mathit{rand}\,(0, 1)$;

Step 4. Calculate the gradient at $x^{(1)}$: $g_1 = \nabla f(x^{(1)})$;

Step 5. $k = 1$;

Step 6. Calculate the search direction: $d^{(k)} = -H_k g_k$;

Step 7. $err = \| g_k \|$;

Step 8. While $err \geq \varepsilon$do;

Step 9. $x^{(k+1)} = x^{(k)} + \lambda d^{(k)}$;

Step 10. $g_{k+1} = \nabla f(x^{(k+1)})$;

Step 11. $q_k^{(k)} = g_{k+1} - g_k$;

Step 12 $p^k = x^{k+1} - x^k$;

Step13. $\boldsymbol{H}_{k+1} = \boldsymbol{H}_k + \frac{p^{(k)} p^{(k)\mathrm{T}}}{p^{(k)\mathrm{T}} q^{(k)}} - \frac{\boldsymbol{H}_k q^{(k)} q^{(k)\mathrm{T}} \boldsymbol{H}_k}{q^{(k)\mathrm{T}} \boldsymbol{H}_k q^{(k)}}$;

Step 14. $d^{k+1} = -\boldsymbol{H}_{k+1} g_{k+1}$;

Step 15. $err = \| g_{k+1} \|$;

Step 16. $k = k + 1$;

Step 17. End while

Lemma 4.1 [31] If the gradient $g_i \neq 0$, the matrix $\boldsymbol{H}_i$ (Step 13 in MDFP algorithm) is symmetric positive definite.

Lemma 4.2 [31] For a symmetric positive definite matrix $\boldsymbol{H}_i$, the search direction vector

$$\boldsymbol{d}^{(k)} = -H_k \nabla f\left(x^{(k)}\right) \tag{4.92}$$

is a descent direction. It shows that a small step along $\boldsymbol{d}^{(k)}$ can guarantee that the objective function is reduced. Thus, the global optimum solution can be approached by an appropriate small step in the direction of descent with limited iteration.

In order to examine the basic convergence properties of EM-MDFP algorithm, general ten test functions taken from the website, http://www-optima.amp.i.kyotou.ac.jp/member/student/hedar/Hedarfiles/go.htm, are calculated by EM-MDFP method.

First, we make use of EM to obtain an approximate solution near to the true solution. Though better results can be achieved by EM when a large number of iterations and sample points (m) are used, we have selected a small number of sample points (m) and maximum iterations aiming to reduce the running time and examine whether the MDPF algorithm can converge to the optimum solution when the quality of initial value given by EM is not good enough. In this section, we have selected the sample points (m) to be 10 and the maximum iterations to be 30, except function Trid and Michalewics (4.82). The number of 30 iterations seems to be not large enough to adequately explore a feasible solution for these two problems. Then, we have selected the maximum iterations to be 100 to calculate function Trid and Michalewics (4.82). The running time, the approximate solution ($x(1)$), and the best objective function values (Best $f(x)$) are recorded. The testing procedure continuously runs 10 times. The average running time (Avg. time) and objective function values (Avg. $f(x)$) are also reported.

Second, the MDFP method starts from the initial point given by EM to search the optimum solution. The stopping criterion is set to be 10^{-6}. For the comparison, the number of iterations, running time, and objective function, including their average values, are recorded.

All the computations are conducted on a Celeron(R) CPU 2.80 GHz PC. The algorithm is coded in MATLAB 6.5. Table 4.4 shows the test results of the test function booth. The known optimum solution and minimum value of the test function booth are, respectively, $x^* = (1, 3)$ and $f(x^*) = 0$. In this example, the search step size λ is assumed to be uniformly distributed between 0 and 1. More function test results are given in Table 4.6, and the process of testing is similar to Table 4.4 Without loss of generality, the search step size λ is randomly set to be a small positive value between 0 and 1. Our results show that EM is able to approximate the optimum. Moreover, although the search step size λ is not optimal, the convergence is extremely rapid and the optimum solutions in all cases can be obtained by MDFP. The function Michalewics in Table 4.5 appears to be a special

Table 4.4 Results of EM-MDFP for test function booth

	EM			MDFP			
Solution number	$x^{(1)}$	$f(x^{(1)})$	Time (s)	$f(x^{*})$	Time (s)	Steps	λ
1	(1.3843, 2.5867)	0.3219	0.3130	0.0000	0.0780	130	0.1
2	(1.2745, 2.8634)	0.1701	0.3130	0.0000	0.0160	70	0.2
3	(1.2524, 2.7525)	0.1251	0.3130	0.0000	0.0160	42	0.3
4	(1.0923, 2.9118)	0.0164	0.3430	0.0000	0.0150	28	0.4
5	(1.1791, 2.8806)	0.0606	0.3280	0.0000	0.0320	24	0.5
6	(1.1484, 2.7962)	0.0758	0.3440	0.0000	0.0000	18	0.6
7	(0.9709, 3.1807)	0.1254	0.3120	0.0000	0.0000	15	0.7
8	(0.6474, 3.4850)	0.4297	0.3130	0.0000	0.0160	12	0.8
9	(1.4567, 2.6599)	0.3786	0.3280	0.0000	0.0310	9	0.9
10	(1.1880, 2.9396)	0.1042	0.3440	0.0000	0.0150	5	1.0
Avg.		0.1808	0.3250	0.0000	0.0219	35	

case of taking more iterations by MDFP. It is well known that the global optimum of function Michalewics is hard to be obtained. To get a better convergence property, a given optimal search step size is used instead of choosing randomly in MDFP. The global optimum is obtained successfully, and the running times of MDFP are, respectively, only about 0.08 and 0.6 s. Notice that the running time is inevitable to be disturbed by other programs running in PC, and the actual time consumption of MDFP method is almost close to 0. This is also why some iterative procedures with a smaller number of iterations need more running time in Tables 4.4 and 4.5.

The structure of the manipulator used in this example is based on the PUMA 560 robot. This robot is chosen because a closed-form solution can be obtained, which can be used for checking the accuracy of the numerical solution. The link parameters and the joint limits of this robot are listed in Table 4.6.

To handle the IK problem of this robot with MDFP method, the search direction must be calculated firstly, which can be obtained by using an approximated Hessian matrix and the gradient vector of the objective function. This Hessian matrix can be computed using the formula (Step 13 in MDFP algorithm). For (4.80), the elements of the gradient vector are given by the following equation [34]:

If the joint is a translational joint, then

$$\frac{\partial e(\boldsymbol{\theta})}{\partial \theta_i} = 2z_i \cdot (\boldsymbol{P}_h(\theta) - \boldsymbol{P}_d) \tag{4.93}$$

If the joint is a rotational joint, then

$$\frac{\partial e(\boldsymbol{\theta})}{\partial \theta_i} = 2z_i \cdot \left\{ [\boldsymbol{P}_d - \boldsymbol{P}_h] \times \boldsymbol{P}_{ih} + \sum_{j=1}^{3} (\boldsymbol{d}_j \cdot \boldsymbol{h}_j - 1)(\boldsymbol{h}_j \times \boldsymbol{d}_j) \right\} \tag{4.94}$$

Table 4.5 Results of EM-MDFP for all test functions

		EM			MDFP			
Function name	Known optimum	Best $f(x)$	Avg. $f(x)$	Avg.time (s)	Avg. $f(x)$	Avg.time (s)	Avg. steps	λ
Beale	0.0	0.0121	0.0650	0.3141	0.0000	0.0108	52	Random
Hump	0.0	0.0000	0.0403	0.5311	0.0000	0.0281	34	Random
Matyas	0.0	0.0150	0.1917	0.3263	0.0000	0.0295	60	Random
Sum squares (5)	0.0	0.3935	1.3336	0.3281	0.0000	0.0138	23	Random
Zakharov (2)	0.0	0.0030	0.0686	0.3166	0.0000	0.0266	26	Random
Levy (2)	0.0	0.0004	0.1783	0.3391	0.0000	0.0231	22	Random
Trid (6)	−50.0	−47.2770	−43.7232	0.9302	−50.00	0.0201	28	Random
Michalewics (2)	−1.8013	−1.8003	−1.7571	0.3469	−1.8013	0.0876	289	0.05
Michalewics (5)	−4.6876	−4.2910	−3.7141	1.0453	−4.6863	0.5967	2 776	0.008

Table 4.6 The link parameters of the PUMA 560 robot

Joint	Link length (m)	Twist angle (°)	Offset length (m)	Joint limitations (°)
1	0	−90	0.6604	[−160, 160]
2	0.4320	0	0.2000	[−225, 45]
3	0	90	−0.0505	[−45, 225]
4	0	−90	0.4320	[−110, 170]
5	0	90	0.0	[−100, 100]
6	0	0	0.0565	[−266, 266]

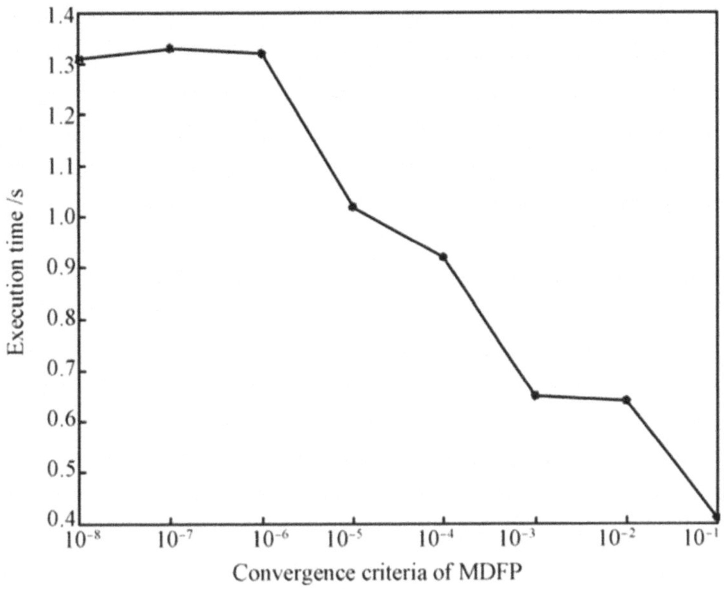

Fig. 4.20 Average execution time for the example

Given $\boldsymbol{\theta}$, z_i, $\boldsymbol{P}_h$, and $\boldsymbol{h}_j (j = 1, 2, 3)$ can be recursively computed by using the forward recursion formulas in [35]. Consequently, the search direction vector can be efficiently evaluated.

4.3.2.4 Example

The convergence criterion of EM in this example is defined by $\varepsilon = 0.1$ instead of the maximum iterations. The convergence of MDFP method is defined by $\varepsilon = 10^{-k}$, where k varies from 1 to 8. For each value of k, ten problems were given to EM-MDFP method to solve. Both the initial and the desired configurations of the problems were randomly generated within the working space of the robot. The average running times of the solved problems are plotted in Fig. 4.20. From this figure, the EM-MDFP method is able to solve all of the given problems at the desired precision. Moreover, even though the required degree of precision is high, the EM-MDFP is also computationally efficient. It should be noted that a C implementation would be much faster than MATLAB. Then, we have reasons to believe that the EM-MDFP is a near real-time algorithm if the evaluations are carried out in C/C++ environment.

References

1. Li Q, Fan Z, Wu Q. Investigation of ice-covered transmission lines and analysis on transmission line failures caused by ice-coating in China. Power system technology. 2008, 32(9):33.
2. Huang X, Liu J, Cai W. Present research situation of icing and snowing of overhead transmission lines in China and foreign countries. Power system technology. 2008,32(4):23.
3. Hydro-QuCbec: "Janvier de cristal: Le QuCbec dans le coeur d'Hydro". [Online]. Available: https://www.hydroquebec.com/history-electricity-in-quebec/great-periods/1998-an-exceptional-response-to-a-crisis.html.
4. Jones K F, Mulherin N D. An evaluation of the severity of the January 1998 ice storm in northern New England. Report for FEMA Region, 1998, 1: 33.
5. Brostrom E. Ice storm modelling in transmission system reliability calculations. KTH, 2007.
6. Laforte J L, Allaire M A, Laflamme J. State-of-the-art on power line de-icing. Atmospheric Research, 1998, 46(1–2): 143–158.
7. Laforte J L, Allaire M A, Laflamme J and Beaudoin P, 1998, An update on overhead cable de-icing technology. In: Proc. 8th Int. Workshop on Atmospheric Icing on Structures (IWAIS). 1998: 215–220.
8. Shan X, Shu N. Discussion on methods of de-icing for overhead transmission lines. Gaodianya Jishu/ High Voltage Engineering, 2006, 32(4): 25–27.
9. Huang Q, Chen Y, Zhang C, Design of an adaptive on-load de-icing scheme for overhead power transmission line. In: 2009 IEEE Power & Energy Society General Meeting. IEEE, 2009: 1–5.
10. Wang Y, Wei S, Yin F, et al. Review on key technology of de-icing robot running on overhead transmission line. Journal of Mechanical Engineering, 2011, 47(23): 30–38.
11. Sullivan C R, Petrenko V F, McCurdy J D et al. Breaking the ice [transmission line icing]. IEEE Industry Applications Magazine, 2003, 9(5): 49–54.
12. Fan S H, Jiang X L, Shu L C, et al. DC ice-melting model for wet-growth icing conductor and its experimental investigation[J]. Science China Technological Sciences, 2010, 53: 3248–3257.
13. Landry M, Beauchemin R, Venne A. De-icing EHV overhead transmission lines using electromagnetic forces generated by moderate short-circuit currents. In: 2000 IEEE ESMO-2000 IEEE 9th International Conference on Transmission and Distribution Construction, Operation and Live-Line Maintenance Proceedings. IEEE, 2000: 94–100.
14. Montambault S, Cote J, Louis M S. Preliminary results on the development of a teleoperated compact trolley for live-line working. In: 2000 IEEE ESMO-2000 IEEE 9th International Conference on Transmission and Distribution Construction, Operation and Live-Line Maintenance Proceedings. IEEE, 2000: 21–27.
15. Debenest P, Guarnieri M, Takita K, et al. Expliner–toward a practical robot for inspection of high-voltage lines. In: Field and Service Robotics: Results of the 7th International Conference. Berlin, Heidelberg: Springer Berlin Heidelberg, 2010: 45–55.
16. Montambault S, Pouliot N. LineScout technology: Development of an inspection robot capable of clearing obstacles while operating on a live line. In: ESMO 2006–2006 IEEE 11th International Conference on Transmission & Distribution Construction, Operation and Live-Line Maintenance. IEEE, 2006.
17. Wu G, Xiao X, Guo Y, et al. Development of a crawling robot for overhead high-voltage transmission line. China Mechanical Engineering, 2006, 17(3): 237–240.

18. Li T, Fang L, Wang H. Development of an inspection robot control system for 500KV extra-high voltage power transmission lines. In: SICE 2004 Annual Conference. IEEE, 2004, 2: 1819–1824.
19. Wang J, Xie Y, Wang F. The study of a mechanical structure for a for power transmission line inspection robot. Machinery Design and Manufacture, 2007, 8: 124–126.
20. Wu G, Xiao X, Xiao H, et al. Development of a mobile inspection robot for high voltage power transmission line. Automation of Electric Power Systems, 2006, 30(13): 90–93.
21. Zhu X, Wang H, Fang L, et al. Analysis of drive dynamic performance and position-pose of autonomous robot for transmission line inspection. Journal of Mechanical Engineering, 2006, 42(12): 143–150.
22. Tang L, Fang L, Wang H. Development of an inspection robot control system for 500KV extra-high voltage power transmission lines. In: SICE 2004 Annual Conference. IEEE, 2004, 2: 1819–1824.
23. Zhu X, Zhou J, Wang H, et al, Single arm running control method of inspection robot based on obliquitous sensor. In: 2006 IEEE International Conference on Robotics and Biomimetics. IEEE, 2006: 187–192.
24. Zhu X, Wang H, Fang L, et al, Dual arms running control method of inspection robot based on obliquitous sensor. In: 2006 IEEE/RSJ International Conference on Intelligent Robots and Systems. IEEE, 2006: 5273–5278.
25. Siciliano B, Khatib O. Springer handbooks of robotics. Springer-Verlag Berlin Heidelberg; 2008.
26. Wang J, Wang S, Lei Y, et al. Design of a novel mechanical structure of deicing robot for power transmission lines. Machinery Design and Manufacture, 2011, 2011(5): 62–64.
27. Manseur R, Doty K L. Structural kinematics of 6-revolute-axis robot manipulators. Mechanism and Machine Theory, 1996, 31(5): 647–657.
28. Zhao J, and Badler N. Inverse kinematics positioning using nonlinear programming for highly articulated figures. ACM Transactions on Graphics, 1994, 13(4): 313–336.
29. Birbil S I, and Fang S. An electromagnetism-like mechanism for global optimization. Journal of global optimization, 2003, 25: 263–282.
30. Birbil S I, Fang S. and Sheu, R. On the convergence of a population-based global optimization algorithm. Journal of global optimization, 2004, 30: 301–318.
31. Chen B. Optimization Theory and Algorithms. Beijing, China: Tsinghua University Press; 2005. pp. 375–376.
32. Angeles J. Fundamentals of Robotic Mechanical Systems: Theory, Methods, and Algorithms. 3rd ed. Berlin: Springer; 2007.
33. Rastegar J, and Fardanesh B. Manipulation workspace analysis using the Monte Carlo method. Mechanism and Machine Theory, 1990, 25(2): 233–239.
34. Wang L, Chen C. A combined optimization method for solving the inverse kinematics problems of mechanical manipulators. IEEE Transactions on Robotics and Automation, 1991, 7(4): 489–499.
35. Wang L, Ravani B. Recursive computations of kinematic and dynamic equations for mechanical manipulators. IEEE Journal on Robotics and Automation, 1985, 1(3): 124–131

Chapter 5
Control Methods for Transmission Line De-icing Robot Manipulator

5.1 Background and Motivation

Due to the dynamic complexity of robot manipulators, many traditional designers of robot manipulators often evade explicitly handling flexibility issues. The motion control problem of these systems is based on the rigid model of the system, and the flexibility problem of the system is considered negligible. To address these issues, adaptive and robust control techniques provide special solutions to cope with uncertainties in their dynamics. The advantages of these techniques are rapidly developing, ensuring global stability and tracking results for controlled systems, and suitable for a relatively large class of nonlinear systems.

Another way to solve this problem is to construct nonlinear functions suitable for the behavior of approximate dynamic systems, which has attracted the attention of many researchers. In recent decades, fuzzy logic systems, neural network systems, and cerebellar model articulation controller (CMAC) systems have been successfully applied to mathematical models that generally approximate dynamic systems. According to the universal approximation theorem, many robust and adaptive fuzzy, neural network, and CMAC control schemes have been developed for handling uncertain nonlinear systems. By combining modern control theory, many robust and adaptive fuzzy, neural network, and CMAC control algorithms based on these technologies have been developed, which are called intelligent adaptive control techniques for handling robust tracking control of disturbed nonlinear systems. Based on the above analysis, this chapter focuses on the design of the manipulator control method to achieve the tracking control performance of the ice removal robot for transmission lines.

Y. Wang et al., *Electric Power Robots*,
https://doi.org/10.1007/978-981-96-2512-3_5

5.2 Control Algorithm for Robot Manipulator

5.2.1 *Self-Organizing Single-Input CMAC Control for Robot Manipulators*

This section presents a self-structured organizing single-input control system based on differentiable cerebellar model articulation controller (CMAC) for an n-link robot manipulator to achieve the high-precision position tracking. In the proposed scheme, the single-input CMAC controller is solely used to control the plant, so the input space dimension of CMAC can be simplified, and no conventional controller is needed. The structure of single-input CMAC will also be self-organized, that is, the layers of single-input CMAC will grow or prune systematically and their receptive functions can be automatically adjusted. The online tuning laws of single-input CMAC parameters are derived in gradient-descent learning method, and the discrete-type Lyapunov function is applied to determine the optimal learning rates of the proposed control system so that the stability of the system can be guaranteed. The simulation results of a two-link robot manipulator are provided to verify the effectiveness of the proposed control methodology.

5.2.1.1 System Description

In general, the dynamic of an n-link robot manipulator may be expressed in the following Lagrange form:

$$\boldsymbol{M}(\boldsymbol{q})\ddot{\boldsymbol{q}} + \boldsymbol{C}(\boldsymbol{q}, \dot{\boldsymbol{q}})\dot{\boldsymbol{q}} + \boldsymbol{G}(\boldsymbol{q}) + \boldsymbol{L} = \boldsymbol{\tau} \tag{5.1}$$

where $\boldsymbol{q}, \dot{\boldsymbol{q}}, \ddot{\boldsymbol{q}} \in \mathbb{R}^n$ are the joint position, velocity, and acceleration vectors, respectively; $\boldsymbol{M}(\boldsymbol{q}) \in \mathbb{R}^{n \times n}$ denotes the inertia matrix; $\boldsymbol{C}(\boldsymbol{q}, \dot{\boldsymbol{q}}) \in \mathbb{R}^{n \times n}$ expresses the matrix of centripetal and Coriolis forces; $\boldsymbol{G}(\boldsymbol{q}) \in \mathbb{R}^{n \times 1}$ is the gravity vector; $\boldsymbol{L} \in \mathbb{R}^{n \times 1}$ represents the vector of external disturbance $\boldsymbol{t}_l$, friction term $\boldsymbol{f}(\dot{\boldsymbol{q}})$, and unmodeled dynamics; and $\boldsymbol{\tau} \in \mathbb{R}^{m \times 1}$ is the torque vectors exerting on joints.

The control problem is to force $q_i(t) \in \mathbb{R}^n$, $i = 1, 2, \cdots, m$, to track a given bounded reference input signal $q_{di}(t) \in \mathbb{R}^n$. Let $\boldsymbol{e}_i(t)$ be the tracking error vector as follows:

$$e_i = q_{di} - q_i, \quad i = 1, 2, m \tag{5.2}$$

and the system tracking error vector is defined as

$$\underline{\mathcal{E}_i} = \begin{pmatrix} k_{1i} & 0 & \cdots & 0 \\ 0 & k_{2i} & \cdots & 0 \\ 0 & 0 & \ddots & 0 \\ 0 & 0 & \cdots & k_m \end{pmatrix} \begin{pmatrix} e_i \\ \dot{e}_i \\ \vdots \\ e_i^{n-1} \end{pmatrix} = \underline{\boldsymbol{K}_{ni}\boldsymbol{e}_i} = \begin{bmatrix} \mathcal{E}_{1i} & \hat{\mathcal{E}}_{2i} & \cdots & \mathcal{E}_{ni}^{n-1} \end{bmatrix}, \quad i = 1, 2, \cdots, m \tag{5.3}$$

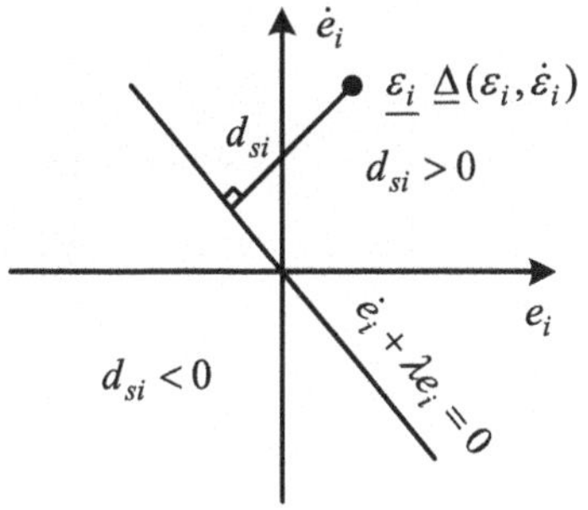

Fig. 5.1 Derivation of a signed distance

where $\boldsymbol{K}_{ni}\underline{\Delta}\text{diag}(k_{ni}) \in \mathbb{R}^{n\times n}$ is the scaling factor matrix for the system tracking vector $\boldsymbol{e}_i\underline{\Delta}\begin{bmatrix} e_i & \dot{e}_i & \cdots & e_i^{n-1}\end{bmatrix}^{\mathrm{T}} \in \mathbb{R}^n$ and $\varepsilon_{ni}^{n-1} = k_{ni}e_i^{n-1}$, $i = 1, 2, \cdots, m$.

Based on [1] and [2], the system tracking error $\underline{\varepsilon}_i \in \mathbb{R}^n$ is transformed into a single variable, termed the signed distance $d_{si} \in \mathbb{R}^m$, which is the distance from an actual state $\varepsilon_i \in R^n$ to the switching line as shown in Fig. 5.1 for a 2-D input. The n-D input switching line is defined as follows:

$$e_i^{n-1} + \lambda_{n-1}e_i^{n-2} + \cdots + \lambda_2 \dot{e}_i + \lambda_1 e_i = 0 \tag{5.4}$$

where $\lambda_{n-1} \in \mathbb{R}^{n-1}$ is a constant. Then, the signed distance between the switching line and operating point $\varepsilon_i \in R^n$ can be expressed by the following equation:

$$d_{si} = \frac{\varepsilon_{ni}^{n-1} + \lambda_{n-1}\varepsilon_{(n-1)i}^{n-2} + \cdots + \lambda_2\varepsilon_{2i} + \lambda_1\varepsilon_{1i}}{\sqrt{1 + \lambda_{n-1}^2 + \cdots + \lambda_2^2 + \lambda_1^2}} \tag{5.5}$$

According to the standalone CMAC control system, the system is shown in Fig. 5.2 for $n = 2$. This control scheme provided better control characteristics due to using the differentiable CMAC in the system. The advantage is that derivative information of input and output variables is preserved in learning process. In addition, the generalization error caused by quantization of input space could be reduced while using the differentiable CMAC.

Based on the standalone CMAC control system, we propose the SOSICM control system as shown in Fig. 5.3, which combines advantages of standalone CMAC, and it does not require prior knowledge of a certain amount of memory space. The self-organizing approach demonstrates the properties of generating and pruning the input layers automatically. The developed self-organizing rule of CMAC is clearly and easily used for real-time systems.

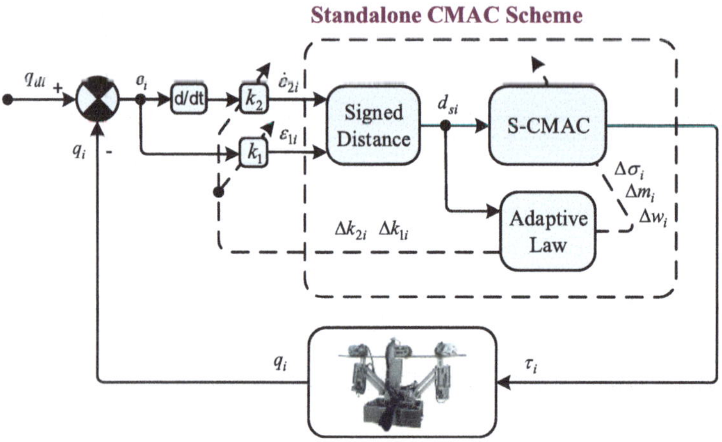

Fig. 5.2 Block diagram of standalone CMAC control system

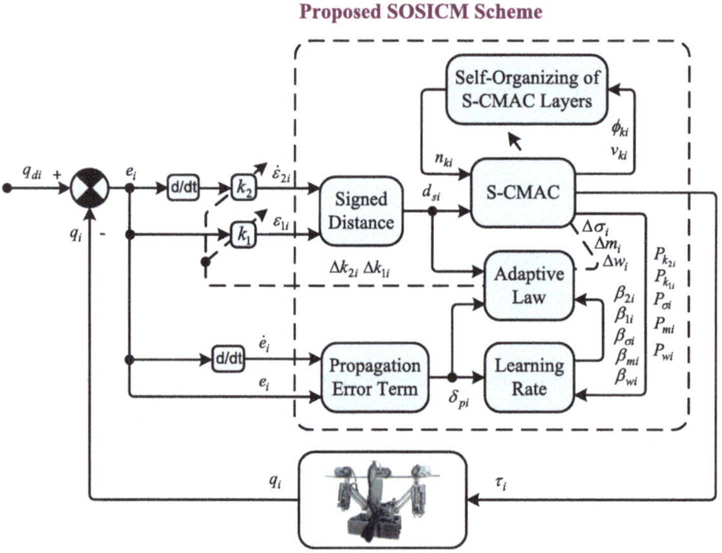

Fig. 5.3 Block diagram of the proposed self-structured organizing single-input CMAC (SOSICM) control system

5.2.1.2 Adaptive SOSICM Control System

An S-CMAC is proposed and shown in Fig. 5.4a. It is composed of an input, association memory, and weight and output spaces. The signal propagation and the basic function in each space are expressed as follows:

Input space D_s: Assume that each input state variable d_{si} can be quantized into N_{si} discrete states and that the information of a quantized state is regarded as region for

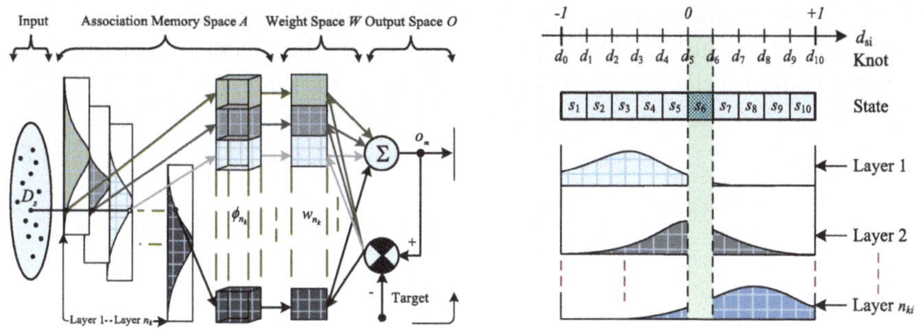

(a)

(b)

Fig. 5.4 Architecture of a single-input CMAC

each layer n_{ki}th. Therefore, there exist N_{si} + 1 individual points on the d_{si} axis. Figure 5.4b shows the case of $N_{si} = 10$. Each activated state in each layer becomes a firing element; thus, the weight of each layer can be obtained. The Gaussian basic function for each layer is given as follows:

$$\phi_{ki}(d_{si}) = \exp\left(-\frac{(d_{si} - m_{ki})^2}{\sigma_{ki}^2}\right), \quad i = 1, 2, \cdots, m, \quad k = 1, 2, \cdots, n_{ki} \tag{5.6}$$

where φ_{ki} represents the k-th layer of the input d_{si} with the mean m_{ki} and the variance σ_{ki}.

Output space O: The output of S-CMAC is the algebraic sum of the firing element with the weight memory and is expressed as

$$\tau_i = \sum_{k=1}^{n_{ki}} a_{ki} w_{ki} \varphi_{ki}(d_{si}) \tag{5.7}$$

where w_{ki} denotes the weight of the k-th layer and $a_{ki} = a_{ki}(d_{si})$, $k = 1, 2$. n_{ki} is the index indicating whether the ith memory element is addressed by the state involving d_{si}. Since each state addressed exactly n_{ki} memory elements, only those addressed a_{ki} are one, and the others are zero.

The block diagram in Fig. 5.2 illustrates a system where only the single-input CMAC (S-CMAC) plays a major role in the control process. To achieve a trade-off between the desired performance and computation loading, we must choose a reasonable number of layers. However, if the number of layers is chosen too small, the learning performance may be insufficient to achieve a desired performance. Otherwise, if the number of layers is chosen too large, the calculation process is too heavy, so it is not suitable for real-time applications. To deal with this problem, a self-structured organizing S-CMAC is proposed that includes structure and parameter learning as shown in Fig. 5.4.

The self-structured organizing S-CMAC can be summarized as follows:
Calculate the firing strength $\phi_{ki} \in R^{n_{ki}}$ of each layer for each input data d_{si} in (5.6). The ***max-min*** method is proposed for layer growing. Find

$$\hat{k}_i = \arg\min_{1 \leq k \leq n_{ki}} \varphi_{ki}(d_{si}), \quad k = 1, 2, n_{ki} \tag{5.8}$$

if

$$\varphi_{\hat{k}_i}(d_{si}) < K_{gi} \tag{5.9}$$

Here, K_{gi} is a threshold value of adaptation with $0 < K_{gi} \leq 1$. In our case, $K_{gi} = 0.1$ and a new layer is generated.

This means that for a new input data, the exciting value of existing basic function is too small. In this case, the number of layers increased as follows:

$$n_{ki}(t+1) = n_{ki}(t) + 1 \tag{5.10}$$

where n_{ki} is the number of layers at time t. Thus, a new layer will be generated, and then the corresponding parameters in the new layer such as the initial mean and variance of Gaussian basic function in association memory space and the weight memory space will be defined as

$$m_{n_{ki}} = d_{si} \tag{5.11}$$

$$\sigma_{n_{ki}} = \sigma_{\hat{k}i} \tag{5.12}$$

$$w_{n_{ki}} = 0 \tag{5.13}$$

Another self-structured organizing learning process is considered to determine whether to delete existing layer, which is inappropriate. A max-min method is proposed for layer pruning. Considering the output of SOSICM in (5.7), the ratio of the k-th component of the output is defined as

$$\mathrm{MM}_{ki} = \frac{v_{ki}}{\tau_i}, \quad k = 1, 2, \cdots, n_{ki} \tag{5.14}$$

where $v_{ki} = \varphi_{ki} w_{ki}$. Then, the minimum ratio of the kth component is defined as follows:

$$\tilde{k}_i = \arg\min_{1 \leq k \leq n_{ki}} \mathrm{MM}_{ki} \tag{5.15}$$

if

$$\mathrm{MM}_{\tilde{k}_i} \leq K_{ci} \tag{5.16}$$

where K_{ci} is a predefined deleting threshold. In our case, $K_{ci} = 0.03$ and the $\tilde{k}_i$-th layer will be deleted. This means that for an output data, if the minimum contribution of a layer is less than the deleting threshold, then this layer will be deleted.

The core aspect of the learning algorithm for a SOSICM involves selecting the parameters of the Gaussian basis function: the weight memory w_{ki}, the mean m_{ki}, and the variance σ_{ki}. Additionally, k_{ni} represents the scaling factors of the error e_i and the change in error. For achieving effective learning, an online learning algorithm, which is derived using the supervised gradient descent method, is introduced so that it can in real time adjust the parameters of SOSICM. The energy function E_i is defined as

$$E_i = \frac{1}{2}(q_{di} - q_i)^2 = \frac{1}{2}e_i^2 \tag{5.17}$$

According to the energy function (5.17) and the system structure in Fig. 5.3, the error term to be propagated is given by

$$\delta_{pi} = -\frac{\partial E_i}{\partial \tau_i} = -\frac{\partial E_i}{\partial q_i}\frac{\partial q_i}{\partial \tau_i} = e_i\frac{\partial q_i}{\partial \tau_i} \tag{5.18}$$

where $\partial q_i/\partial \tau_i$ represent the sensitivity of the plant with respect to its input. With the energy function E_i, the parameters' updating law(s) based on the normalized gradient descent method can be derived as follows:

The updating law for the kth weight memory can be derived according to

$$\begin{aligned}\Delta w_{ki} &= -\beta_{wi}\frac{\partial E_i}{\partial w_{ki}} = -\beta_{wi}\frac{\partial E_i}{\partial \tau_i}\frac{\partial \tau_i}{\partial w_{ki}} \\ &= a_{ki}\beta_{wi}\delta_{pi}\varphi_{ki}(d_{si})\end{aligned} \tag{5.19}$$

where β_{wi} is the positive learning rate for the output weight memory w_{ki}, and the connective weight can be updated according to the following equation:

$$w_{ki}(t+1) = w_{ki}(t) + \Delta w_{ki} \tag{5.20}$$

The mean and variance of the kth Gaussian basic function can be also updated according to

$$\begin{aligned}\Delta m_{ki} &= -\beta_{mi}\frac{\partial E_i}{\partial m_{ki}} = -\beta_{wi}\frac{\partial E_i}{\partial \tau_i}\frac{\partial \tau_i}{\partial m_{ki}} \\ &= a_{ki}\beta_{mi}\delta_{pi}w_{ki}\varphi_{ki}(d_{si})\frac{2(d_{si} - m_{ki})^2}{\sigma_{ki}^2}\end{aligned} \tag{5.21}$$

$$\begin{aligned}\Delta\sigma_{ki} &= -\beta_{\sigma i}\frac{\partial E_i}{\partial \sigma_{ki}} = -\beta_{wi}\frac{\partial E_i}{\partial \tau_i}\frac{\partial \tau_i}{\partial \sigma_{ki}} \\ &= a_{ki}\beta_{\sigma i}\delta_{pi}w_{ki}\varphi_{ki}(d_{si})\frac{2(d_{si}-m_{ki})^2}{\sigma_{ki}^3}\end{aligned} \tag{5.22}$$

where $\beta_{mi}, \beta_{\sigma i}$ are positive learning rates for the mean and variance, respectively. The mean and variance can be updated as follows:

$$m_{ki}(t+1) = m_{ki}(t) + \Delta m_{ki} \tag{5.23}$$

$$\sigma_{ki}(t+1) = \sigma_{ki}(t) + \Delta\sigma_{ki} \tag{5.24}$$

Finally, the updating law for scaling factors can be derived as follows:

$$\begin{aligned}\Delta k_{ni} &= -\beta_{ni}\frac{\partial E_i}{\partial k_{ni}} = -\beta_{ni}\frac{\partial E_i}{\partial \tau_i}\frac{\partial \tau_i}{\partial d_{si}}\frac{d_{si}}{k_{ni}} \\ &= \beta_{ni}\delta_{pi}\left[\sum_{k=1}^{n_{ki}} a_{ki}w_{ki}\varphi_{ki}(d_{si})\frac{-2(d_{si}-m_{ki})}{\sigma_{ki}^2}\right] \times \frac{\lambda_n e_{ni}^{n-1}}{\sqrt{1+\lambda_{n-1}^2+\cdots+\lambda_2^2+\lambda_1^2}}\end{aligned} \tag{5.25}$$

where β_{ni} is the learning rate, and it can be updated by the following:

$$k_{ni}(t+1) = k_{ni}(t) + \Delta k_{ni} \tag{5.26}$$

The plant sensitivity $\partial q_i/\partial \tau_i$ in (5.18) can be calculated if the plant model is exactly known. However, the plant model is unknown, so $\partial q_i/\partial \tau_i$ cannot be obtained in advance. To deal with this problem, in [3], a simple approximation of the error term of the system can be used as follows:

$$\delta_{pi} \cong \dot{e}_i + \boldsymbol{e}_i \tag{5.27}$$

The updating laws of Eqs. (5.19), (5.21), (5.22), and (5.25) require a proper choice of the learning rates β_{wi}, β_{mi}, $\beta_{\sigma i}$, and β_{ni} in order to guarantee the convergence of the output error; however, this is not easy, which depends on each person's experience. To train the S-CMAC effectively, the variable learning rates that guarantee the convergence of the output error are derived in the following:

A discrete-type Lyapunov function can be defined and is given by the following expression:

$$V_i(k) = \frac{1}{2}e_i^2(k) \tag{5.28}$$

Thus, the change of the Lyapunov due to the training process is obtained as

$$\Delta V_i(k) = V_i(k+1) - V_i(k) = \frac{1}{2}\left[e_i^2(k+1) - e_i^2(k)\right] \tag{5.29}$$

where $e_i(k + 1)$ is represented by [3]

$$e_i(k+1) = e_i(k) + \Delta e_i(k) = e_i(k) + \left(\frac{\partial e_i(k)}{\partial P_i}\right)^{\mathrm{T}} \Delta P_i \tag{5.30}$$

where Δe_i represents the change in the tracking error during the learning process, and ΔP_i denotes a change of an adjustable parameter. Using (5.18), we have $\partial e_i/\partial P_i = -\delta_{pi}\partial\tau_i/e_i(k)\partial P_i$ and $\Delta P_i = -\beta_{pi}\partial E_i/\partial P_i = \beta_{pi}\delta_{pi}\partial\tau_i/\partial P_i$, where β_{pi} is the learning rate for the parameter P_i.

Thus

$$\begin{aligned}\Delta V_i(k) &= \Delta e_i(k)\left[e_i(k) + \frac{1}{2}\Delta e_i(k)\right] = -\frac{\beta_{pi}\delta_{pi}^2}{e_i(k)}\left\|\frac{\partial \tau_i}{\partial P_i}\right\|^2 \\ &\quad \left[e_i(k) - \frac{1}{2}\frac{\beta_{pi}\delta_{pi}^2}{e_i(k)}\left\|\frac{\partial \tau_i}{\partial P_i}\right\|^2\right] \\ &= \frac{1}{2}\beta_{pi}\delta_{pi}^2\left\|\frac{\partial \tau_i}{\partial P_i}\right\|^2\left[\beta_{pi}\left(\frac{\delta_{pi}}{e_i(k)}\right)^2\left\|\frac{\partial \tau_i}{\partial P_i}\right\|^2 - 2\right]\end{aligned} \tag{5.31}$$

If the learning rate β_{pi} is selected as

$$0 < \beta_{pi} < \frac{2}{\left[\frac{\delta_{pi}}{e_i(k)}\right]^2\left\|\frac{\partial \tau_i}{\partial P_i}\right\|^2} \tag{5.32}$$

then $\Delta V_i(k) \le 0$; therefore, $V_i(k + 1) \le V_i(k)$, and the Lyapunov stability (system stability) and the convergence of the tracking error could be guaranteed. In addition, the optimal learning rate can be found for achieving faster convergence by taking the differential eq. (5.31) with respect to β_{pi} and equals to zero. Finally, the optimal learning rate can be determined as follows:

$$\beta_{pi}^* = \frac{1}{\left[\frac{\delta_{pi}^2}{e_i(k)}\right]^2\left\|\frac{\partial \tau_i}{\partial P_i'}\right\|^2} \tag{5.33}$$

where $\partial\tau_i/\partial P_i$ for $P_i = w_{ki}$, m_{ki}, σ_{ki}, and k_{ni}, which can be obtained as follows:

$$P_{wi}(k) = \frac{\partial \tau_i}{\partial w_{ki}} = a_{ki}\varphi_{ki},$$

$$P_{mi}(k) = \frac{\partial \tau_i}{\partial m_{ki}} = a_{ki}w_{ki}\varphi_{ki}\frac{2(d_{si} - m_{ki})}{\sigma_{ki}^2}$$

$$P_{\sigma i}(k) = \frac{\partial \tau_i}{\partial \sigma_{ki}} = a_{ki}w_{ki}\varphi_{ki}\frac{2(d_{si} - m_{ki})^2}{\sigma_{ki}^3}$$

$$P_{k_{ni}}(k) = \frac{\partial \tau_i}{\partial k_{ni}} = \left[\sum_{k=1}^{n_{ki}} a_{ki}w_{ki}\varphi_{ki}(d_{si})\frac{-2(d_{si} - m_{ki})}{\sigma_{ki}^2}\right]\frac{\lambda_n e_{ni}^{n-1}}{\sqrt{1 + \lambda_{n-1}^2 + \cdots + \lambda_2^2 + \lambda_1^2}} \tag{5.34}$$

5.2.1.3 Simulation Results

A two-link robot manipulator as shown in Fig. 5.14 is utilized in this section to verify the effectiveness of the proposed control scheme. The detailed system parameters of this robot manipulator are given as $m_1 = 4.6$ kg, $m_2 = 2.3$ kg, $l_1 = 0.5$ m, $l_2 = 0.2$ m, and $g = 9.8\text{m/s}^2$. The initial conditions are $q_1(0) = 0.5$, $q_2(0) = 0.5$, and $\dot{q}_1(0) = 0$. The desired reference trajectories are $q_{d1}(t) = \sin(t)$ and $q_{d2}(t) = \cos(t)$.

The most important parameters that affect the control performance of the robotic system are the external disturbance t_l and the friction term $f(\dot{q})$; in simulation, parameter variation situation and disturbance situation occurring at 5 s are considered, which are injected into the robotic system, and their shapes are expressed as follows:

$$t_l(t) = [5\sin(5t) \quad 0.5\sin(5t)]^{\mathrm{T}} \tag{5.35}$$

In addition, friction forces are also considered in this simulation and given as

$$f(\dot{q}) = [20\dot{q}_1 + 0.8\,\mathrm{sgn}(\dot{q}_1) \quad 4\dot{q}_2 + 0.1\,\mathrm{sgn}(\dot{q}_2)]^{\mathrm{T}} \tag{5.36}$$

In order to exhibit the superior control performance of the proposed SOSICM control system, the control system standalone CMAC introduced in Fig. 5.1 is examined in the meantime [3]. They are applied to control a two-link robot manipulator, and the same setting of SOSICM and standalone CMAC control system are chosen in the following: the input space of S-CMAC are d_{s1} and d_{s2}, and the mean and variance of Gaussian basic functions are selected to cover the input space$\{[-1 \quad 1][-1 \quad 1]\}$; all initial weights are set to zero, i.e., $w_{k1} = w_{k2} = 0$, $k = 1, 2, \cdots, n_{ki}$. The parameter λ in the switching line is one. For recording respective control performance, the mean square error of the position tracking response is defined as

$$mse_i = \frac{1}{T}\sum_{j=1}^{T}[q_{di}(j) - q_i(j)]^2, \quad i = 1, 2 \tag{5.37}$$

where T is the total sampling instant and q_i and q_{di} are the elements. In this section, the numerical simulation results are carried out by using MATLAB software.

Example 5.1 Consider the standalone CMAC control system shown in Fig. 5.2.

For the standalone CMAC control system, the parameters are chosen in the following: $\beta_{wi} = 0.05$, $\beta_{mi} = 0.05$, $\beta_{\sigma i} = 0.05$, and $\beta_{ni} = 0.02$. The initial value of the Gaussian basic functions and scaling factors are chosen as $m_{1i} = -1.0$, $m_{2i} = -0.8$, $m_{3i} = -0.6$, $m_{4i} = -0.4$, $m_{5i} = -0.2$, $m_{6i} = 0$, $m_{7i} = 0.2$, $m_{8i} = 0.4$, $m_{9i} = 0.6$, $m_{10i} = 0.8$, $m_{11i} = 1.0$, $\sigma_{ki} = 0.15$, $k_{1i} = 0.5$, and $k_{2i} = 0.2$ for $k = 1, 2, \cdots, 11$, $i = 1, 2$. The simulation results of the standalone CMAC system, the responses of joint position, *MSE*, and tracking error are depicted in Fig. 5.5a–f, respectively.

Example 5.2 Consider the proposed SOSICM control system shown in Fig. 5.3. For the proposed SOSICM control system, the parameters are chosen in the following: $\beta^*_{pi} = 1/\left[\delta_{pi}/e_i(k)\right]^2 \|\partial\tau_i/\partial P_i\|^2$ for $P_i = w_{ki}, m_{ki}, \sigma_{ki}$, and k_{ni}, and the initial values of the system parameters are given as $n_{ki} = 2$, the inputs of S-CMAC are d_{s1} and d_{s2}, and the mean and variance of Gaussian basic functions are selected to cover the input space $\{[-1 \quad 1][-1 \quad 1]\}$. The threshold value of K_{gi} is set as 0.1; K_{ci} is set as 0.03 for $i = 1, 2$. The simulation results of the proposed SOSICM system, the responses of joint position, MSE, tracking error, and layer number are depicted in Fig. 5.6a–h, respectively.

According to the simulation results shown in Figs. 5.5 and 5.6, the joint position tracking responses of the SOSICM system can be controlled to more closely follow the desired reference trajectories than the standalone CMAC as shown in Fig. 5.5a and b. In Figs. 5.5 and 5.6c and d, the MSE of the proposed control system for each joint reduced faster than that of the standalone CMAC control system and finally converges to 0.0003 and 0.0006; meanwhile the MSE of the standalone CMAC is 0.004 and 0.003, and the number of layers of S-CMACs converges to four and six layers as shown in Fig. 5.6g and h.

5.2.2 Adaptive Iterative Learning Control for De-icing Robot Manipulators

5.2.2.1 Robot Manipulator Dynamics

In general, the dynamic of an n-link robot manipulator may be expressed in the following Lagrange form:

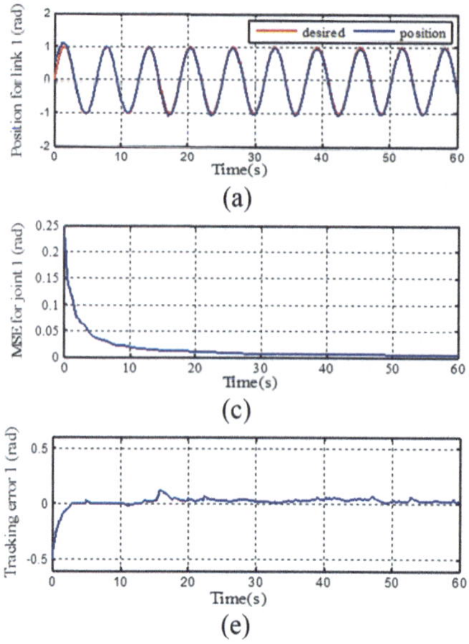

Fig. 5.5 Simulated position responses, MSEs, and tracking errors of the standalone CMAC control system at joints 1 and 2

$$\boldsymbol{M}\big(\boldsymbol{q}^i(t)\big)\ddot{\boldsymbol{q}}^i(t)+\boldsymbol{C}\big(\boldsymbol{q}^i(t),\dot{\boldsymbol{q}}^i(t)\big)\dot{\boldsymbol{q}}^i(t)+\boldsymbol{G}\big(\boldsymbol{q}^i(t),\dot{\boldsymbol{q}}^i(t)\big)+\boldsymbol{\tau}_a(t)=\tau^i(t) \tag{5.38}$$

where $t \in [0, t_f]$ denotes the time and $i \in N$ denotes the iteration; $\boldsymbol{q}^i(t) \in \mathbb{R}^n$, $\dot{\boldsymbol{q}}^i(t) \in \mathbb{R}^n$, and $\ddot{\boldsymbol{q}}^i(t) \in \mathbb{R}^n$ are the joint position, joint velocity, and joint acceleration variable vectors, respectively; $\boldsymbol{M}(\boldsymbol{q}^i(t)) \in \mathbb{R}^{n\times n}$ is the inertia matrix; $\boldsymbol{C}\big(\boldsymbol{q}^i(t),\dot{\boldsymbol{q}}^i(t)\big) \in \mathbb{R}^n$ is the Coriolis-centripetal matrix; $\boldsymbol{G}\big(\boldsymbol{q}^i(t),\dot{\boldsymbol{q}}^i(t)\big) \in \mathbb{R}^n$ is the gravity vector plus friction force vector; $\tau_a(t) \in \mathbb{R}^n$ is the bounded unknown disturbances; and $\tau^i(t) \in \mathbb{R}^n$ is the control input torque. For convenience, a two-link robot manipulator, as shown in Fig. 5.7, is utilized to verify the dynamic properties given in this section.

Property 5.1 The inertia matrix $\boldsymbol{M}(\boldsymbol{q}^i(t))$ is symmetric and positive definite. It is also bounded as a function of $q : m_1\boldsymbol{I} \le \boldsymbol{M}(\boldsymbol{q}^i(t)) \le m_2\boldsymbol{I}$, $m_1, m_2 > 0$.

Property 5.2 $\dot{\boldsymbol{M}}\big(\boldsymbol{q}^i(t)\big) - 2\boldsymbol{C}\big(\boldsymbol{q}^i(t),\dot{\boldsymbol{q}}^i(t)\big)$ is a skew symmetric matrix. Therefore, $\boldsymbol{y}^{\top}\big[\dot{\boldsymbol{M}}\big(\boldsymbol{q}^i(t)\big) - 2\boldsymbol{C}\big(\boldsymbol{q}^i(t),\dot{\boldsymbol{q}}^i(t)\big)\big]\boldsymbol{y}=0$, where $\boldsymbol{y}$ is an $n \times 1$ nonzero vector.

Assumption 5.1 The given desired joint trajectory $\boldsymbol{q}_d(t)$ belongs to $C^3[0, t_f]$, where $C^3[0, t_f]$ is the set of twice continuously differentiable functions on $t \in [0, t_f]$.

Assumption 5.2 We also assume that time t is reset to zero at the starting point of each iterative. We assume that $q^i(0) = q_d(0)$ and $\dot{q}^i(0) = \dot{q}_d(0)$ for all $i \ge 1$.

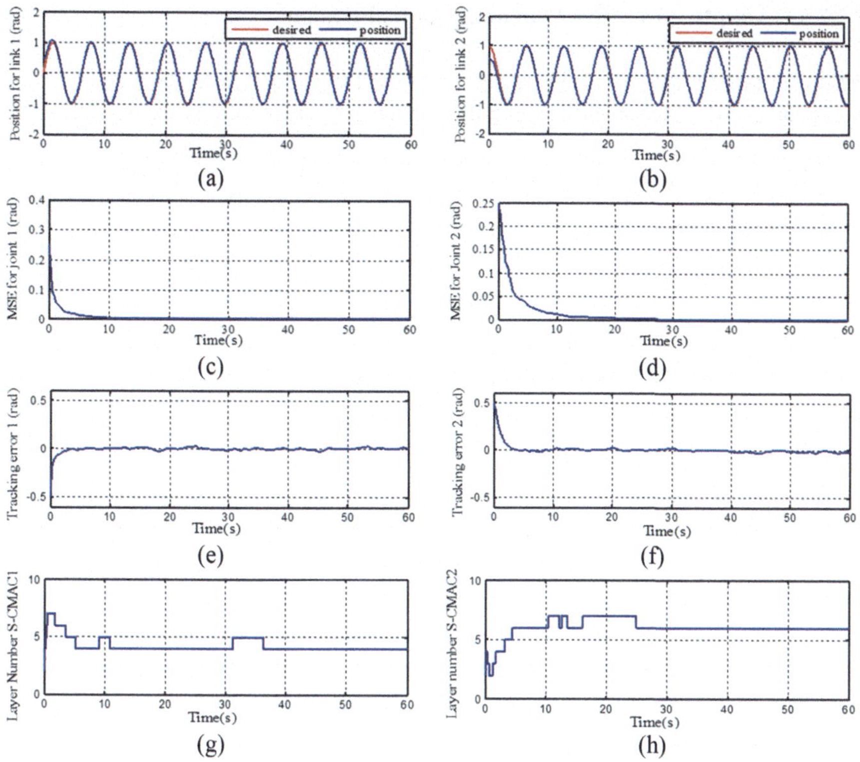

Fig. 5.6 Simulated position responses, MSEs, layer number, and tracking errors of the SOSICM control system at joints 1 and 2

Fig. 5.7 Architecture of a two-link robot manipulator

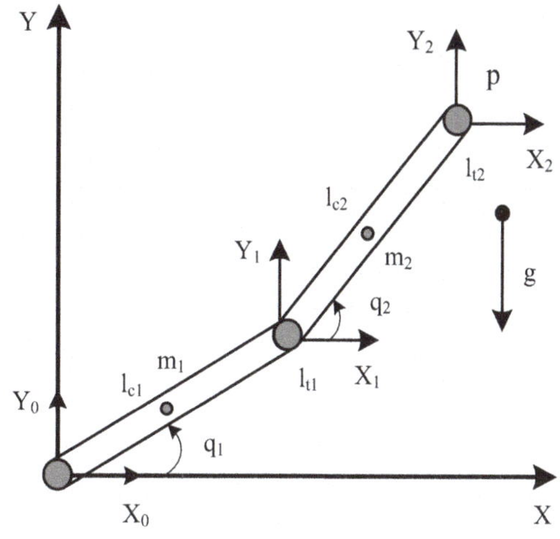

5.2.2.2 Adaptive Iterative Learning Control Design

By linearizing the system (5.38) along the desired trajectory $q_d(t), \dot{q}_d(t), \ddot{q}_d(t)$ at the i-th iterative, we obtain the following linear time-varying system according to [4, 5]:

$$\begin{aligned} &M(t)\ddot{e}^i(t) + [C(t) + C_1(t)]\dot{e}^i(t) + R(t)e^i(t) + n\left(\ddot{e}^i, \dot{e}^i, e^i, t\right) - \tau_a(t) \\ &= S(t) - \tau^i(t) \end{aligned} \tag{5.39}$$

where $M(t) = M(q_d(t))$, $C(t) = C(q_d(t), \dot{q}_d(t))$, $C_1(t) = \frac{\partial C}{\partial \dot{q}}\Big|_{q_d(t), \dot{q}_d(t)} \dot{q}_d(t) + \frac{\partial G}{\partial \dot{q}}\Big|_{q_d(t), \dot{q}_d(t)}$, $R(t) = \frac{\partial M}{\partial q}\Big|_{q_d(t)} \ddot{q}_d(t) + \frac{\partial C}{\partial q}\Big|_{q_d(t), \dot{q}_d(t)} \dot{q}_d(t) + \frac{\partial G}{\partial q}\Big|_{q_d(t)}$, $S(t) = M(q_d(t))\ddot{q}_d(t) + C(q_d(t), \dot{q}_d(t))q_d(t) + G(q_d(t)) + \tau_a(t)$, and $e(t) = q_d(t) - q(t)$.

The term $n\left(\ddot{e}^i, \dot{e}^i, e^i, t\right)$contains the higher order of terms $\ddot{e}^i(t)$, $\dot{e}^i(t)$, and $e^i(t)$, and it can be negligible.

The control problem is to find a control law so that the position $q(t)$ can track specific commands $q_d(t)$. We construct controller as follows:

$$\tau^i = \tau_e^i + H^i \tag{5.40}$$

where the first term $\tau_e^i = K_p^i(e^i(t)) + K_d^i\left(\dot{e}^i(t)\right)$ is the feedback proportional-derivative (PD) control law with the following gain switching rule in [42]:

$$\begin{cases} K_p^{i+1} = \beta(i)K_p^i \\ K_d^{i+1} = \beta(i)K_d^i \qquad i = 0, 1, 2, \cdots, N \\ \beta(i+1) > \beta(i) \end{cases} \tag{5.41}$$

where K_p^i and K_d^i are the initial proportional and derivative control gain matrices that are diagonal and positive definite. K_p^{i+1} and K_d^{i+1} are the control gains of the i-th iterative. $\beta(i) > 1$ is the gain switching factor. The gains of the adaptive laws in (5.41) are used to adjust the PD gains from iterative to iterative. H^i is the initial predicted feedforward control input to be computed at each iterative by a learning rule.

As demonstrated in [6], the system described by (5.39) is guaranteed to converge when employing a feedback PD control law with the gain switching factor as given in (5.41), in conjunction with a feedforward learning control law that utilizes an input force profile. However, while the trajectory tracking converges quickly in some initial iterations, we cannot arbitrarily increase the switching factor due to the limitations of actuator forces, especially when the system has modeling errors or nonlinearities. Hence, to deal with this problem, we propose feedforward control input $H^i(t)$ with a learning rule so that $H^i(t)$ converges to $R(t)$ for all $t \in [0, t_f]$ as follows:

$$\boldsymbol{H}^{i+1}(t) = \boldsymbol{H}^{i}(t) + \alpha \boldsymbol{\tau}_{e}^{i} \tag{5.42}$$

At the initial stage of learning, $\boldsymbol{H}^{i}(t)$ are set to zero and α is a positive constant often called a training factor.

Therefore, for i-th and $(i + 1)$-th iterations, applying the input (5.40) and (5.42) to system (5.39), we obtain an error equation as follows:

$$\boldsymbol{M}(t)\ddot{\boldsymbol{e}}^{i}(t) + [\boldsymbol{C}(t) + \boldsymbol{C}_{1}(t)]\dot{\boldsymbol{e}}^{i}(t) + \boldsymbol{R}(t)\boldsymbol{e}^{i}(t) = \boldsymbol{S}(t) - \boldsymbol{\tau}_{e}^{i} - \boldsymbol{H}^{i} \tag{5.43}$$

$$\boldsymbol{M}(t)\ddot{\boldsymbol{e}}^{i+1}(t) + [\boldsymbol{C}(t) + \boldsymbol{C}_{1}(t)]\dot{\boldsymbol{e}}^{i+1}(t) + \boldsymbol{R}(t)\boldsymbol{e}^{i+1}(t) = \boldsymbol{S}(t) - \boldsymbol{\tau}_{e}^{i+1} - \boldsymbol{H}^{i} - \alpha\boldsymbol{\tau}_{e}^{i} \tag{5.44}$$

To simplify the proof of stability, let $K_{p}^{i} = aK_{d}^{i}$ for the initial iteration, and define the filter errors as follows:

$$\tilde{\boldsymbol{x}}^{i}(t) = \dot{\boldsymbol{e}}^{i}(t) + a\boldsymbol{e}^{i}(t) \tag{5.45}$$

Also, define $\delta\tilde{\boldsymbol{x}}^{i} = \tilde{\boldsymbol{x}}^{i+1} - \tilde{\boldsymbol{x}}^{i}$ and $\delta\boldsymbol{e}^{i} = \boldsymbol{e}^{i+1} - \boldsymbol{e}^{i}$. Then, from (5.45)

$$\delta\tilde{\boldsymbol{x}}^{i} = \delta\dot{\boldsymbol{e}}^{i} + a\delta\boldsymbol{e}^{i} \tag{5.46}$$

From (5.41)–(5.46), one can obtain the following equation:

$$\begin{aligned} &\boldsymbol{M}\delta\dot{\tilde{\boldsymbol{x}}}^{i} + \left(\boldsymbol{C} + \boldsymbol{C}_{1} - a\boldsymbol{M} + \boldsymbol{K}_{d}^{i+1}\right)\delta\tilde{\boldsymbol{x}}^{i} + (\boldsymbol{R} - a(\boldsymbol{C} + \boldsymbol{C}_{1} - a\boldsymbol{M}))\delta\boldsymbol{e}^{i} \\ &= -\left(\boldsymbol{K}_{d}^{i+1} + (\alpha - 1)\boldsymbol{K}_{d}^{i}\right)\tilde{\boldsymbol{x}}^{i} \end{aligned} \tag{5.47}$$

The following theorem can be proved:

Theorem 5.1 Consider an n-link robot manipulator dynamic represented by (5.38) that satisfies Properties 5.1 and 5.2 and Assumptions 5.1 and 5.2. If the control laws of the AILC system are designed according to (5.40), with the gain switching rule given by (5.41) and the learning rule by (5.42), then the following conditions should hold for all $t \in [0, t_f]$:

$$\boldsymbol{q}^{i}(t) \overset{i \to \infty}{\rightarrow} \boldsymbol{q}_{d}(t)$$

$$\dot{\boldsymbol{q}}^{i}(t) \overset{i \to \infty}{\rightarrow} \dot{\boldsymbol{q}}_{d}(t)$$

If the controller gains are selected, the following relationships hold:

$$l_{p} = \lambda_{d_{1}\,min}^{i} \tag{5.48}$$

$$l_r = \lambda^i_{d_{1\,min}} \tag{5.49}$$

$$l_p l_r \geq \| \boldsymbol{R}/a - (\boldsymbol{C} + \boldsymbol{C}_1 - a\boldsymbol{D}) \|^2_{max} \tag{5.50}$$

where $\lambda_{\min}(\boldsymbol{A})$ is the minimum eigenvalue of matrix $\boldsymbol{A}$ and $\|\boldsymbol{D}\|_{\max} = \max \|\boldsymbol{D}(t)\|$ for $t \in [0, t_f]$. Here, $\|\boldsymbol{D}\|$ represents the Euclidean norm of M.

Proof We select a performance index $V^i(t)$ as follows:

$$V^i = \int_0^t \mathrm{e}^{-\rho\tau} \tilde{\boldsymbol{x}}^{i\mathrm{T}} \boldsymbol{Q} \tilde{x}^i \mathrm{d}\tau \geq 0 \tag{5.51}$$

Thus, $\beta(i) > 1$ according to (5.41), so we have $\boldsymbol{K}^{i+1}_d > \boldsymbol{K}^i_d$ and $\alpha \leq 1$ is a positive constant, so $\boldsymbol{Q} = \boldsymbol{K}^{i+1}_d + (\alpha - 1)\boldsymbol{K}^i_d) > 0$.

From the definition of V^i, for the $(i + 1)$-th iteration, we can get

$$V^{i+1} = \int_0^t \mathrm{e}^{-\rho\tau} \tilde{\boldsymbol{x}}^{i+1\mathrm{T}} \boldsymbol{Q} \tilde{\boldsymbol{x}}^{i+1} \mathrm{d}\tau \tag{5.52}$$

Let $\Delta V^i = V^{i+1} - V^i$; then from (5.51), (5.52), and (5.47), we obtain

$$\begin{aligned}
\Delta V^i &= \int_0^t \mathrm{e}^{-\rho\tau} \left(\delta\tilde{\boldsymbol{x}}^{i\mathrm{T}} \boldsymbol{Q} \delta\tilde{\boldsymbol{x}}^i + 2\delta\tilde{\boldsymbol{x}}^{i\mathrm{T}} \boldsymbol{Q} \tilde{\boldsymbol{x}}^i \right) \mathrm{d}\tau \\
&= \int_0^t \mathrm{e}^{-\rho\tau} \delta\tilde{\boldsymbol{x}}^{i\mathrm{T}} \boldsymbol{Q} \delta\tilde{\boldsymbol{x}}^i \mathrm{d}\tau - 2 \int_0^t \mathrm{e}^{-\rho\tau} \delta\tilde{\boldsymbol{x}}^{i\mathrm{T}} \boldsymbol{M} \delta\dot{\tilde{\boldsymbol{x}}}^i \mathrm{d}\tau \\
&\quad - 2 \int_0^t \mathrm{e}^{-\rho\tau} \delta\tilde{\boldsymbol{x}}^{i\mathrm{T}} \left((\boldsymbol{C} + \boldsymbol{C}_1 - a\boldsymbol{M} + \boldsymbol{K}^{i+1}_d) \delta\tilde{\boldsymbol{x}}^i \right. \\
&\quad \left. + (\boldsymbol{R} - a(\boldsymbol{C} + \boldsymbol{C}_1 - a\boldsymbol{M}))\delta\boldsymbol{e}^i \right) \mathrm{d}\tau
\end{aligned} \tag{5.53}$$

By applying the partial integration, from Assumption 5.2 and Property 5.2, we have

$$\int_0^t e^{-\rho\tau}\delta\tilde{x}^{iT}M\delta\dot{\tilde{x}}^i d\tau = e^{-\rho\tau}\delta\tilde{x}^{iT}M\delta\tilde{x}^i\Big|_0^t - \int_0^t \left(e^{-\rho\tau}\delta\tilde{x}^{iT}M\right)'\delta\tilde{x}^i d\tau$$
$$= e^{-\rho\tau}\delta\tilde{x}^{iT}(t)M(t)\delta\tilde{x}^i(t) + \rho\int_0^t e^{-\rho\tau}\delta\tilde{x}^{iT}M\delta\tilde{x}^i d\tau$$
$$-\int_0^t e^{-\rho\tau}\delta\tilde{x}^{iT}M\delta\dot{\tilde{x}}^i d\tau - 2\int_0^t e^{-\rho\tau}\delta\tilde{x}^{iT}C\delta\tilde{x}^i d\tau \tag{5.54}$$

Substituting (5.54) into (5.53) and $Q = K_d^{i+1} + (\alpha - 1)K_d^i$ yields

$$\Delta V^i = -e^{-\rho\tau}\delta\tilde{x}^{iT}(t)M(t)\delta\tilde{x}^i(t) - \rho\int_0^t e^{-\rho\tau}\delta\tilde{x}^{iT}M\delta\tilde{x}^i d\tau - 2\int_0^t e^{-\rho\tau}\delta\tilde{x}^{iT}\big(R - a\big(C$$
$$+C_1 - aM\big)\big)\delta e^i d\tau - \int_0^t e^{-\rho\tau}\delta\tilde{x}^{iT}\left(K_d^{i+1} - (\alpha-1)K_d^i + 2C_1 - 2aM\right)\delta\tilde{x}^i d\tau \tag{5.55}$$

From (5.41), we have

$$\int_0^t e^{-\rho\tau}\delta\tilde{x}^{iT}K_d^{i+1}\delta\tilde{x}^i d\tau = \beta(i+1)\int_0^t e^{-\rho\tau}\delta\tilde{x}^{iT}K_d^i\delta\tilde{x}^i d\tau$$
$$\geq \int_0^t e^{-\rho\tau}\delta\tilde{x}^{iT}K_d^i\delta\tilde{x}^i d\tau \tag{5.56}$$

Substituting (5.46) into (5.55) and noticing (5.56), we obtain

$$
\begin{aligned}
\Delta V^i \leq & -\mathrm{e}^{-\rho\tau}\delta\tilde{\boldsymbol{x}}^{i\mathrm{T}}(t)\boldsymbol{M}(t)\delta\tilde{\boldsymbol{x}}^i(t)-\rho\int_0^t \mathrm{e}^{-\rho\tau}\delta\tilde{\boldsymbol{x}}^{i\mathrm{T}}\boldsymbol{M}\delta\tilde{\boldsymbol{x}}^i\mathrm{d}\boldsymbol{\tau} \\
& -\int_0^t \mathrm{e}^{-\rho\tau}\delta\dot{\boldsymbol{e}}^{i\mathrm{T}}\left((2-\alpha)\boldsymbol{K}_d^i+2\boldsymbol{C}_1-2a\boldsymbol{M}\right)\delta\dot{\boldsymbol{e}}^i\mathrm{d}\boldsymbol{\tau} \\
& -2a\int_0^t \mathrm{e}^{-\rho\tau}\delta\boldsymbol{e}^{i\mathrm{T}}\left((2-\alpha)\boldsymbol{K}_d^i+2\boldsymbol{C}_1-2a\boldsymbol{M}\right)\delta\dot{\boldsymbol{e}}^i\mathrm{d}\boldsymbol{\tau} \\
& -2\int_0^t \mathrm{e}^{-\rho\tau}\delta\dot{\boldsymbol{e}}^{i\mathrm{T}}\left(\boldsymbol{R}-a(\boldsymbol{C}+\boldsymbol{C}_1-a\boldsymbol{M}\right)\delta\boldsymbol{e}^i\mathrm{d}\boldsymbol{\tau}-a^2\int_0^t \mathrm{e}^{-\rho\tau}\delta\boldsymbol{e}^{i\mathrm{T}}\left((2-\alpha)\boldsymbol{K}_d^i\right. \\
& \left.+2\boldsymbol{C}_1-2a\boldsymbol{M}\right)\delta\boldsymbol{e}^i\mathrm{d}\boldsymbol{\tau}-2a\int_0^t \mathrm{e}^{-\rho\tau}\delta\boldsymbol{e}^{i\mathrm{T}}\left(\boldsymbol{R}-a(\boldsymbol{C}+\boldsymbol{C}_1-a\boldsymbol{M}\right)\delta\boldsymbol{e}^i\mathrm{d}\boldsymbol{\tau}
\end{aligned}
\tag{5.57}
$$

By applying the method of partial integration, we obtain the following result

$$
\begin{aligned}
& \int_0^t \mathrm{e}^{-\rho\tau}\delta\boldsymbol{e}^{i\mathrm{T}}\left((2-\alpha)\boldsymbol{K}_d^i+2\boldsymbol{C}_1-2a\boldsymbol{M}\right)\delta\dot{\boldsymbol{e}}^i\mathrm{d}\boldsymbol{\tau} \\
& =\mathrm{e}^{-\rho\tau}\delta\boldsymbol{e}^{i\mathrm{T}}\left((2-\alpha)\boldsymbol{K}_d^i+2\boldsymbol{C}_1-2a\boldsymbol{M}\right)\delta\boldsymbol{e}^i\Big|_0^t+\rho\int_0^t \mathrm{e}^{-\rho\tau}\delta\boldsymbol{e}^{i\mathrm{T}}\left((2-\alpha)\boldsymbol{K}_d^i\right. \\
& \left.+2\boldsymbol{C}_1-2a\boldsymbol{M}\right)\delta\boldsymbol{e}^i\mathrm{d}\boldsymbol{\tau}-\int_0^t \mathrm{e}^{-\rho\tau}\delta\dot{\boldsymbol{e}}^{i\mathrm{T}}\left((2-\alpha)\boldsymbol{K}_d^i+2\boldsymbol{C}_1-2a\boldsymbol{M}\right)\delta\boldsymbol{e}^i\mathrm{d}\boldsymbol{\tau} \\
& +2\int_0^t \mathrm{e}^{-\rho\tau}\delta\boldsymbol{e}^{i\mathrm{T}}\left(a\dot{\boldsymbol{M}}-\dot{\boldsymbol{C}}_1\right)\delta\boldsymbol{e}^i\mathrm{d}\boldsymbol{\tau}
\end{aligned}
\tag{5.58}
$$

Therefore,

$$\Delta V^i \leq -\mathrm{e}^{-\rho\tau}\delta\tilde{\boldsymbol{x}}^{i\mathrm{T}}\boldsymbol{M}\delta\tilde{\boldsymbol{x}}^i - \rho\int_0^t \mathrm{e}^{-\rho\tau}\delta\tilde{\boldsymbol{x}}^{i\mathrm{T}}\boldsymbol{M}\delta\tilde{\boldsymbol{x}}^i d\tau - a\mathrm{e}^{-\rho\tau}\delta\boldsymbol{e}^{i\mathrm{T}}\left((2-\alpha)\boldsymbol{K}_d^i + 2\boldsymbol{C}_1\right.$$
$$\left. - 2a\boldsymbol{M}\right)\delta\boldsymbol{e}^i - \rho a\int_0^t \mathrm{e}^{-\rho\tau}\delta\boldsymbol{e}^{i\mathrm{T}}\left((2-\alpha)\boldsymbol{K}_d^i + 2\boldsymbol{C}_1 - 2a\boldsymbol{M}\right)\delta\boldsymbol{e}^i d\boldsymbol{\tau} - \int_0^t \mathrm{e}^{-\rho\tau}\boldsymbol{w}d\boldsymbol{\tau}$$
$$\leq -\mathrm{e}^{-\rho\tau}\delta\tilde{\boldsymbol{x}}^{i\mathrm{T}}\boldsymbol{M}\delta\tilde{\boldsymbol{x}}^i - a\mathrm{e}^{-\rho\tau}\delta\boldsymbol{e}^{i\mathrm{T}}l_p\delta\boldsymbol{e}^i - \rho\int_0^t \mathrm{e}^{-\rho\tau}\delta\tilde{\boldsymbol{x}}^{i\mathrm{T}}\boldsymbol{M}\delta\tilde{\boldsymbol{x}}^i d\boldsymbol{\tau}$$
$$-\rho a\int_0^t \mathrm{e}^{-\rho\tau}\delta\boldsymbol{e}^{i\mathrm{T}}l_p\delta\boldsymbol{e}^i d\boldsymbol{\tau} - \int_0^t \mathrm{e}^{-\rho\tau}\boldsymbol{w}d\boldsymbol{\tau} \tag{5.59}$$

where

$$\begin{aligned}\boldsymbol{w} &= \delta\dot{\boldsymbol{e}}^{i\mathrm{T}}\left((2-\alpha)\boldsymbol{K}_d^i + 2\boldsymbol{C}_1 - 2a\boldsymbol{M}\right)\delta\dot{\boldsymbol{e}}^i + 2a\delta\dot{\boldsymbol{e}}^{i\mathrm{T}}(\boldsymbol{R}/a - (\boldsymbol{C}+\boldsymbol{C}_1 - a\boldsymbol{M}))\delta\boldsymbol{e}^i \\ &+a^2\delta\boldsymbol{e}^{i\mathrm{T}}\left((2-\alpha)\boldsymbol{K}_d^i + 2\boldsymbol{R}/a + 2\boldsymbol{C} - 2\dot{\boldsymbol{C}}_1/a\right)\delta\boldsymbol{e}^i\end{aligned} \tag{5.60}$$

Let $\boldsymbol{P} = \boldsymbol{R}/a - (\boldsymbol{C} + \boldsymbol{C}_1 - a\boldsymbol{M})$. Then, from (5.48) and (5.49), we obtain

$$\boldsymbol{w} \geq l_p\|\delta\dot{\boldsymbol{e}}\|^2 + 2a\delta\dot{\boldsymbol{e}}^{\mathrm{T}}\boldsymbol{P}\delta\boldsymbol{e} + a^2 l_r\|\delta\boldsymbol{e}\|^2 \tag{5.61}$$

By applying the Cauchy-Schwarz inequality, we obtain

$$\delta\dot{\boldsymbol{e}}^{\mathrm{T}}\boldsymbol{P}\delta\boldsymbol{e} \geq -\|\delta\dot{\boldsymbol{e}}\|\,\|\boldsymbol{P}\|_{\max}\|\delta\boldsymbol{e}\| \tag{5.62}$$

From (5.48)–(5.50),

$$\begin{aligned} w &= l_p\|\delta\dot{\boldsymbol{e}}\|^2 - 2a\|\delta\dot{\boldsymbol{e}}\|\ \|\boldsymbol{P}\|_{\max}\|\delta\boldsymbol{e}\| + a^2 l_r\|\delta\boldsymbol{e}\|^2 \\ &= l_p\left(\|\delta\dot{\boldsymbol{e}}\| - \frac{a}{l_p}\|\boldsymbol{P}\|_{\max}\|\delta\boldsymbol{e}\|\right)^2 + a^2\left(l_p - \frac{1}{l_r}\|\boldsymbol{P}\|_{\max}^2\right)\|\delta\boldsymbol{e}\|^2 \geq 0 \end{aligned} \tag{5.63}$$

According to properties (5.1) and (5.63), based on (5.59), it can be ensured that $\Delta V^i \leq 0$; therefore, $V^{i+1} \leq V^i$. From the definition, $\boldsymbol{K}_d^i$ is a positive definite matrix. From the definition of V^i, $V^i \geq 0$, and V^i is bounded. As a result, $\tilde{\boldsymbol{x}}(t) \to 0$ when $i \to \infty$. Since $\boldsymbol{e}^i(t)$ and $\dot{\boldsymbol{e}}^i(t)$ are two independent variables and a is a positive constant, it follows that if $i \to \infty$, then $\dot{\boldsymbol{e}}^i(t) \to 0$ for all $t \in [0, t_f]$.

Finally, the following conclusions hold for $t \in [0, t_f]$:

$$q^i(t) \overset{i \to \infty}{\rightarrow} q_d(t),$$

$$\dot{q}^i(t) \overset{i \to \infty}{\rightarrow} \dot{q}_d(t)$$

According to the analysis, it can be seen that the adaptive iterative learning control (AILC) method guarantees that the tracking errors converge arbitrarily close to zero as the number of iterations increases. The following case is demonstrated based on simulation results for this conclusion.

5.2.2.3 Numerical Simulation

A two-link robot manipulator as shown in Fig. 5.7 is utilized in this section to verify the effectiveness of the proposed control scheme. The dynamic equation of the robot manipulator is adopted in [7]. Usually, the manipulator task specification is given relative to the end effector. Thus, it is natural to attempt to derive the control algorithm directly in task space, better than in the joint space. Denote the end-effector position and orientation in task space by $x \in R^n$. The task space dynamic can be rewritten as follows:

$$D_x\left(q^i(t)\right)\ddot{x}^i(t) + C_x\left(q^i(t), \dot{q}^i(t)\right)\dot{x}^i(t) + G_x\left(q^i(t), \dot{q}^i(t)\right) = \tau_a(t) + F_x{}^i(t) \quad (5.64)$$

where $D_x(q^i(t)) = J^{-\mathrm{T}}(q^i(t))D(q^i(t))J^{-1}(q^i(t))$, $C_x\left(q^i(t), \dot{q}^i(t)\right) = J^{-\mathrm{T}}\left(q^i(t)\right)\left(C\left(q^i(t), \dot{q}^i(t)\right) - D\left(q^i(t)\right)J^{-1}\left(q^i(t)\right)\dot{J}\left(q^i(t)\right)\right)J^{-1}\left(q^i(t)\right)$, $G_x\left(q^i(t), \dot{q}^i(t)\right) = J^{-\mathrm{T}}\left(q^i(t)\right)G\left(q^i(t), \dot{q}^i(t)\right)$, and $F_x^i = J^{-\mathrm{T}}\left(q^i(t)\right)\tau^i(t)$.where J-$(q^i(t)) \in \mathbb{R}^{n \times n}$ is the configuration-dependent Jacobian matrix, which is assumed to be nonsingular in the finite workspace Ω. The Jacobian matrix $J(q)$ is known as

$$J(q) = \begin{pmatrix} -l_1 \sin(q_1) - l_2 \sin(q_1 + q_2) & -l_2 \sin(q_1 + q_2) \\ l_1 \cos(q_1) + l_2 \cos(q_1 + q_2) & l_2 \cos(q_1 + q_2) \end{pmatrix} \quad (5.65)$$

The desired reference trajectories in the Cartesian space are $x_{d1} = 1.0 + 0.2 \cos(\pi t)$ and $x_{d2} = 1.0 + 0.2 \sin(\pi t)$, which represent a circle of radius 0.2 m, and its centers are located at $(x_1, x_2) = (1.0, 1.0)$. The robot is initially rested with its end effector positioned at the center of the circle. The initial conditions are$x_1(0) = x_{d1}(0) = 1.2$, $x_2(0) = x_{d2}(0) = 1.0$, $\dot{x}_1(0) = \dot{x}_{d1}(0) = 1.0$, $\dot{x}_2(0) = \dot{x}_{d2}(0) = 1.2$, and $t = [0, 2]$.

To demonstrate the superior control performance of the AILC system, we compare it with two additional control systems: an adaptive switching learning PD control system (ASL-PD), which is also described in [6], and an iterative learning control (ILC) system, which is represented in [8].

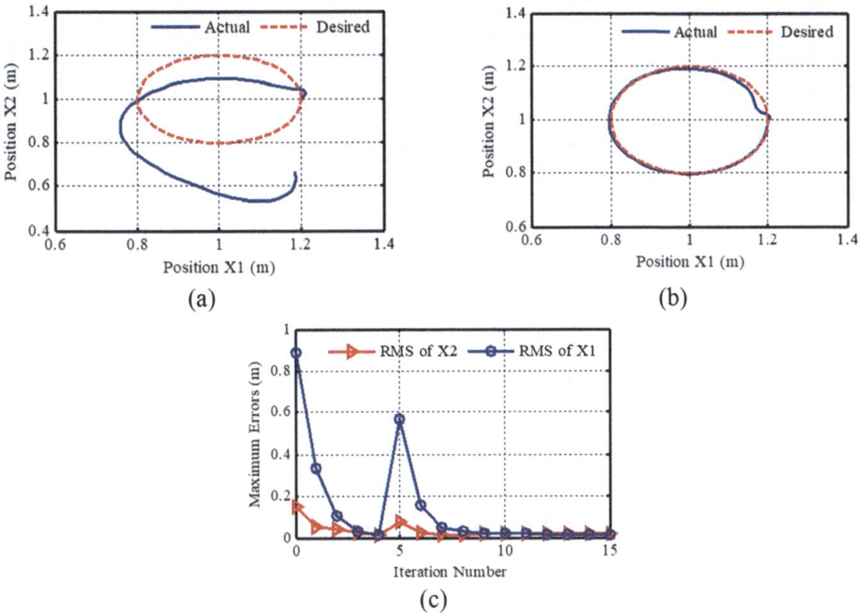

Fig. 5.8 Reference tracking and actual trajectory for end effector at different iterations under AILC adopted in [8]. (**a**) Position tracking of end effector at the first iteration. (**b**) Position tracking of end effector after the 15th iteration. (**c**) Profile of RMS position errors and payload and disturbance variation after the fifth iteration

The simulation results of the ILC system, the responses of end-effector position at the 1st and 15th iteration, and the tracking error from iteration to iteration are depicted in Fig. 5.8a–c, respectively. From the simulation results, it can be seen that the tracking performance was not acceptable because the errors were too large compared to the ASL-PD and the proposed controllers; especially at the fifth iteration, poor tracking error responses resulted due to the occurrence of payload variation situation and external disturbance. The details of the ASL-PD system are depicted in [6]. The end-effector position responses and tracking error from iteration to iteration are depicted in Fig. 5.9a–c, respectively. The end-effector tracking performance is obvious under the occurrence of payload variation and external disturbance. The convergence rate increased greatly compared with the ILC method.

Now, the AILC system depicted in Fig. 5.10 is applied to control the robot manipulator for comparison. The simulation results of end-effector position responses and tracking error from iteration to iteration are depicted in Fig. 5.11a–c, respectively. Table 5.1 shows the tracking performance of the proposed system from the initial to the 15th iteration is obvious. Therefore, the comparison of their method and our method demonstrated a bit fast convergence rate with the proposed control method. Specially, the control torque of the proposed system is much smaller than the ASL-PD controller. Detailed results are depicted in Table 5.2.

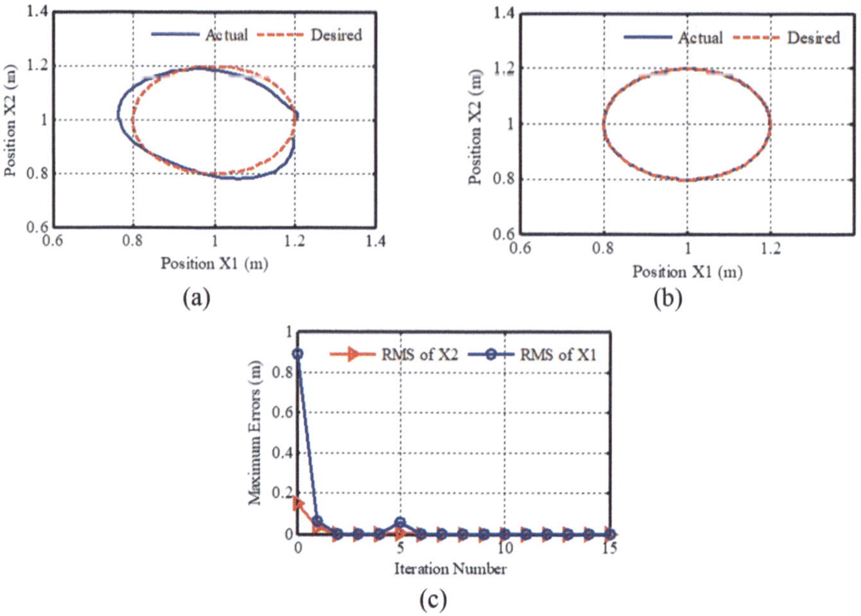

Fig. 5.9 Reference tracking and actual trajectory for end effector at different iterations under ASL-PD adopted in [6]. (**a**) Position tracking of end effector at the first iteration. (**b**) Position tracking of end effector after the 15th iteration. (**c**) Profile of RMS position errors and payload and disturbance variation after the fifth iteration

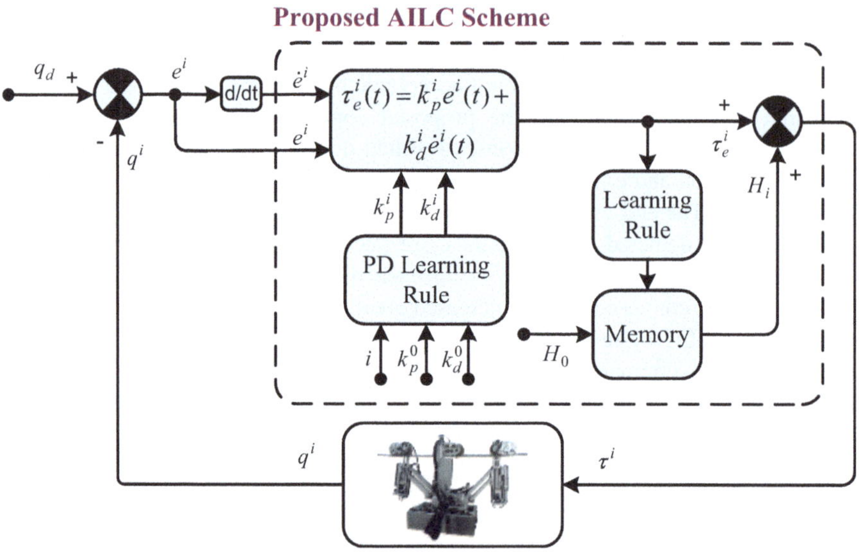

Fig. 5.10 Block diagram of the proposed AILC system

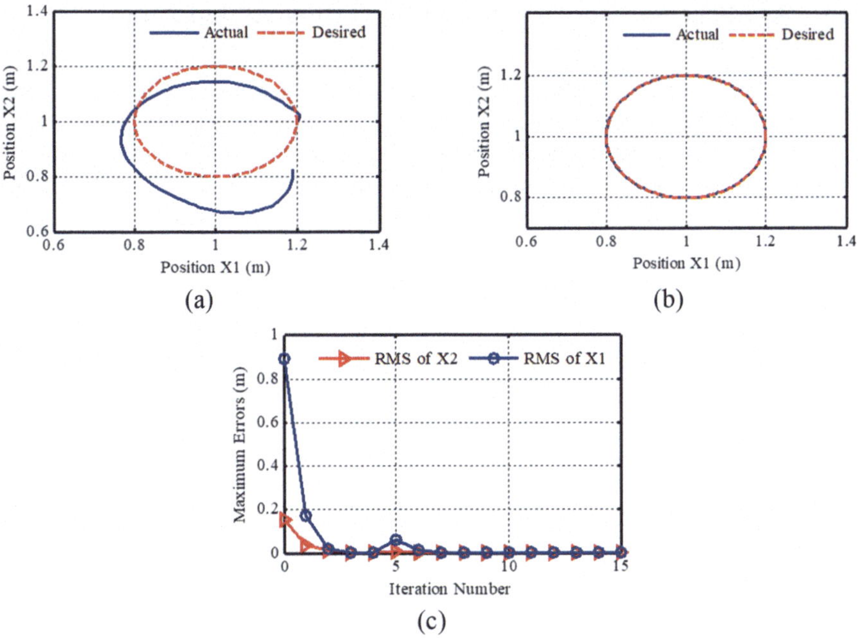

Fig. 5.11 Reference tracking and actual trajectory for end effector at different iterations under AILC controller. (**a**) Position tracking of end effector at the first iteration. (**b**) Position tracking of end effector after the 15th iteration. (**c**) Profile of RMS position errors and payload and disturbance variation after the fifth iteration

Table 5.1 Maximum tracking errors from iteration to iteration

Iterative		0	5	10	15
ILC	$\text{Max}\lvert e_1^i\rvert$	0.0822	0.0773	0.0168	0.0140
	$\text{Max}\lvert e_2^i\rvert$	0.1998	0.5634	0.0130	0.0105
ASL-PD	$\text{Max}\lvert e_1^i\rvert$	0.0398	0.0076	0.0012	0.0010
	$\text{Max}\lvert e_2^i\rvert$	0.0646	0.0593	0.0008	0.0006
AILC	$\text{Max}\lvert e_1^i\rvert$	0.0377	0.0076	0.0011	0.0009
	$\text{Max}\lvert e_2^i\rvert$	0.1741	0.0590	0.0008	0.0006

Table 5.2 Maximum control forces from iteration to iteration

Iteration		0	5	10	15
ASL-PD	$\text{Max}\lvert\tau_1^i\rvert$(N.m)	90	930	3330	7230
	$\text{Max}\lvert\tau_2^i\rvert$(N.m)	51.45	531.6	1903	3618
AILC	$\text{Max}\lvert\tau_1^i\rvert$(N.m)	82.50	772.5	2647	5647
	$\text{Max}\lvert\tau_2^i\rvert$(N.m)	47.14	441.6	1513	3228

5.3 Application: De-icing Robot for Transmission Lines

The de-icing robot for transmission lines is a product that involves multiple disciplines such as mechanical engineering, electronics, computer science, communication, and automation. It operates in harsh environments and has a weak research foundation both domestically and internationally. Currently, there are no mature technologies available for reference. Compared to transmission line inspection robots, the de-icing robot faces greater challenges in terms of implementation. During the development process, the main technical difficulties can be summarized as follows:

1. High-reliability motion mechanism on complex and flexible transmission lines
2. Continuous power supply for long-term operation in large-scale outdoor work
3. Development of non-damaging and efficient de-icing mechanism
4. Autonomous behavior control and navigation of robots in complex unstructured environments
5. Remote communication, monitoring, and teleoperation of robots under long-distance strong electromagnetic interference conditions
6. Special protection mechanisms in harsh working environments

5.3.1 Mechanical Design of De-icing Robot Manipulator

As shown in Fig. 5.12, the three-arm robot consists of the front, middle, and rear arms, as well as the control box (center-of-gravity adjustment mechanism). The front and rear arms of the robot are equipped with walking devices, de-icing mechanisms, and brake mechanisms. The middle arm is equipped with a walking mechanism and a brake device. Each of the three arms is designed with a telescopic device, allowing the arm length to be flexibly extended or shortened. Additionally, each arm is

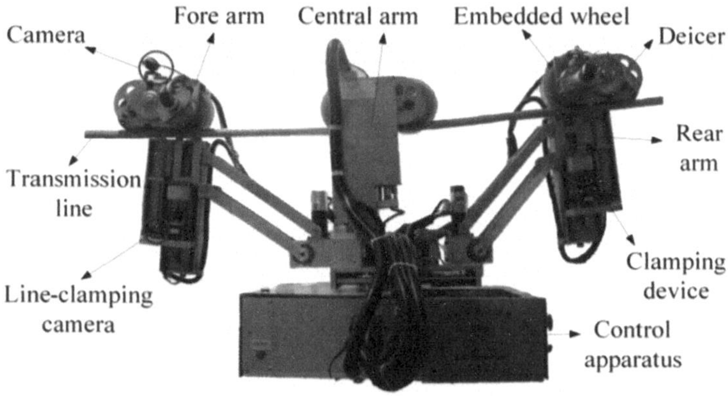

Fig. 5.12 Mechanical structure of power transmission line de-icing robot

equipped with a vertical rotation joint and a horizontal rotation joint. The center-of-gravity adjustment mechanism includes the robot's center-of-gravity adjustment device, control, and power box. The center-of-gravity adjustment equipment moves the control box forward or backward to achieve front-to-back center-of-gravity adjustment for the robot.

5.3.1.1 Design of Robot Manipulator

The forearm and rear arm components have identical structures, with the only difference being the installation method. They are symmetrically installed relative to the middle arm. Both the forearm and rear arm are equipped with de-icing devices (including cutting de-icing mechanisms and impact de-icing mechanisms), walking mechanisms, braking devices, telescopic mechanisms, vertical plane rotation joints, and horizontal plane rotation joints, as shown in Fig. 5.13. The structure of the middle arm and the front and rear arms is essentially the same, with the only difference being that the middle arm is not equipped with a de-icing mechanism but instead modified with a two-wheel drive structure, which stabilizes the robot and improves driving power.

De-icing device: The de-icing action of the de-icing mechanism consists of two parts: the cutting motion for removing the upper layer of ice and the striking motion for breaking the lower layer of ice. The cutting motion is achieved by the combined cutting blades driven by a pancake motor, while the striking de-icing motion is achieved by the rotation of the striking rod driven by a motor.

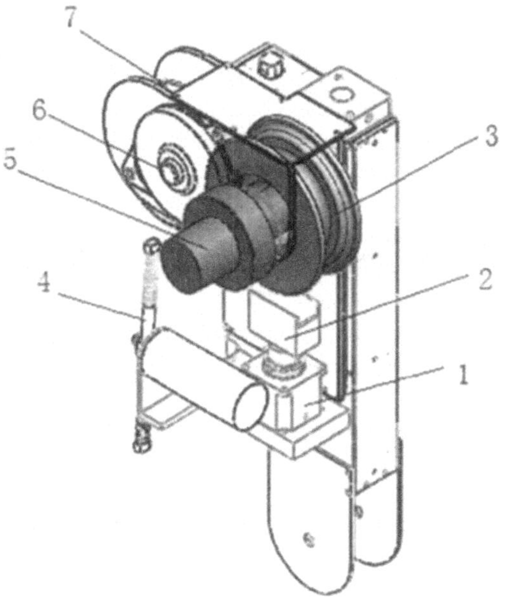

Fig. 5.13 Schematic diagram of obstacle avoidance arm structure for de-icing robot. (1) Brake motor; (2) brake pad; (3) drive wheel; (4) impact de-icing mechanism; (5) drive motor; (6) de-icing motor; (7) de-icing mechanism

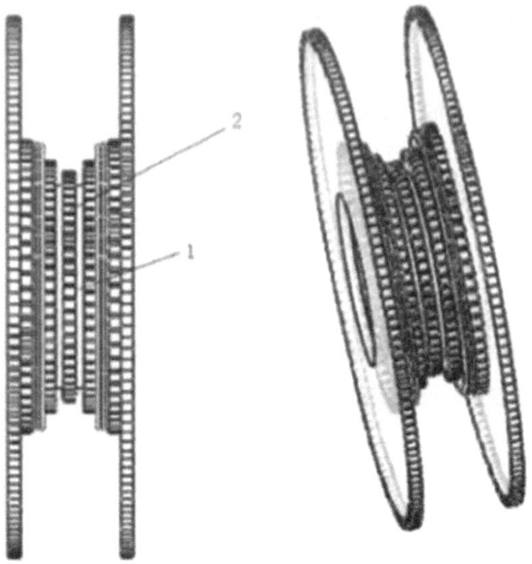

Fig. 5.14 Combination cutting de-icing knife structure: (1) cutting blade; (2) anti-cutting gasket

As shown in Fig. 5.14, the combination cutting tool consists of six cutting blades, four spacers, and two anti-cutting plates. In addition to its de-icing function, it also serves the purpose of preventing the cutting of high-voltage power lines. The protection mechanism works as follows: after the upper part of the ice is removed, the symmetrical protective plates will clamp onto the high-voltage transmission line, ensuring that the cutting blades do not damage the power line. The combination cutting tool is driven by a pancake motor with a power of 120 W and weighs only 0.7 kg.

As shown in Fig. 5.15, the percussion rod consists of motor connecting components, connecting rods, elastic connections, and percussion hammerheads. The elastic connection can be achieved using a tension spring, with the elasticity being moderate. If the elasticity is too small, it may cause the motor to jam, while if it is too large, the impact force of the percussion hammerhead on the ice cover will be too small. The percussion hammerhead has cutting edges and can achieve a certain cutting motion while percussing.

Walking mechanism: The walking mechanism mainly consists of a driving motor and walking wheels. The driving motor adopts Maxon motor, equipped with an encoder and a brake, which allows it to stop when necessary. The walking wheels are made of polyester composite material to increase friction with the track.

The structure of the walking mechanism is shown in Fig. 5.16. When one arm is performing de-icing work, the walking mechanism of that arm has a minimal driving effect on the robot. Instead, it relies on the walking mechanisms of the two rear arms to propel the robot forward and achieve the feeding de-icing motion of the forearm.

Braking system: As shown in Fig. 5.17, the brake device consists of brake pads with a certain degree of flexibility, brake pad mounting blocks, telescopic rods, linear stepping motor, and motor fixed parts. Linear stepping motor is the driving

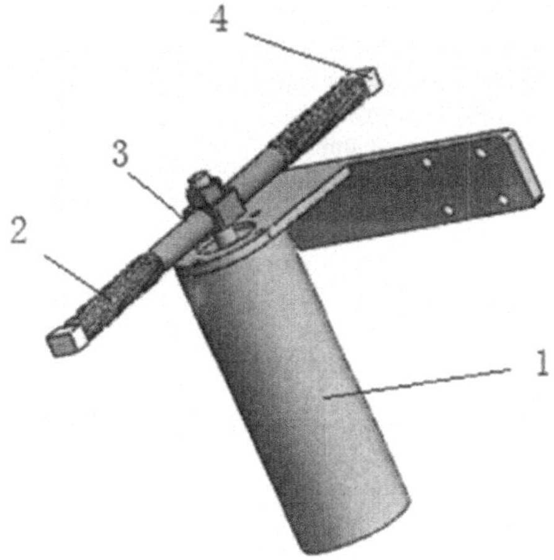

Fig. 5.15 Structure of impact de-icing mechanism: (1) drive motor; (2) elastic connector; (3) connecting rod; (4) impact head

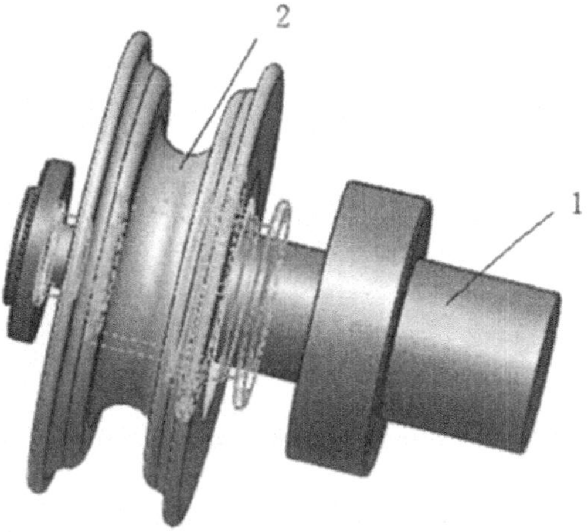

Fig. 5.16 Main drive wheel structure: (1) drive motor; (2) wire groove for carding wheel

component; the linear motor model 57H4A-3.25-078 is selected, which is characterized by small power consumption, low weight, thrust, small size, and high precision advantages; and the output axis is for linear motion without rotary motion, so it is suitable for brake motion.

Telescopic mechanism: The telescopic mechanism consists of three main components: two large and small telescopic rods and an arm frame. The large and small telescopic rods are fixedly installed on the arm frame. The large telescopic rod is connected to the cutting motion mechanism of the walking mechanism and de-icing mechanism, with an extension distance of approximately 200 mm. The small telescopic rod is connected to the striking motion device of the brake and de-icing device, with an extension distance of approximately 200 mm. The telescopic rods have a self-locking function and a pushing force of approximately 75 kgf, which fully meets the requirements for the normal operation of the de-icing robot.

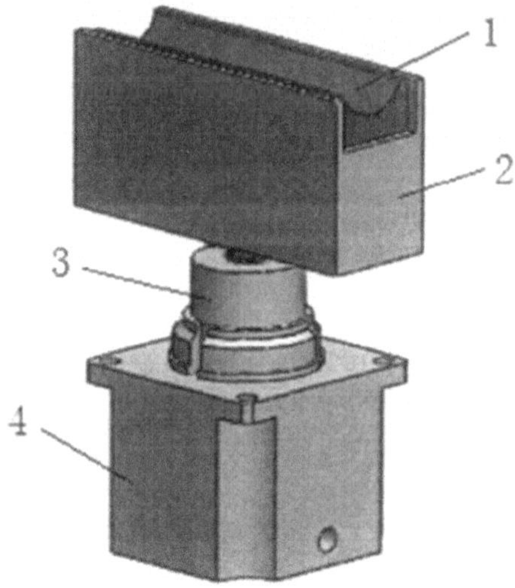

Fig. 5.17 De-icing robot manipulator brake system: (1) brake pad; (2) mounting component; (3) telescopic rod; (4) motor mounting component

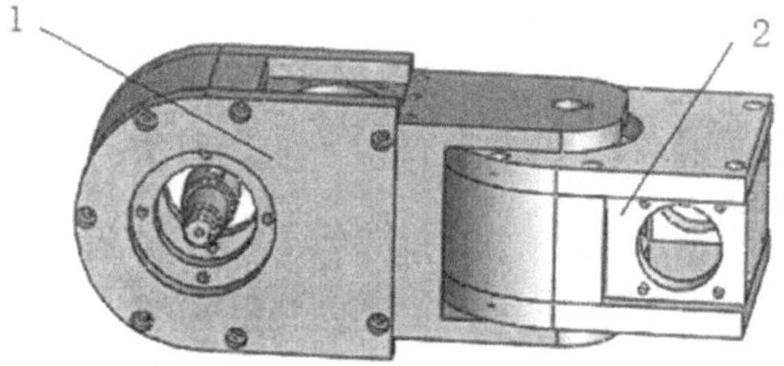

Fig. 5.18 Ice removal robot arm mechanism: (1) vertical rotation joint; (2) horizontal rotation joint

5.3.1.2 Swing Arm Mechanism

The swing arm mechanism is used to connect the robotic arm of the de-icing robot to the chassis, allowing the robotic arm to have degrees of freedom for rotation in both the vertical and horizontal planes. This ensures the de-icing robot can operate normally and smoothly overcome various obstacles. As shown in Fig. 5.18, the swing arm mechanism of the de-icing robot consists of a horizontal rotation mechanism and a vertical rotation mechanism, which are used to adjust the robot's arm posture in the vertical and horizontal planes when overcoming obstacles. They are driven by Maxon motors with encoders, through a worm gear mechanism. The worm gear mechanism has a self-locking function, which allows the arm to be stopped at the desired posture, enabling the robot to maintain a specific posture.

5.3.1.3 Adjusting the Balance Box of the Center of Gravity

The main function of the center-of-gravity adjustment balance box is to ensure the balance of the center of gravity during the obstacle-crossing process, as the movements of various arms of the robot may cause the center of gravity to deviate. The center-of-gravity adjustment mechanism is used to maintain the balance of the center of gravity during the movement of the de-icing robot. Additionally, it can be used as a load mechanism for various control hardware settings. As shown in Fig. 5.19, the center-of-gravity adjustment balance box consists of the robot center-of-gravity adjustment mechanism and the control box. The control box houses the control system and power system. The center-of-gravity adjustment mechanism moves the control box forward or backward to adjust the front-to-back center of gravity of the robot. The center-of-gravity adjustment mechanism is mainly composed of a screw and two guide rods, with a movement distance of approximately ±100 mm.

The physical structure of the main components of the three-arm de-icing robot is shown in Fig. 5.20. This robot is designed to perform online de-icing work while ensuring the integrity of the transmission lines. Its main features include separating the de-icing power from the robot's locomotion power, ensuring the effectiveness of the de-icing work, and adjustable de-icing speed. The arms adopt an eccentric integral structure, simplifying the obstacle-crossing process. Additionally, the unique design of the parallel four-bar linkage arm structure not only reduces the weight of the robot's arms but also decreases the motor load and energy consumption. It allows the robot to maintain a consistent arm posture during obstacle crossing, always remaining perpendicular to the horizontal plane where the transmission lines are located. This greatly reduces the difficulty of controlling the arm posture during obstacle crossing and ensures the reliability of the de-icing robot's obstacle-crossing operation. Furthermore, by adjusting the center-of-gravity balance mechanism, the stability of the robot's operation can be further improved.

5.3.2 *Adaptive Robust Sliding Tracking Control for Transmission Line De-icing Robot Manipulator*

The practical equipment system for the de-icing robot is shown in Fig. 5.21, and the hardware block chart of the power line de-icing robot manipulator (PLDRM) system

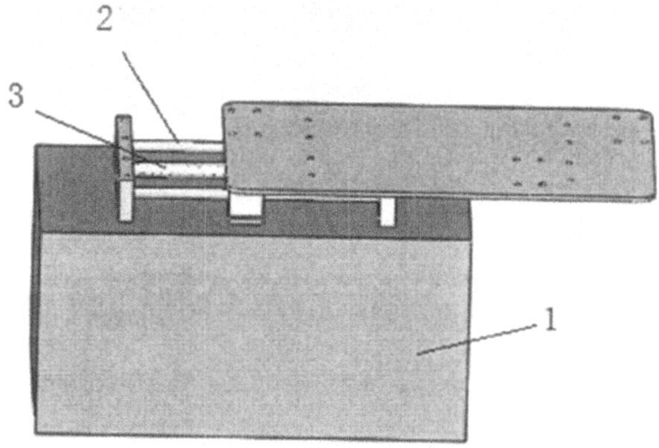

Fig. 5.19 Ice removal robot control and center-of-gravity balance box: (1) control and power box; (2) guide rod; (3) lead screw

Fig. 5.20 Physical diagram of the de-icing robot's main components: (1) combination ice blade; (2) de-icing drive motor; (3) walking drive motor; (4) walking wheel; (5) camera; (6) upper arm linkage mechanism; (7) center-of-gravity balancing box; (8) horizontal rotational joint; (9) vertical rotational joint; (10) braking mechanism

Fig. 5.21 The practical equipment system of the PLDRM

is shown in Fig. 5.22. This PLDRM is experimented under indoor experimental condition. The parts of the PLDRM system comprise robot control apparatus, de-icing device, manipulators, motion structure, and brake system. The left and right manipulators have three links (i.e., two revolute joints and one prismatic joint) and work during normal operating conditions. The middle manipulator has

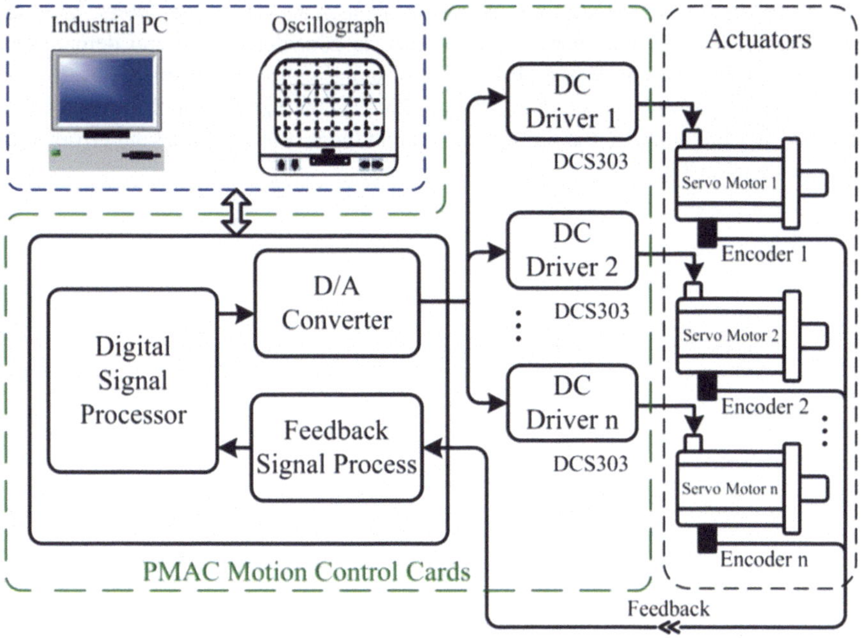

Fig. 5.22 The hardware block chart of PLDRM system

only two links (i.e., one revolute joint and one prismatic joint), and it only works when the PLDRM voids obstacles on power transmission line. The end effectors of the manipulators are attached to the de-icing device and the motion structure, which is designed to move the PLDRM along the power transmission line. The parts of the manipulator system contain joint motors, gearboxes, worm gears, motor drivers, and encoder sensors. The EC-45-type Maxon DC servo motor is used for each joint of manipulator, and this motor is attached with an encoder. The programmable multi-axis controller (PMAC) cards with peripheral component interconnect (PCI) interfaces connected to the industrial personal computer (IPC) are used to control the motion of DC servo motors. Each PMAC motion control card involves the interface circuits of the encoders and the multi-channels of digital/analog (D/A). A TMS320F2812-type digital signal processor (DSP) designed by Texas Instruments is applied to perform the proposed control program and to carry out the real-time processing. The DSP uses external pulse counters to read the outputs of encoder sensors and produces digital output signals. The digital output signals are sent to the DCS303-type digital DC servo drivers to control the DC servo motors.

In fact, the operation of the PLDRM is so complex. In this experiment, the aim of the control is that the joint positions of a three-link PLDRM can track periodic cosine or sinusoidal trajectory by using the proposed adaptive robust sliding tracking control (ARSTC) scheme. The experimental parameters and the desired trajectories of the ARSTC system are selected to be the same as in the simulation, and the initial conditions of joint positions are given as follows: $q_1(0) = 1$ rad, $q_2(0) = 0$, and $q_3(0) = 0$.

The external disturbance term is added by increasing friction forces in joint connections, and the parameter variation also is considered. It is too demanding for the hardware implementation if the control sampling rate is considered by $T_s - 1$ ms. Herein, T_s is thus set by $T_s = 20$ ms. Finally, the experimental results of the ARSTC system comprise the joint positions and the response of tracking errors that are illustrated in Fig. 5.23.

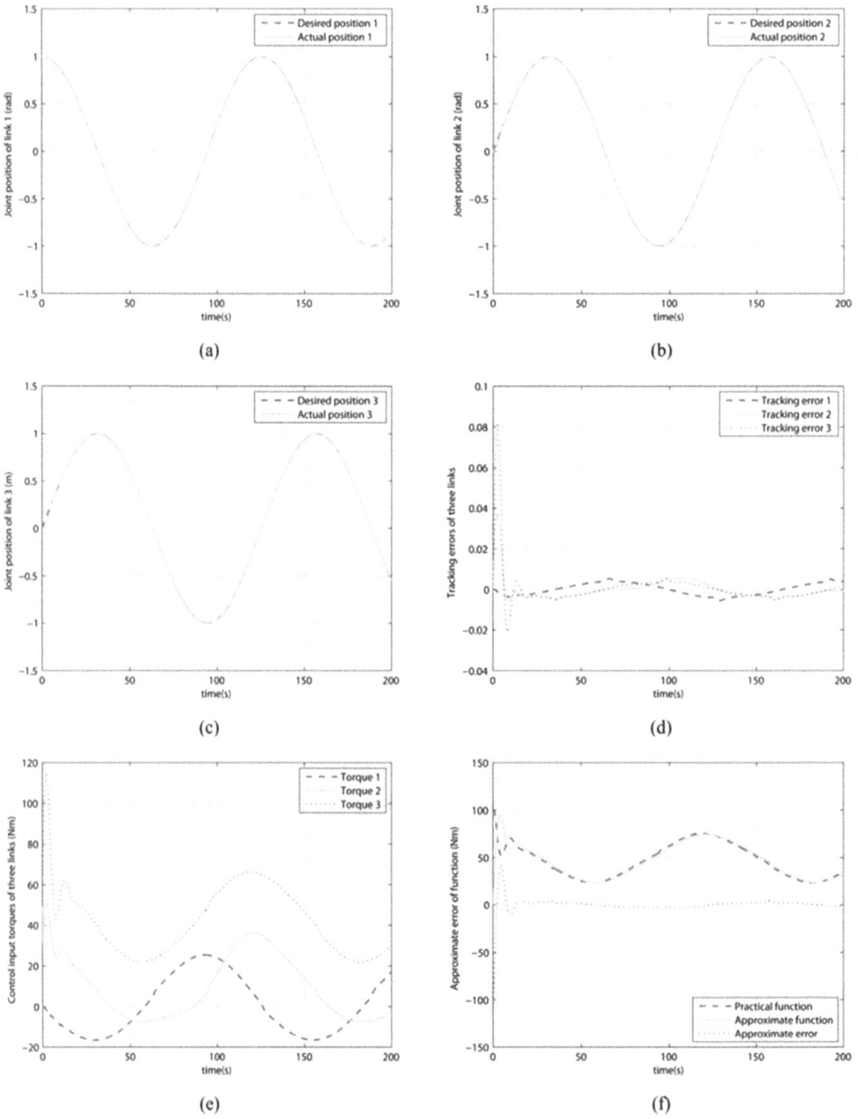

Fig. 5.23 Experimental responses for ARSTC system. (**a**) Joint position of link 1. (**b**) Joint position of link 2. (**c**) Joint position of link 3. (**d**) Tracking errors of three links. (**e**) Control torques of three links. (**f**) Approximate error of function

The experimental results in Fig. 5.23 show that in the appearance of both the uncertainties and the inescapable disturbances, the proposed ARSTC system for a three-link PLDRM achieves the high-accuracy position tracking without the chattering.

References

1. Choi BJ, Kwak SW, Kim BK. Design of single-input fuzzy logic controller and its properties. Fuzzy Sets Syst. 1999;106:299–308.
2. Choi B J, Kwak S W, Kim B K. Design and stability analysis of single-input fuzzy logic controller. IEEE Transactions on Systems, Man, and Cybernetics, Part B, 2000, 30(2): 303–309.
3. Yeh M F, Tsai C H. Standalone CMAC control system with online learning ability. IEEE Transactions on Systems, Man, and Cybernetics, Part B, 2009, 40(1): 43–53.
4. Xu J, Tan Y. Linear and nonlinear iterative learning control. Springer; 2003.
5. Bristow D A, Tharayil M, Alleyne A G. A survey of iterative learning control. IEEE control systems magazine, 2006, 26(3): 96–114.
6. Ouyang PR, Zhang W, Gupta MM. An adaptive switching learning control method for trajectory tracking of robot manipulators. Mechatronics. 2006;16(1):51–61.
7. Debenest P, Guarnieri M, Takita K, et al. Expliner—robot for inspection of transmission lines. In: 2008 IEEE international conference on robotics and automation. IEEE; 2008. pp. 3978–3984.
8. Kuc T, Nam K, Lee J S, et al. An iterative learning control of robot Manipulators. IEEE Trans Robot Automat. 1991;7(6):835–42.

Chapter 6
Unmanned Aerial Vehicle with Stable Flight for Power Inspection

6.1 Background and Motivation

In recent years, rapid technological advancements in computers, artificial intelligence, and microelectronics have led to an increased use of robots in various fields such as commerce, healthcare, industry, and national defense. In particular, the rapid deployment of artificial intelligence algorithms and computing technology has enabled robots to exhibit a certain level of intelligence in specific scenarios. The most typical examples are dialogue interaction robots, such as Pepper from Japan's SoftBank [1], and floor-sweeping robots [2]. It is evident that robotic technology and products will play a significantly important role in our lives and environment in the future, eventually becoming an integral part of our lives and even coexisting with us. In the field of industry, the emergence of human-machine collaborative robots [3] and swarm robots [4] has gradually changed the existing industrial manufacturing mode, leading to significant improvements in efficiency and flexibility. In terms of national defense, robotic technology is becoming the development direction of future weapons, especially in unmanned ground vehicles, unmanned underwater combat vessels, and unmanned combat aircraft. In summary, the level of robotic technology reflects the overall scientific and technological level of a country to a certain extent. Robotics is "the pearl at the top of the crown of manufacturing," with its research, development, manufacture, and application serving as critical indicators of a country's scientific and technological innovation and high-end manufacturing proficiency. Consequently, various notions have introduced relevant policies and plans for robotics, such as the "EU Robotics 2020" [5], the US DARPA Plan, and the "New Robotics Strategy" in Japan [6]. China has also outlined a series of plans for robotics in scientific research and policies, such as the National Medium- and Long-Term Plan for Science and Technology Development (2006–2020), which proposes the development of industrial robots and specialized operation robots through advanced applicable technologies and aims to promote the industrialization of

Y. Wang et al., *Electric Power Robots*,
https://doi.org/10.1007/978-981-96-2512-3_6

robotic science and technology. It also calls for the development of modular robotic units and the enhancement of the international competitiveness of China's intelligent manufacturing industry. With a focus on this, the Ministry of Science and Technology issued the "12th Five-Year Plan for the Development of Intelligent Manufacturing Science and Technology" in April 2012, which calls for tackling the core common technologies of industrial robots, independently developing industrial robot engineering products, and achieving technological breakthroughs and industrialization of industrial robots and their core components. Therefore, research on the fundamental theories and technologies related to robotics is of great significance in promoting the advancement of the social economy.

Mobile robots, as an important category of robots, have broad practical applications. Unlike traditional industrial robots that operate in structured environments and perform repetitive tasks through programming, mobile robots mainly operate in large-scale unstructured environments [7]. As a result, research on mobile robots primarily focuses on their autonomy, enabling them to complete specific tasks independently such as navigation, positioning, path planning, and control. Typical examples of mobile robots include unmanned ground vehicles (UGVs), unmanned aerial vehicles (UAVs), unmanned surface vessels (USVs), and autonomous underwater vehicles (AUVs). Currently, these robots are generally used to carry specific sensor equipment for passive observation and monitoring of environmental information. According to the subjective initiative of the robots, mobile robots can be categorized as information observation-type robots, which typically consist of specific sensor devices added to the mobile platforms to carry out specific tasks.

With the rapid development of micro-electromechanical system (MEMS) technology and high-power density dynamic system technology, UAVs, especially rotary-wing UAVs, have made significant breakthroughs and achieved substantial progress in various applications over the past two decades, including aerial photography [8], map creation and measurement [9], disaster rescue [10], power inspection [11], battlefield surveillance [12], and others, as shown in Fig. 6.1. Many research institutions around the world have conducted studies on modeling [13], planning [14], autonomous control [15], and other aspects of this type of robot, achieving phased results that enable fixed-point hovering, maneuvering flight, and autonomous takeoff and landing flight. The recent development stage is similar to the early development stages of ground, underwater, deep space, and other mobile robots, and they are all in the research period of platform autonomous control, which has achieved successful demonstrations of information acquisition applications.

6.2 Kinematic and Dynamic Modeling for UAV

Euler's theorem describes the relationship between displacement and rotational motion of a rigid body, namely, the displacement of a rigid body around a fixed point can also be the composition of several finite rotations around that point. The abovementioned geometric relationship also exists correspondingly between the

(a) Map creation and measurement [9]. (b) Disaster rescue [10].

(c) Power inspection [11]. (d) Battlefield surveillance [12].

Fig. 6.1 Application cases of UAVs. (**a**) Map creation and measurement [9]. (**b**) Disaster rescue [10]. (**c**) Power inspection [11]. (**d**) Battlefield surveillance [12]

inertial coordinate system and the body coordinate system, and the two coordinate systems can be transformed through the corresponding transformation matrix. Firstly, vectors $\boldsymbol{\xi} = [x, y, z]^\top$ and $\boldsymbol{\nu} = [v_x, v_y, v_z]^\top$ are the linear position and velocity defined in the body coordinate system, respectively. The vector $\boldsymbol{\eta} = [\phi, \theta, \psi]^\top$ is the Euler angles. The vector $\boldsymbol{\omega} = [\omega_x, \omega_y, \omega_z]^\top$ is the angular velocity defined in the body coordinate system.

The translation matrix $\boldsymbol{R}_{BE}$ for the linear velocity conversion from the body coordinate system to the inertial coordinate system is expressed as

$$\boldsymbol{R}_{BE} = \begin{bmatrix} C\psi C\theta & C\psi S\theta S\phi - S\psi C\phi & C\psi S\theta C\phi + S\psi S\phi \\ S\psi C\theta & S\psi S\theta S\phi + C\psi C\phi & S\psi S\theta C\phi - C\psi S\phi \\ -S\theta & C\theta S\phi & C\theta C\phi \end{bmatrix} \tag{6.1}$$

where $\boldsymbol{C}\cdot = \cos(\cdot)$ and $\boldsymbol{S}\cdot = \sin(\cdot)$.

The angular velocity transformation matrix $\boldsymbol{W}_{BE}$ from the body coordinate system to the inertial coordinate system is given as

$$\boldsymbol{W}_{BE} = \begin{bmatrix} 1 & \sin\phi\tan\theta & \cos\phi\tan\theta \\ 0 & \cos\phi & -\sin\phi \\ 0 & \sin\phi\sec\theta & \cos\phi\sec\theta \end{bmatrix} \tag{6.2}$$

Using the linear velocity transformation matrix (6.1) and angular velocity transformation matrix (6.2), the kinematic model of a quadrotor drone can be obtained:

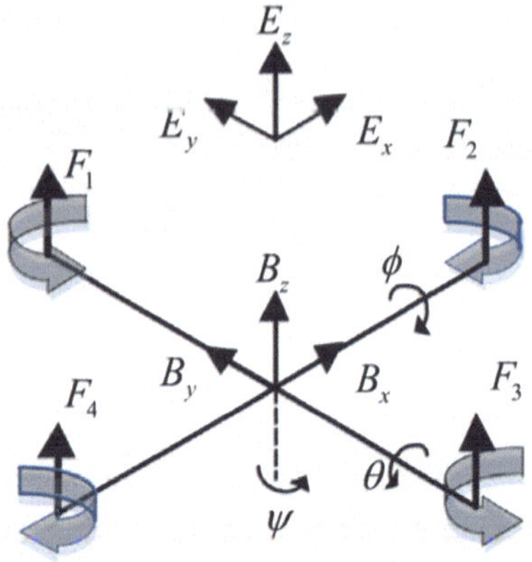

Fig. 6.2 Quadrotor model

$$\begin{bmatrix} \boldsymbol{\nu} \\ \omega \end{bmatrix} = \begin{bmatrix} R_{\text{BE}}^T & 0_{3\times 3} \\ \boldsymbol{0}_{3\times 3} & W_{\text{BE}}^{-1} \end{bmatrix} \begin{bmatrix} \dot{\boldsymbol{\xi}} \\ \dot{\eta} \end{bmatrix} \tag{6.3}$$

where $\boldsymbol{0}_{3\times 3}$ is the zero matrix.

A quadrotor aircraft is considered, which is subjected to parameter uncertainties, aerodynamic drag, and wind gusts. This kind of quadrotor aircraft can be modeled by the Euler method. As shown in Fig. 6.2, a quadrotor is a cross rigid frame with four rotors to produce the driving forces.

In the quadrotor system, Ω_i stands for the angular speed of the i-th propeller. Based on the simplified rotor model, the control inputs T, τ_1, τ_2, τ_3 can be obtained using the rotational speed $\boldsymbol{\Omega} \in R^4$, which is defined as [16]

$$\begin{cases} T = b\sum_{i=1}^{4}\Omega_i^2 \\ \tau_1 = db\left(\Omega_1^2 - \Omega_3^2\right) \\ \tau_2 = db\left(\Omega_2^2 - \Omega_4^2\right) \\ \tau_3 = \kappa\left(\Omega_1^2 + \Omega_3^2 - \Omega_2^2 - \Omega_4^2\right) \end{cases} \tag{6.4}$$

where $T \in \mathbb{R}$ is the thrust magnitude; $\boldsymbol{\tau} = [\tau_1, \tau_2, \tau_3]^\top \in \mathbb{R}^3$ are the roll, pitch, and yaw moments; d represents the distance from the mass center to each rotor; and $\kappa > 0$ and $b > 0$ are the thrust and drag coefficients, respectively. Finally, the thrust magnitude T and the items of the moment vector $\boldsymbol{\tau}$ are regarded as four control inputs of the quadrotor aircraft system.

Let $B = [B_x, B_y, B_z]$ be the body-frame coordinates shown in Fig. 6.2, and let $E = [E_x, E_y, E_z]$ be the earth-frame coordinates. A quadrotor system has six state variables, expressed by three translational motion variables $\xi = [x, y, z]^\top$ and three rotational motion variables $\eta = [\phi, \theta, \psi]^\top$. By sequentially rotating aircrafts around the three axes in body coordinates, rotation matrix $\boldsymbol{R}_{BE}$ can be obtained to describe the linear velocity relationship between the body frame and the inertial frame [5]. The rotation matrix is given by (6.1).

The matrix $\boldsymbol{W}_{BE}$ can be obtained to describe the angular velocity relationship between the body frame and the inertial frame [17], which is defined as (6.2). Then, the dynamic model of the quadrotor can be expressed as

$$L(q,\dot{q}) = E_T + E_R - E_P, \tag{6.5}$$

$$\frac{\mathrm{d}}{\mathrm{d}t}\frac{\partial L}{\partial \dot{q}} - \frac{\partial L}{\partial q} = \begin{bmatrix} \zeta \\ \tau \end{bmatrix}, \tag{6.6}$$

where L is the Lagrangian of the model, $E_T = (m/2)\dot{\xi}^\top\dot{\xi}$ is the translational kinetic energy, $E_R = (1/2)\dot{\omega}^\top \boldsymbol{I}\dot{\omega}$ is the rotational kinetic energy, ω is the angular velocity resolved in the body frame, $\boldsymbol{I}$ is the inertial matrix, $E_P = [0,0,mgz]^\top$ is the potential energy, m is the mass of the aircraft, g is the gravitational acceleration, and $\zeta = \boldsymbol{R}_{BE}[0,0,T]^\top$ is the translational force due to thrust magnitude T.

According to (6.6), the aircraft model can be divided into two subsystems. For translational subsystem, the model equation can be written as

$$m\ddot{\xi} + mg[0,0,1]^\top = \zeta. \tag{6.7}$$

From (6.2), the angular velocity correlation is defined as

$$\dot{\eta} = W_{BE}\omega. \tag{6.8}$$

To represent the rotational energy E_R as a function of the generalized coordinate η, substituting (6.8) into the rotational energy equation, the following equation is obtained:

$$E_R = \frac{1}{2}\dot{\eta}^\top W_{\tilde{BE}}^\top I W_{\tilde{BE}}^{-1}\dot{\eta}. \tag{6.9}$$

The model equation of rotational subsystem can be obtained from (6.5), (6.6), and (6.9), and it is expressed as

$$\boldsymbol{M}(\boldsymbol{\eta})\ddot{\eta} + \boldsymbol{C}(\boldsymbol{\eta},\dot{\boldsymbol{\eta}})\dot{\boldsymbol{\eta}} = \tau, \tag{6.10}$$

where $\boldsymbol{M}(\boldsymbol{\eta}) = \boldsymbol{W}_{\bar{BE}}^{-\top}\boldsymbol{I}\boldsymbol{W}_{\bar{BE}}^{-1}$, and the matrix $\boldsymbol{M}(\boldsymbol{\eta})$ is written as

$$
\boldsymbol{M}(\boldsymbol{\eta}) = \begin{bmatrix} \boldsymbol{I}_{xx} & 0 & -\boldsymbol{I}_{xx}S\theta \\ 0 & \boldsymbol{I}_{yy}C^2\phi + \boldsymbol{I}_{zz}S^2\phi & (\boldsymbol{I}_{yy} - \boldsymbol{I}_{zz})C\phi S\phi C\theta \\ -\boldsymbol{I}_{xx}S\theta & (\boldsymbol{I}_{yy} - \boldsymbol{I}_{zz})C\phi S\phi C\theta & \boldsymbol{I}_{xx}S^2\theta + \boldsymbol{I}_{yy}S^2\phi C^2\theta + \boldsymbol{I}_{zz}C^2\phi C^2\theta \end{bmatrix}. \tag{6.11}
$$

The specific expression for the matrix $\boldsymbol{C}(\boldsymbol{\eta}, \dot{\boldsymbol{\eta}})$ is complex, and it has not been provided here. For detailed information, it can be referenced in [18].

6.3 Adaptive UAV Controller Under Disturbance and Input Saturation

6.3.1 Quadrotor Aircraft Control with Adaptive RBFNNs/Integral Sliding Mode

Kinematic and dynamic modeling can be referred to Sect. 6.2. Therefore, translational subsystem equation and rotational subsystem equation are given by (6.7) and (6.10), respectively. The objective of this section is to design the control input signals of T and τ to guarantee the trajectory tracking for nonlinear coupling and underactuated quadrotors while rejecting the effect of disturbances and parameter uncertainties. The overall control strategy is illustrated in Fig. 6.3.

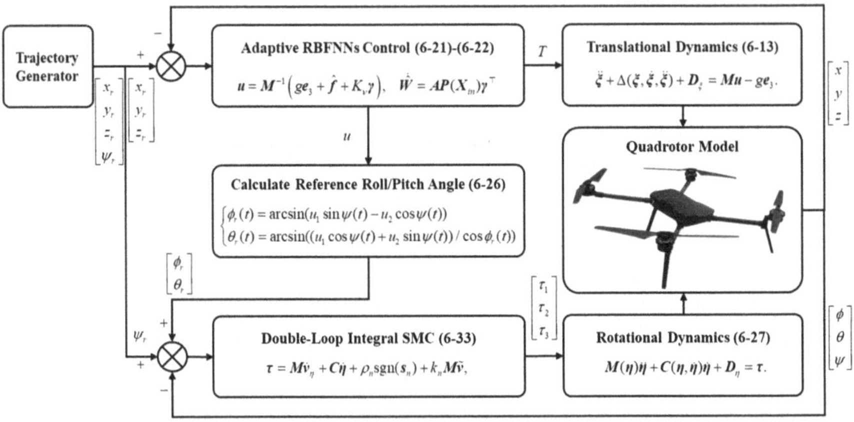

Fig. 6.3 Adaptive RBFNNs/integral sliding-mode control diagram for a quadrotor aircraft

6.3.1.1 Controller Design

The overall control objective for the UAV is to track a desired trajectory $\xi_r = [x_r, y_r, z_r]^\top$ and a desired yaw angle ψ_r. In translational subsystem, an adaptive radial basis function neural network (RBFNN) controller is designed to calculate T, ϕ_r, and θ_r, which enables the aircraft to track the desired trajectory ψ_r, and the neural network compensation helps eliminate the effects caused by parameter uncertainties and disturbances. Therefore, the desired angles $\eta_r = [\phi_r, \theta_r, \psi_r]^\top$ are derived. In rotational subsystem, an integral sliding-mode controller is used to obtain proper values of rotor moment $\tau = [\tau_1, \tau_2, \tau_3]^\top$ to make the aircraft track the desired attitude angles η_r, and the integral item helps eliminate state errors. Finally, stability analysis is proved by using the Lyapunov theory.

The translational dynamic equation is described in (6.7). To begin, define new control inputs u_1 and u_2, satisfying

$$\begin{cases} u_1(t) = \cos\psi(t)\sin\theta(t)\cos\phi(t) + \sin\psi(t)\sin\phi(t) \\ u_2(t) = \sin\psi(t)\sin\theta(t)\cos\phi(t) - \cos\psi(t)\sin\phi(t). \end{cases} \tag{6.12}$$

In view of (6.12), considering the parameter uncertainties and disturbances, the translational dynamics can be rewritten as follows:

$$\ddot{\xi} + \Delta\left(\xi, \dot{\xi}, \ddot{\xi}\right) + D_\xi = Mu - ge_3. \tag{6.13}$$

where $\boldsymbol{\Delta}\left(\xi, \dot{\xi}, \ddot{\xi}\right) \in \mathbb{R}^3$ is the vector of model uncertainties, $\boldsymbol{M} =$ diag $[1/m, 1/m, \cos\theta \cos\phi/m]$ is the inertial matrix, $D_\xi = [D_{\xi x}, D_{\xi y}, D_{\xi z}]^\top$ is the disturbance force like wind, $u = [u_1 T, u_2 T, T]^\top$ is the control input, and $\boldsymbol{e}_3 = [0, 0, 1]^\top$.

The displacement tracking error can be defined as

$$\tilde{\xi}(t) = \xi_r(t) - \xi(t). \tag{6.14}$$

Assumption 6.1 The real time ξ, the desired ξ_r and its first-order derivative $\dot{\xi}_r$, and its second-order derivative $\ddot{\xi}_r$ are all measurable.

The following auxiliary state error vector $\gamma(t)$ is introduced to facilitate the control formulation and stability analysis:

$$\gamma(t) = \dot{\tilde{\xi}}(t) + \Lambda\tilde{\xi}(t) \tag{6.15}$$

where $\boldsymbol{\Lambda} = \boldsymbol{\Lambda}^\top > 0$. Substituting the derivative of (6.14) into (6.15) reveals

$$\dot{\xi} = \dot{\xi}_r - \gamma + \Lambda\tilde{\xi}. \tag{6.16}$$

Substituting (6.13) and (6.16) into the derivative of (6.15) yields

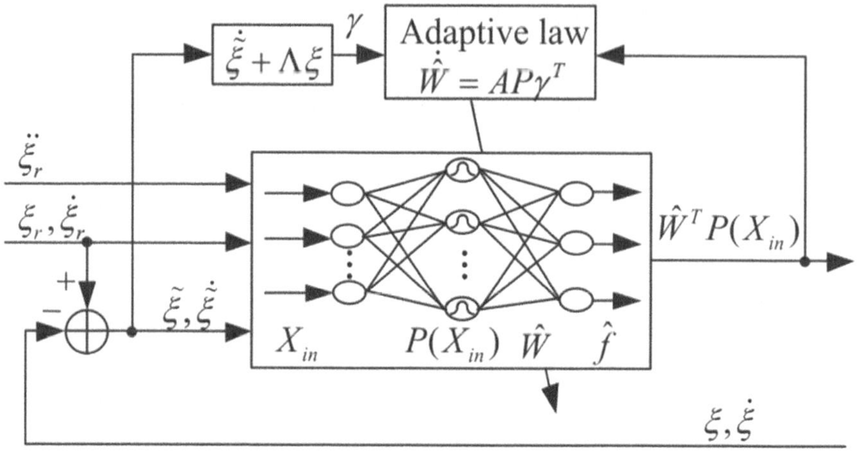

Fig. 6.4 Adaptive RBFNN control

$$\begin{aligned} \dot{\gamma} &= \ddot{\xi}_r - \xi + \Lambda\dot{\tilde{\xi}} \\ &= \ddot{\xi}_r + \Lambda\dot{\xi} - Mu + ge_3 + \Delta(\xi,\xi,\xi) + D_\xi \\ &= -Mu + ge_3 + f, \end{aligned} \tag{6.17}$$

where $f = \ddot{\xi}_r + \Lambda\dot{\tilde{\xi}} + \Delta(\xi,\dot{\xi},\ddot{\xi}) + D_\xi$ represents the whole dynamic model including nominal model and unknown parts.

The adaptive RBFNN scheme is proposed for the unknown MIMO nonlinear system, since RBFNNs can estimate the unknown continuous functions f and ensure that the tracking error ultimately converges to an adequately small compact set [19]. Illustrated in Fig. 6.4, the adaptive RBFNNs can be written as

$$\hat{\boldsymbol{f}}(\boldsymbol{X}_{in}) = \hat{\boldsymbol{W}}^{\top}\boldsymbol{P}(\boldsymbol{X}_{in}), \tag{6.18}$$

where $\hat{\boldsymbol{f}} \in \mathbb{R}^O$ is the output vector of the neural network, $\boldsymbol{X}_{in} \in \mathbb{R}^I$ is the input vector of the RBFNNs, $\boldsymbol{W} \in \mathbb{R}^{J \times O}$ is the weight vector, $J > 1$ is the node number of middle hidden layer, and $\boldsymbol{P}(\cdot) : R^I \rightarrow R^J$ is the basis function vector with p_i being the commonly used Gaussian functions, which is simplified as

$$p_i(\boldsymbol{X}_{in}) = \exp\left(-\frac{\|\boldsymbol{X}_{in} - c_i\|^2}{\sigma_i^2}\right), \tag{6.19}$$

where $\mathbf{c}_i$ is the center of the receptive field and σ_i is the width of the Gaussian function. There exists an optimal vector $\boldsymbol{W}^*$ such that [20]

$$f(X_{in}) = (W^*)^\top P(X_{in}) + \epsilon, \tag{6.20}$$

where $\epsilon_M > 0$ denotes the maximum value of the neural approximation error ϵ. In this case, the inputs for the neural networks are taken as $X_{in} = \left[\tilde{\xi}, \dot{\tilde{\xi}}, \xi_r, \dot{\xi}_r, \ddot{\xi}_r\right]$.

Theorem 6.1 Consider the quadrotor translational dynamics described in (6.13) or (6.17)–(6.20). If the control input and the adaptive weight update law are designed as

$$u = M^{-1}\left(ge_3 + \hat{f} + K_\nu \gamma\right) \tag{6.21}$$

$$\dot{\hat{W}} = AP(X_{in})\gamma^\top \tag{6.22}$$

where K_ν is a positive constant and A is a positive definite symmetric matrix of gains, then the speed tracking and position tracking can be guaranteed regardless of uncertain systematic parameters and disturbances.

Proof Consider the following Lyapunov function candidate:

$$V = \frac{1}{2}\gamma^\top \gamma + \frac{1}{2}\operatorname{tr}\left(\tilde{W}^\top A^{-1}\tilde{W}\right), \tag{6.23}$$

where $\tilde{W} = W^* - \hat{W}$ denotes the vector of weight errors and A^{-1} is the adaptive gain. Differentiating (6.23) using (6.17) produces

$$\dot{V} = \gamma^T(-Mu + ge_3 + f) + \operatorname{tr}\left(\tilde{W}^\top A^{-1}\dot{\tilde{W}}\right). \tag{6.24}$$

From the adaptive law (6.22), it can be deduced that $\dot{\tilde{W}} = -AP(X_{\text{in}})\gamma^\top$. In view of (6.21), the Lyapunov derivative (6.24) becomes

$$\begin{aligned}\dot{V} &= \gamma^\top\left(f - \hat{f} - K_V\gamma\right) + \operatorname{tr}\left(\tilde{W}^\top A^{-1}\dot{\tilde{W}}\right)\\ &= -\gamma^\top K_\nu\gamma + \operatorname{tr}\tilde{W}^\top\left(A^{-1}\dot{\tilde{W}} + P\gamma^\top\right) + \gamma^\top\epsilon\\ &= -\gamma^\top K_\nu\gamma + \gamma^\top\epsilon,\end{aligned} \tag{6.25}$$

where $\|\gamma\| > \|\epsilon\|/K_\nu$ ensures $\dot{V} < 0$. Thus, by choosing an appropriate value of K_ν, $\dot{V} < 0$ can be obtained. Summing up, V is positive definite and $\dot{V}$ is negative, so the tracking error will be uniformly ultimately bounded.

Furthermore, the control signal T is u_3, and from (6.12), two desired attitude angles can be obtained, which will be used as the desired trajectory in the rotational subsystem, that is

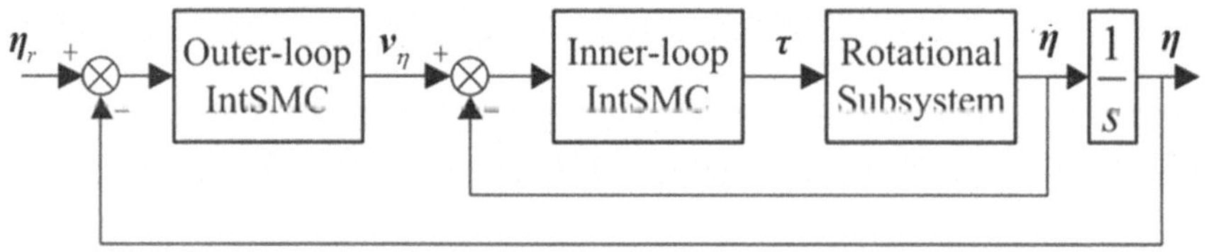

Fig. 6.5 Double-loop integral sliding-mode control

$$\begin{cases} \phi_r(t) = \arcsin(u_1 \sin\psi(t) - u_2 \cos\psi(t)) \\ \theta_r(t) = \arcsin((u_1 \cos\psi(t) + u_2 \sin\psi(t))/\cos\phi_r(t)) \end{cases} \tag{6.26}$$

In traditional sliding-mode control, it is hard to solve the derivatives of $\boldsymbol{M}(\boldsymbol{\eta})$ in (6.10), because there is strong coupling between ϕ, θ, and ψ. Thus, the double-loop integral sliding-mode control (SMC) is introduced to avoid the above problem, while the integral function helps to guarantee system's stability and zero errors under disturbances. The control diagram is given in Fig. 6.5.

Adding the disturbance item $D_\eta = [D_{\eta 1}, D_{\eta 2}, D_{\eta 3}]^\top$ into the rotational model (6.10), the rotational model equation can be rewritten as

$$\boldsymbol{M}(\boldsymbol{\eta})\ddot{\boldsymbol{\eta}} + \boldsymbol{C}(\boldsymbol{\eta}, \dot{\boldsymbol{\eta}})\dot{\boldsymbol{\eta}} + D_\eta = \tau. \tag{6.27}$$

For the control of attitude angles in the outer loop, the integral sliding-mode surface is defined as

$$s_w = \widetilde{\boldsymbol{\eta}} + k_w \int_0^t \widetilde{\boldsymbol{\eta}} \mathrm{d}t, \tag{6.28}$$

where k_w is a positive constant and $\widetilde{\boldsymbol{\eta}} = \eta_r - \eta$ is the error vector of the aircraft attitude. Differentiating (6.28) produces

$$\dot{s}_w = \dot{\eta}_r - \boldsymbol{\nu}_\eta + k_w \widetilde{\eta} \tag{6.29}$$

where $\boldsymbol{\nu}_\eta$ represents the control inputs of outer-loop SMC, which can be designed using the Lyapunov theory and can be achieved by inner-loop SMC.

For the control of attitude angles' velocity in the inner loop, the sliding-mode surface is defined as

$$s_n = \widetilde{\boldsymbol{\nu}} + k_n \int_0^t \widetilde{\boldsymbol{\nu}} \mathrm{d}t \tag{6.30}$$

where k_n is a positive constant and $\widetilde{\boldsymbol{\nu}} = \nu_\eta - \dot{\eta}$ is the error vector of angle velocity. According to (6.27), the differential form of (6.30) can be written as

$$\dot{s}_n = \dot{\nu}_\eta + \boldsymbol{M}^{-1}\boldsymbol{C}\dot{\eta} - \boldsymbol{M}^{-1}\tau - \boldsymbol{M}^{-1}D_\eta + k_n\widetilde{\boldsymbol{\nu}}. \tag{6.31}$$

To achieve $\boldsymbol{\nu}_\eta \to \dot{\eta}$, the control law τ is designed for tracking the angle velocity.

Theorem 6.2 In the outer-loop integral SMC controller, the control law of $\boldsymbol{\nu}_\eta$ is designed as

$$\boldsymbol{\nu}_\eta = \dot{\eta}_r + k_w\widetilde{\eta} + \rho_w \operatorname{sgn}(s_w), \tag{6.32}$$

where ρ_w is a positive constant, which determines the convergence rate of attitude angles. The control law of the inner loop is designed as

$$\tau = \boldsymbol{M}\dot{\boldsymbol{\nu}}_\eta + \boldsymbol{C}\dot{\eta} + \boldsymbol{\rho}_n \operatorname{sgn}(s_n) + k_n\boldsymbol{M}\widetilde{\boldsymbol{\nu}}, \tag{6.33}$$

where ρ_n is a positive constant. Designing these two laws with appropriate feedback gains ensures that the aircraft's attitude position and velocity achieve asymptotic tracking convergence, even in the presence of disturbances.

Proof Consider the following Lyapunov function:

$$V_w = \frac{1}{2}s_w^\top s_w \tag{6.34}$$

From (6.29), the differential form of (6.34) is

$$\dot{V}_W = s_w^\top\left(\dot{\eta}_r - \boldsymbol{\nu}_\eta + k_\eta\widetilde{\eta}\right) \tag{6.35}$$

Substituting the control law (6.32) into (6.35) results in

$$\begin{aligned}\dot{V}_w &= -\rho_w s_w^\top \operatorname{sgn}(s_w)\\ &= -\rho_w \sum_{i=1}^{3} |s_{wi}|\\ &\leq 0.\end{aligned} \tag{6.36}$$

Thus, attitude angle errors will converge to zero along the desired integral sliding-mode surface.

In the inner loop, the Lyapunov function is chosen as

$$V_n = \frac{1}{2}s_n^\top s_n. \tag{6.37}$$

According to (6.31) and the control law (6.32), the differential form of (6.37) is given as follows:

$$
\begin{aligned}
\dot{V}_n &= s_n^\top\left(\dot{\boldsymbol{\nu}}_\eta + \boldsymbol{M}^{-1}\boldsymbol{C}\dot{\eta} - \boldsymbol{M}^{-1}\tau - \boldsymbol{M}^{-1}D_\eta + k_n\widetilde{\boldsymbol{\nu}}\right)\\
&= s_n^\top\boldsymbol{M}^{-1}\left(-\rho_n\,\mathrm{sgn}(s_n) - D_\eta\right)\\
&\leq -\lambda_{\max}\left(\boldsymbol{M}^{-1}\right)|s_n|\left(\rho_n - \|D_\eta\|\right),
\end{aligned}
\tag{6.38}
$$

with $-\lambda_{\max}(\boldsymbol{M}^{-1})$ denoting the maximum eigenvalue of $\boldsymbol{M}^{-1}$. Thus, by choosing$\rho_n \geq \|D_\eta\|$, the Lyapunov equation $\dot{V}_n \leq 0$ can ensure that attitude angles' velocity errors will converge to zero along the above integral sliding-mode surface. In practical applications, to avoid the chattering phenomenon, sigmoid functions are introduced to replace sign functions, and they are 2/(1 + exp $(-x/500)$) − 1 and 2/(1 + exp $(-x/10)$) − 1 in outer loop and inner loop, respectively.

6.3.1.2 Simulation Results

In this section, simulation examples are provided to demonstrate the effectiveness of the proposed control strategy for path tracking problems. The parameters of the quadrotor used in the simulation are shown in Table 6.1, which were selected from tests [18]. In addition, the simulation compares the control structure developed in this section with traditional proportional-derivative (PD) strategies to show its improvement. The simulation considers external disturbances and uncertainties in the system model across six degrees of freedom (DOFs).

Example In this example, the reference path is set as a circle evolving in the Cartesian space defined by

$$
x_r = \frac{1}{2}\cos\left(\frac{\pi t}{20}\right)\mathrm{m},\quad y_r = \frac{1}{2}\sin\left(\frac{\pi t}{20}\right)\mathrm{m}
$$
$$
z_r = 3 - 2\cos\left(\frac{\pi t}{20}\right)\mathrm{m},\quad \psi_r = 0\mathrm{rad}
$$

Sustained disturbances and parameter uncertainties have been imposed on the quadrotor to indicate the ability of anti-disturbance and robustness. Also, the result is compared with PD-InitSMC control method. The initial conditions of the aircraft are set as $\boldsymbol{\xi}_0 = [0.45, 0.05, 0.45]^\top$ m and $\boldsymbol{\eta}_0 = [0.1, 0.1, 0.1]^\top$ rad. The parameters of the adaptive RBFNN controller are adjusted as $J = 7$, $\mathbf{c}_i = [-1.5, -1, -0.5, 0, 0.5, 1, 1.5]$, $\sigma = 5$, $\Lambda = 3$, $A = 100$, and $K_v = 15$. The sliding-mode controller gains are tuned with the following values: $k_w = 100$, $\rho_w = 5$, $k_n = 20$, and $\rho_n = 25$.

Table 6.1 Parameters of the quadrotor aircraft

Symbol	Value	Unit
m	0.74	kg
g	9.81	m/s^2
d	0.2	m
I_{xx}, I_{yy}	0.004	kg · m^2
I_{zz}	0.0084	kg · m^2

Simulation results of the path following and control inputs are shown in Figs. 6.6, 6.7, 6.8, 6.9, and 6.10, and the performance is noticeable in these figures especially at the time when force and moment disturbances occur. It can be seen that starting from an initial position far from the reference, the proposed control strategy is able to make the vehicle follow the reference trajectory in a small delay. Position tracing results in 3-D space and separate space are present in Figs. 6.6 and 6.7, respectively. Figure 6.8 shows the translational tracking errors, and it can be observed that translational tracking is well achieved, even if structural uncertainty is considered in the vehicle.

In contrast, simulation results using PD-InitSMC controller without adaptive RBFNN compensation is considered, and the results are also presented in Figs. 6.6, 6.7, and 6.8. It can be seen the quadrotor leaves the predetermined trajectory when a disturbance is introduced, and it never reaches the reference trajectory again. Besides, the curve is affected when parameter uncertainties of Λ changed.

Figure 6.9 shows that the inner integral sliding-mode controller makes the vehicle track its rotational reference trajectory when each degree of freedom is affected by the moment disturbances. The reference attitude angles of ϕ_r and θ_r generated by the translational controller vary with time, but the integral sliding-mode controller achieves null steady-state error. Finally, four control inputs are presented in Fig. 6.10, including three torque inputs and one force input. Due to the replacement of the sign function by the sigmoid function, the control inputs avoid the chattering problem so that the obtained input control signals are acceptable and physically

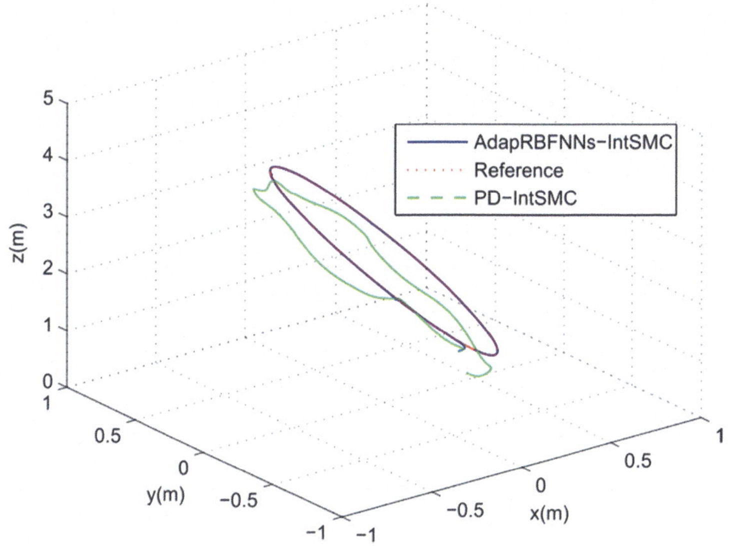

Fig. 6.6 Path following

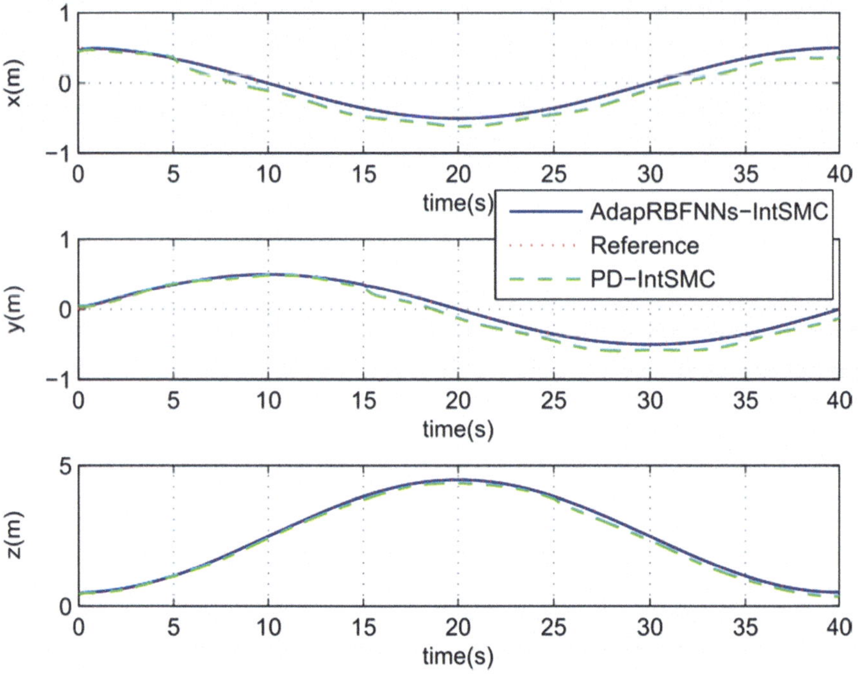

Fig. 6.7 Position tracing

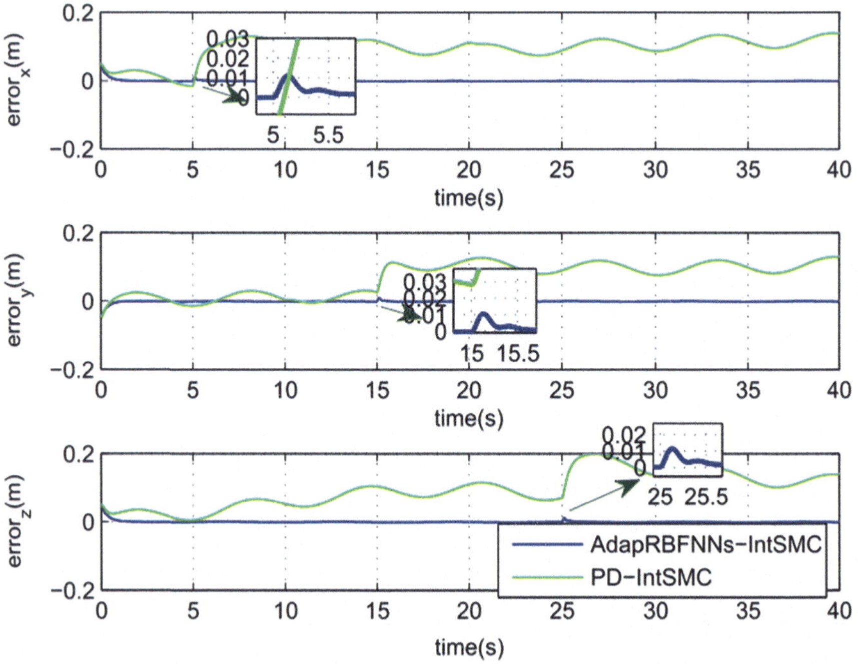

Fig. 6.8 Position error

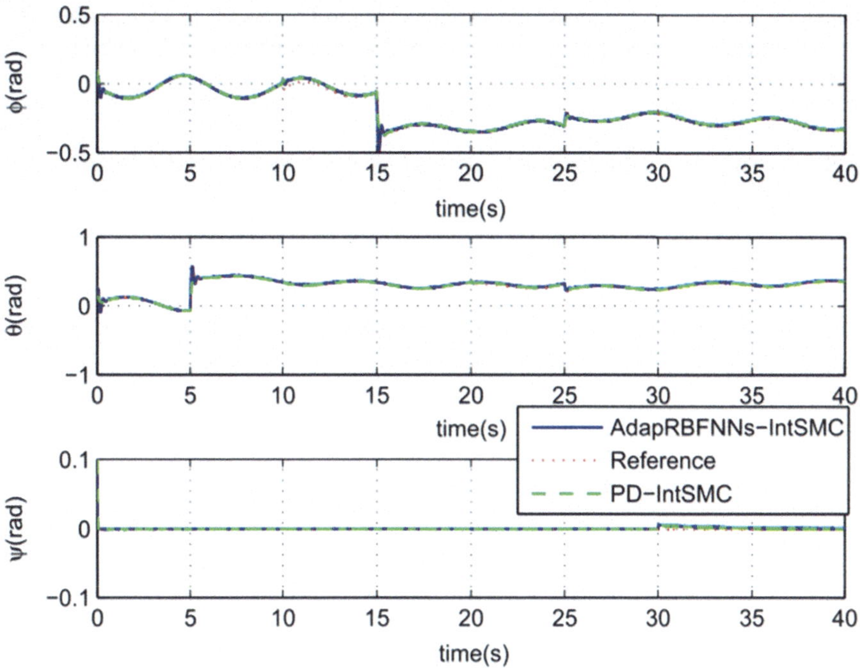

Fig. 6.9 Orientation tracing

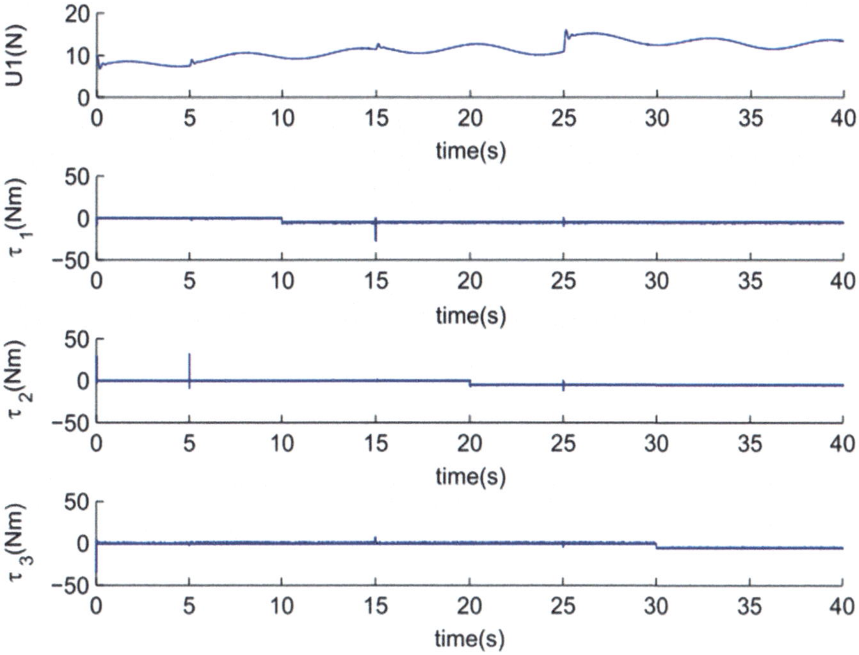

Fig. 6.10 Control inputs

realizable. From these figures, the designed controller succeeds in reaching the desired trajectory, while disturbances and model uncertainty are well compensated.

6.3.2 *Adaptive and Robust Control of Quadrotors with Input Saturation*

6.3.2.1 Problem Formulation

Based on the simplified rotor model in Fig. 6.2, the rotational speed vector $\mathbf{\Omega} \in \mathbb{R}^4$ can be utilized to obtain mapped control inputs u_T, u_ϕ, u_θ, and u_ψ, defined as [16]

$$\begin{cases} u_T = b \sum_{i=1}^{4} \Omega_i^2 \\ u_\phi = lb\left(\Omega_1^2 - \Omega_3^2\right) \\ u_\theta = lb\left(\Omega_2^2 - \Omega_4^2\right) \\ u_\psi = \kappa\left(-\Omega_1^2 + \Omega_2^2 - \Omega_3^2 + \Omega_4^2\right), \end{cases} \tag{6.39}$$

where $u_T \in \mathbb{R}$ is the thrust magnitude; u_ϕ, u_θ, and u_ψ represent roll, pitch, and yaw moments, respectively; l denotes the distance from the mass center to each rotor; and $\kappa > 0$ and $b > 0$ are the thrust and drag coefficients, respectively. Finally, the thrust magnitude u_T and three rotational moments u_ϕ, u_θ, and u_ψ are considered as the real control inputs to the dynamical system.

Let $B = [B_x, B_y, B_z]$ be the body-fixed frame and $E = [E_x, E_y, E_z]$ be the earth-fixed inertial frame. Define the vector $[\boldsymbol{p}, \boldsymbol{q}, \boldsymbol{r}]^\top$ to represent the quadrotor's angular velocity in the body frame. A quadrotor system has six state variables, which are three translational motion variables $\xi = [x, y, z]^\top$ and three rotational motion variables $\eta = [\phi, \theta, \psi]^\top$. The orientation of the quadrotor from the inertial frame to the body-fixed frame can be obtained via three successive rotations about the three axes, and the transformation matrix from $[\boldsymbol{p}, \boldsymbol{q}, \boldsymbol{r}]^\top$ to $[\boldsymbol{\phi}, \boldsymbol{\theta}, \boldsymbol{\psi}]^\top$ is given by [21]

$$\begin{bmatrix} \dot{\boldsymbol{\phi}} \\ \dot{\boldsymbol{\theta}} \\ \dot{\boldsymbol{\psi}} \end{bmatrix} = \begin{bmatrix} 1 & \sin\phi\tan\theta & \cos\phi tan\theta \\ 0 & \cos\phi & -\sin\phi \\ 0 & \sin\phi\sec\theta & \cos\phi\sec\theta \end{bmatrix} \begin{bmatrix} \boldsymbol{p} \\ \boldsymbol{q} \\ \boldsymbol{r} \end{bmatrix} \tag{6.40}$$

In view of the above discussion and [22], the dynamical model of the quadrotor can be described by the following equations:

$$
\begin{cases}
m\ddot{x} = (\cos\phi\sin\theta\cos\psi + \sin\phi\sin\psi)u_T - k_X\dot{x} - A_X \\
m\ddot{y} = (\cos\phi\sin\theta\sin\psi - \sin\phi\cos\psi)u_T - k_y\dot{y} - A_y \\
m\ddot{z} = (\cos\phi\cos\theta)u_T - mg - k_z\dot{z} - A_z \\
I_\phi\ddot{\phi} = \dot{\theta}\dot{\psi}(I_\theta - I_\psi) + lu_\phi - k_\phi l\dot{\phi} - A_\phi \\
I_\theta\ddot{\theta} = \dot{\psi}\dot{\phi}(I_\psi - I_\phi) + lu_\theta - k_\theta l\dot{\theta} - A_\theta \\
I_\psi\ddot{\psi} = \dot{\phi}\dot{\theta}(I_\phi - I_\theta) + u_\psi - k_\psi\dot{\psi} - A_\psi
\end{cases}
\tag{6.41}
$$

where k_i is the drag coefficient, which is a positive constant; I_ϕ, I_θ, and I_ψ represent the inertias of the aircrafts, respectively; m stands for the mass; g is the gravitational acceleration; and A_i is the disturbance force or disturbance moment. This model is presented and is tested experimentally in [23].

Remark 6.1 It should be noted that the model (6.41) considers the parametric uncertainties of the quadrotor aircraft, and external disturbances on the six degrees of freedom $A_i(i = x, y, z, \phi, \theta, \psi)$ when sustaining winds disturb the whole system. In practice, it is extremely difficult to accurately obtain external disturbances such as wind gusts for aircraft. Therefore, the sustained disturbance signal model for quadrotor aircrafts, presented in [18], is adopted. The following adaptive control will be designed to compensate disturbance influence by adaptively identifying this unknown sustained disturbance vector where its derivative is regarded as zero. Some recent literature also handles the quadrotor aircraft disturbance in the same way, where persistent light gusts are considered as external disturbances on the aerodynamic forces and moments [24].The overall control objective is to design a nonlinear adaptive controller for the quadrotor such that the UAVs can track a desired trajectory $\xi_r = [x_r, y_r, z_r]^\top$ and a desired yaw angle ψ_r. In translational subsystem, an adaptive controller is designed to obtain u_T, ϕ_r, and θ_r, which enables the aircraft to track the desired trajectory ξ_r. The adaptive law is employed to compensate the uncertain system parameters as well as eliminate the effects caused by wind-gust disturbances. Based on the derived desired angles η_r, another adaptive controller is utilized in rotational subsystem to obtain proper rotor moment values $[u_\phi, u_\theta, u_\psi]^\top$ to keep the aircraft tracking the desired attitude angles η_r. Finally, stability analysis is proved by using the Lyapunov theory. To streamline the technical proof of the main results, the following lemma is needed, where Lemma 6.1 is brought from [25].

Lemma 6.1 For any uniformly continuous function $f(t)$, if $\lim_{t\to\infty}\int_0^t f(\tau)\mathrm{d}\tau$ exists and is finite, then $\lim_{t\to\infty} f(t) = 0$.

6.3.2.2 Adaptive Control Without Input Saturation

In this section, an attitude controller and a position controller are designed based on the adaptive control approach, respectively. Then, a novel adaptive and robust controller is presented with an emendatory compensation, which is introduced to attenuate the saturation influence involved in translational dynamics and rotational dynamics.

The translational dynamic equations of quadrotors can be obtained from (6.41). To begin, define new control inputs u_x, u_y, and u_z, satisfying

$$\begin{cases} u_X(t) = (\cos\phi(t)\sin\theta(t)\cos\psi(t) + \sin\phi(t)\sin\psi(t))u_T \\ u_y(t) = (\cos\phi(t)\sin\theta(t)\sin\psi(t) - \sin\phi(t)\cos\psi(t))u_T \\ u_z(t) = (\cos\phi(t)\cos\theta(t))u_T \end{cases} \tag{6.42}$$

The translational and rotational position and velocity tracking errors can be defined as follows:

$$e_{\xi i} = \xi_i - \xi_{ir}, \quad \dot{e}_{\xi i} = \dot{\xi}_i - \dot{\xi}_{ir}, \quad i = x, y, z \tag{6.43}$$

$$e_{\eta i} = \eta_i - \eta_{ir}, \quad \dot{e}_{\eta i} = \dot{\eta}_i - \dot{\eta}_{ir}, \quad i = \phi, \theta, \psi \tag{6.44}$$

where ξ_{ir} and $\dot{\xi}_{ir}$ are the desired translational trajectory and the corresponding velocity, respectively. The following auxiliary state error $\gamma_{\xi i}$ is introduced to facilitate the control formulation and stability analysis:

$$\gamma_{\xi i} = \dot{e}_{\xi i} + \alpha_i e_{\xi i}, \quad i = x, y, z \tag{6.45}$$

$$\gamma_{\eta i} = \dot{e}_{\eta i} + \alpha_i e_{\eta i}, \quad i = \phi, \theta, \psi \tag{6.46}$$

where $\alpha_i > 0$ is an arbitrarily chosen constant to weigh the two terms.

For the translational subsystem, differentiating (6.43) with respect to time, one obtains

$$\dot{\gamma}_{\xi i} = \ddot{e}_{\xi i} + \alpha_i \dot{e}_{\xi i} = \ddot{\xi}_i - \left(\ddot{\xi}_{ir} - \alpha_i \dot{e}_{\xi i}\right). \tag{6.47}$$

Using (6.41) and (6.47), the height model in z direction can be considered as

$$m\dot{\gamma}_{\xi z} = u_z - mg - k_z\dot{z} - A_z - m\left(\ddot{\xi}_{zr} - \alpha_z \dot{e}_{\xi z}\right). \tag{6.48}$$

For technical simplicity, (6.48) can be written in the following compact form:

$$m\dot{\gamma}_{\xi z} = u_z - \boldsymbol{H}_z \boldsymbol{\rho}_z, \tag{6.49}$$

with the known regression matrix $\boldsymbol{H}_z \in \mathbb{R}^{1 \times 3}$ given by

$$\boldsymbol{H}_z = \left[\dot{z}, 1, \ddot{\xi}_{zr} - \alpha_z \dot{e}_{\xi z} + g\right], \tag{6.50}$$

and the unknown parameter vector $\boldsymbol{\rho}_z \in \mathbb{R}^{3 \times 1}$ given by

$$\boldsymbol{\rho}_z = [k_z, A_z, m]^\top. \tag{6.51}$$

Similarly, according to (6.41) and (6.47), the regression matrix and parameter vector for x and y channels are defined as

$$\boldsymbol{H}_i = \left[\dot{\xi}_i\ , 1, \ddot{\xi}_{ir}\ - \alpha_i \dot{e}_{\xi i}\right], \quad i = x, y \tag{6.52}$$

$$\boldsymbol{\rho}_i = [k_i, A_i, m]^\top, \quad i = x, y \tag{6.53}$$

Therefore, the translational subsystem model with error signals can be described as

$$m\dot{\gamma}_{\xi i} = u_i - \boldsymbol{H}_i \boldsymbol{\rho}_i, \quad i = x, y, z \tag{6.54}$$

As for the rotational subsystem, substituting the derivative of (6.46) into (6.41) leads to

$$I_i \dot{\gamma}_{\eta i} = f_i(\dot{\phi}, \dot{\theta}, \dot{\psi})\widetilde{I}_i + l_i u_i - k_i l_i \dot{\eta}_i - A_i - I_i(\ddot{\eta}_{ir} - \alpha_i \dot{e}_{\eta i}), \quad i = \phi, \theta, \psi \tag{6.55}$$

where $f_\phi = \dot{\theta}\dot{\psi}$, $f_\theta = \dot{\psi}\dot{\phi}$, $f_\psi = \dot{\phi}\dot{\theta}$, $\widetilde{I}_\phi = I_\theta - I_\psi$, $\widetilde{I}_\theta = I_\psi - I_\phi$, $\widetilde{I}_\psi = I_\phi - I_\theta$, $l_\phi = l$, $l_\theta = l$, and $l_\psi = 1$.

For technical simplicity, (6.55) can also be written in the following compact form:

$$I_i \dot{\gamma}_{\eta i} = l_i u_i - \boldsymbol{H}_i \boldsymbol{\rho}_i, \quad i = \phi, \theta, \psi \tag{6.56}$$

where the regression matrix and corresponding unknown parameter vector are defined as

$$\boldsymbol{H}_i = \left[-f_i(\dot{\phi}, \dot{\theta}, \dot{\psi}), l_i \dot{\eta}_i, 1, \ddot{\eta}_{ir} - \alpha_i \dot{e}_{\eta i}\right], \quad i = \phi, \theta, \psi \tag{6.57}$$

$$\boldsymbol{\rho}_i = \left[\widetilde{I}_i, k_i, A_i, I_i\right]^\top, \quad i = \phi, \theta, \psi \tag{6.58}$$

Theorem 6.3 Consider the position subsystem in (6.54) and rotation subsystem in (6.56). If the adaptive control laws are defined by

$$u_i = \boldsymbol{H}_i \widehat{\boldsymbol{\rho}}_i - \mu_i \gamma_{\xi i}, \quad \dot{\widehat{\boldsymbol{\rho}}}_i = -\Gamma_i \boldsymbol{H}_i^\top \gamma_{\xi i}, \quad i = x, y, z \tag{6.59}$$

$$u_i = \left(\boldsymbol{H}_i \widehat{\boldsymbol{\rho}}_i - \mu_i \gamma_{\eta i}\right)/l_i, \quad \dot{\widehat{\boldsymbol{\rho}}}_i = -\Gamma_i \boldsymbol{H}_i^\top \gamma_{\eta i}, \quad i = \phi, \theta, \psi \tag{6.60}$$

where $\widehat{\boldsymbol{\rho}}_i$ represents the estimated value of the unknown parameter vector, then the position tracking error, attitude tracking error, and their speed tracking error will decay to zero regardless of unknown system dynamics, and all closed-loop signals are bounded.

Proof At first, consider the following Lyapunov candidate for position subsystem:

$$V = \frac{1}{2}\gamma_{\xi i} m \gamma_{\xi i} + \frac{1}{2}\widetilde{\rho}_i^{\top}\Gamma_i^{-1}\widetilde{\rho}_i, \quad i = x, y, z \tag{6.61}$$

where $\widetilde{\boldsymbol{\rho}}_i = \boldsymbol{\rho}_i - \widehat{\boldsymbol{\rho}}_i$ denotes the estimated error. As stated in Remark 6.1, the sustained disturbance is an unknown constant vector that needs to be estimated in an adaption process. In fact, even if it varies slightly with time, the disturbance term A_i can still be regarded as a constant vector in a short time. Then, the tracking error bound regarding disturbance can be made arbitrarily small by selecting appropriate adaptive gains.

Using this notation, the adaptation law in (6.59) yields

$$\dot{\widetilde{\boldsymbol{\rho}}}_i = \Gamma_i \boldsymbol{H}_i^{\top}\gamma_{\xi i}, \quad i = x, y, z \tag{6.62}$$

Then, substituting (6.59) and (6.62) into (6.44), one obtains

$$m\dot{\gamma}_{\xi i} = -\boldsymbol{H}_i\widetilde{\rho}_i - \mu_i\gamma_{\xi i}, \quad i = x, y, z \tag{6.63}$$

Differentiating (6.61) by using (6.62) and (6.63), the derivative of (6.61) is then obtained by

$$\begin{aligned}\dot{V} &= \gamma_{\xi i}\left(-\boldsymbol{H}_i\widetilde{\rho}_i - \mu_i\gamma_{\xi i}\right) + \widetilde{\boldsymbol{\rho}}_i^{\top}\Gamma_i^{-1}\Gamma_i\boldsymbol{H}_i^{\top}\gamma_{\xi i} \\ &= -\mu_i\gamma_{\xi i}^2, \quad i = x, y, z\end{aligned} \tag{6.64}$$

Similarly, the Lyapunov function adopted for rotation subsystem is given as follows:

$$V = \frac{1}{2}\gamma_{\eta i} I_i \gamma_{\eta i} + \frac{1}{2}\widetilde{\rho}_i^{\top}\Gamma_i^{-1}\widetilde{\boldsymbol{\rho}}_i, \quad i = \phi, \theta, \psi \tag{6.65}$$

Considering $\widetilde{\rho}_i = \boldsymbol{\rho}_i - \widehat{\boldsymbol{\rho}}_i$, (6.56), and (6.60), the differential form of (6.56) satisfies

$$\dot{V} = -\mu_i\gamma_{\eta i}^2, \quad i = \phi, \theta, \psi \tag{6.66}$$

Since $\dot{V}$ is semi-negative definite, V has an upper bound $V(0)$, which guarantees the boundedness of $\gamma_{\xi i}$, $\gamma_{\eta i}$, $\widetilde{\rho}_i$, and u_i. According to (6.56) and (6.66), one can obtain that e, $\dot{e}$, ξ_i, $\dot{\xi}_i$, η_i, and $\dot{\eta}_i$ are all bounded based on standard linear control arguments. Then, from (6.63), one can get that $\dot{\gamma}$ is bounded, which further ensures $\ddot{V}$ obtained by differentiating (6.64) and (6.66) to be bounded. Thus, $\dot{V}$ is uniformly continuous. Overall, V is positive definite and bounded, and $\dot{V}$ is semi-negative definite and uniformly continuous, so Lemma 6.1 implies that

$$\lim_{t \to \infty} \dot{V}(t) = 0. \tag{6.67}$$

In view of (6.64), (6.66), and (6.67), the limitation of γ is

$$\lim_{t \to \infty} \gamma(t) = 0. \tag{6.68}$$

Recall (6.56) and (6.66), suppose that the initial error value is $e(0) = e_0$, and then the error can be solved as follows:

$$e(t) = e(0)e^{-\alpha t} + \frac{\gamma(t)}{\alpha}(1 - e^{-\alpha t}), \tag{6.69}$$

which further implies that $e(t)$ and $\dot{e}(t)$ will converge to zero as $t \to \infty$ based on (6.68) and (6.69). Therefore, the position tracking error, attitude tracking error, and their velocity errors of the quadrotor aircrafts will decay to zero even in the presence of the external disturbances and unknown dynamic parameters.

Remark 6.2 If the adaptive parameter Γ_i is sufficiently large, one can conclude that $\widehat{\rho}_i$ will approximate the unknown parameter vector ρ_i, and thereby, the effects caused by the external disturbances and unknown dynamic parameters can be attenuated.

6.3.2.3 Adaptive Robust Control with Input Saturation (Fig. 6.11)

The control functions u_z, u_ϕ, u_θ, and u_ψ cannot be put into use due to the rotational speed limitations in (6.39). In fact, four control inputs are influenced by the following saturation function:

$$u_i = \text{sat}(u_{iL}, \overline{u}_i, u_{iU}) = \begin{cases} u_{iL} & \overline{u}_i \quad\quad\quad < u_{iL} \\ \overline{u}_i & u_{iL} < \overline{u}_i \quad < u_{iU} \\ u_{iU} & \overline{u}_i \quad\quad\quad > u_{iU} \end{cases} \tag{6.70}$$

where u_{iL} and u_{iU} $(i = z, \phi, \theta, \psi)$ are the lower and upper limit bounds of the four control inputs, respectively.

Once the saturation occurs, the tracking errors of e and $\dot{e}$ will increase such that the filtered error signal γ will be increased, which also leads to an oscillation in the adaptive laws. Thus, the control performance under control laws (6.59) and (6.60) will be ruined. The basic idea underlying the anti-windup designs with input saturating actuators is to introduce modifications and compensations in order to reduce the influence of saturation nonlinearity.

In translational subsystem with actuator saturation $u_{zL} < u_z < u_{zU}$, define a new auxiliary error $\overline{\gamma}_{\xi i} = \gamma_{\xi i} - \chi_{\xi i}$, which satisfies

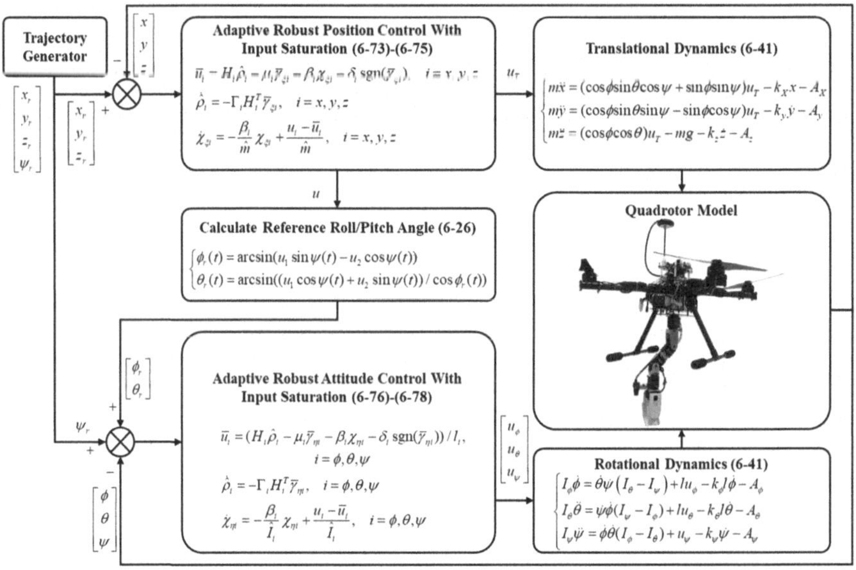

Fig. 6.11 Adaptive robust control of quadrotors with input saturation

$$\begin{aligned} m\dot{\bar{\gamma}}_{\xi i} &= m\dot{\gamma}_{\xi i} - m\dot{\chi}_{\xi i} \\ &= u_i - \boldsymbol{H}_i\boldsymbol{\rho}_i - m\dot{\chi}_{\xi i}, \quad i = x, y, z \end{aligned} \tag{6.71}$$

with the regression matrix $\boldsymbol{H}_i$ and the unknown parameter vector $\boldsymbol{\rho}_i$. Then, the newly introduced signal of $\chi_{\xi i}$ can change adaptively to attenuate the influence from actuator saturation, and its adaptive law will be discussed later.

When the saturation is considered, the adaptive control law given by (6.59) can be modified, and the new control law is designed in the following form:

$$u_i = \text{sat}(u_{iL}, \overline{u}_i, u_{iU}), \quad i = x, y, z \tag{6.72}$$

$$\overline{u}_i = \boldsymbol{H}_i\widehat{\boldsymbol{\rho}}_i - \mu_i\overline{\gamma}_{\xi i} - \beta_i\chi_{\xi i} - \delta_i\,\text{sgn}(\overline{\gamma}_{\xi i}), \quad i = x, y, z \tag{6.73}$$

with the unknown parameter vector and auxiliary error being updated as follows:

$$\dot{\widehat{\boldsymbol{\rho}}}_i = -\Gamma_i\boldsymbol{H}_i^{\mathrm{T}}\overline{\gamma}_{\xi i}, \quad i = x, y, z \tag{6.74}$$

$$\dot{\chi}_{\xi i} = -\frac{\beta_i}{\widehat{m}}\chi_{\xi i} + \frac{u_i - \overline{u}_i}{\widehat{m}}, \quad i = x, y, z \tag{6.75}$$

where the function $\text{sat}(u_{iL}, \overline{u}_i, u_{iU})$ is defined in (6-70); u_i and β_i represent freely chosen positive constants, respectively; and the robust signal term $\delta_i\,\text{sgn}(\overline{\gamma}_{\xi i})$ will be discussed later.

Consider that the precise physical parameters of quadrotor UAVs cannot be obtained in practical environment, for example, the mass m and inertia moments I_ϕ, I_θ, and I_ψ. Based on the invariance of these parameters and bounds of the inputs, the following definitions are made:

Definition 6.1 The difference proportion ϵ_m of aircraft mass is defined as $m/\widehat{m} = 1 + \epsilon_m$, and the known bound value ϵ_m^* guarantees $\epsilon_m^* > |\epsilon_m| = |m/\widehat{m} - 1|$.

According to (6.75), there exists bound u_i^* such that $u_i^* > |\beta_i \chi_{\xi i} - u_i + \overline{u}_i|$, $i = x, y, z$. $\dot{\beta}_i \chi_{\xi i}$ varies with the change of $u_i - \overline{u}_i$ until the adaptation error becomes zero; thus, the upper bound of their subtraction will exist as u_i^*.

Following a similar procedure, the robust adaptive control function for attitude motion of the vehicle body under torque saturations of (6.70) can be designed as follows:

$$\overline{u}_i = \left(\boldsymbol{H}_i \widehat{\boldsymbol{\rho}}_i - \mu_i \overline{\gamma}_{\eta i} - \beta_i \chi_{\eta i} - \delta_i \operatorname{sgn}\left(\overline{\gamma}_{\eta i}\right)\right)/l_i, \quad i = \phi, \theta, \psi \tag{6.76}$$

with the unknown parameter vector and auxiliary error being updated as follows:

$$\dot{\widehat{\boldsymbol{\rho}}}_i = -\Gamma_i \boldsymbol{H}_i^\top \overline{\gamma}_{\eta i}, \quad i = \phi, \theta, \psi \tag{6.77}$$

$$\dot{\chi}_{\eta i} = -\frac{\beta_i}{\widehat{I}_i} \chi_{\eta i} + \frac{u_i - \overline{u}_i}{\widehat{I}_i}, \quad i = \phi, \theta, \psi \tag{6.78}$$

To cope with the inertial parameter uncertainty in aircraft body and the input torque saturation, some definitions are expressed as follows:

Definition 6.2 The difference proportion of inertial values in three axes can be written as $\epsilon_i (i = \phi, \theta, \psi)$, which is further defined as $I_i/\widehat{I}_i = 1 + \epsilon_i$, and the known bound value ϵ_i^* guarantees $\epsilon_i^* > |\epsilon_i| = \left|I_i/\widehat{I}_i - 1\right|$. According to (6.78), there exists the bound u_i^* such that $u_i^* > \left|\beta \chi_{\eta i} - u_i + \overline{u}_i\right|$.

The following theorem will provide our main results on the adaptive robust control for quadrotor aircraft movements with input saturation:

Theorem 4.4 Consider the position subsystem with control signals in (6.54) and rotation subsystem in (6.56) both in the presence of finite-time saturation (6.70) under Definitions 6.1 and 6.2. If the modified control inputs are designed as (6.72)–(6.75) and (6.76)–(6.78) with the robust term satisfying $\delta_i \geq \epsilon_i^* u_i^*$, then the stability of the position tracking and attitude tracking can be guaranteed even in the presence of input saturation, and the control laws are robust to the parameter uncertainties including m and I. Also, all the signals in the closed-loop systems are bounded.

Proof For position subsystem, select the quadratic Lyapunov function candidate as

$$V=\frac{1}{2}\overline{\gamma}_{\xi i}m\overline{\gamma}_{\xi i}+\frac{1}{2}\widetilde{\rho}_i^{\mathrm{T}}\Gamma_i^{-1}\widetilde{\rho}_i,\quad i=x,y,z \tag{6.79}$$

Considering (6.71) and (6.74), the derivative of (6.79) can be rewritten as

$$\begin{aligned}\dot{V}&=\overline{\gamma}_{\xi i}m\dot{\overline{\gamma}}_{\xi i}+\widetilde{\boldsymbol{\rho}}_i^{\mathrm{T}}\Gamma_i^{-1}\dot{\widetilde{\boldsymbol{\rho}}}_i\\&=\overline{\gamma}_{\xi i}\left(u_i-\boldsymbol{H}_i\boldsymbol{\rho}_i-m\dot{\chi}_{\xi i}\right)+\widetilde{\boldsymbol{\rho}}_i^{\mathrm{T}}\Gamma_i^{-1}\Gamma_i\boldsymbol{H}_i^{\mathrm{T}}\overline{\gamma}_{\xi i}\\&=\overline{\gamma}_{\xi i}\left(u_i-\boldsymbol{H}_i\widehat{\boldsymbol{\rho}}_i-m\dot{\chi}_{\xi i}\right).\end{aligned} \tag{6.80}$$

Arranging (6.73) and (6.75) together obtains the further derivative of (6.80) as follows:

$$\begin{aligned}\dot{V}&=\overline{\gamma}_{\xi i}\left(u_i-\boldsymbol{H}_i\widehat{\boldsymbol{\rho}}_i+\frac{m}{\widehat{m}}\left(\beta_i\chi_{\xi i}-(u_i-\overline{u}_i)\right)\right)\\&=\overline{\gamma}_{\xi i}\left(u_i-\boldsymbol{H}_i\widehat{\boldsymbol{\rho}}_i+(1+\epsilon_m)\left(\beta_i\chi_{\xi i}-(u_i-\overline{u}_i)\right)\right)\\&=\overline{\gamma}_{\xi i}\left(\overline{u}_i-\boldsymbol{H}_i\widehat{\boldsymbol{\rho}}_i+\beta_i\chi_{\xi i}+\epsilon_m\left(\beta_i\chi_{\xi i}-u_i+\overline{u}_i\right)\right)\\&=-\mu_i\overline{\gamma}_{\xi i}^2-\overline{\gamma}_{\xi i}\delta_i\operatorname{sgn}\left(\overline{\gamma}_{\xi i}\right)+\overline{\gamma}_{\xi i}\epsilon_m\left(\beta_i\chi_{\xi i}-u_i+\overline{u}_i\right)\\&\le-\mu_i\overline{\gamma}_{\xi i}^2-\mid\overline{\gamma}_{\xi i}\mid\delta_i+\mid\overline{\gamma}_{\xi i}\mid\epsilon_m^*u_i^*\\&\le-\mu_i\overline{\gamma}_{\xi i}^2\le 0,\quad i=x,y,z\end{aligned} \tag{6.81}$$

As for the rotational subsystem, define the following Lyapunov function:

$$V=\frac{1}{2}\overline{\gamma}_{\eta i}I_i\overline{\gamma}_{\eta i}+\frac{1}{2}\widetilde{\boldsymbol{\rho}}_i^{\mathrm{T}}\Gamma_i^{-1}\widetilde{\boldsymbol{\rho}}_i,\quad i=\phi,\theta,\psi \tag{6.82}$$

Substituting (6.76)–(6.78) to the derivative of (6.82), the similar result can be obtained as follows:

$$\dot{V}\le-\mu_i\overline{\gamma}_{\eta i}^2-\mid\overline{\gamma}_{\eta i}\mid\delta_i+\mid\overline{\gamma}_{\eta i}\mid\epsilon_i^*u_i^*\le 0,\quad i=\phi,\theta,\psi \tag{6.83}$$

Therefore, the proof of this theorem can be completed by using the same procedures in the proof of Theorem 6.3.

Remark 6.3 When the saturation occurs, the overall closed-loop quadrotor systems are stable in the sense of Lyapunov due to the modified adaptive robust controller. The signal χ helps attenuate the windup influence caused by input saturation. Specifically, signal value χ will increase along with error γ increasing, which in turn guarantees that there is no sudden rise in the modified tracking error $\overline{\gamma}$ and all the signals are bounded despite the parametric uncertainties.

6.3.2.4 Simulation Results

Several simulation examples are provided to demonstrate the effectiveness of the proposed control strategy for path tracking problems. The parameters of the quadrotor used in the simulation are shown in Table 6.2 and were selected from the experiment [18]. All simulations consider external disturbances and uncertainties in system parameters across six degrees of freedom, demonstrating the robustness of the control method.

The desired reference path used is a circle evolving in the Cartesian space, where the yaw angle is required to be stabilized at zero. The circle is defined by $x_r = 0.5\cos(\pi t/20)$ m, $y_r = 0.5\sin(\pi t/20)$ m, $z_r = 3 - 2\cos(\pi t/20)$ m, and $\psi_r = 0$.

Suppose that the initial state values of the aircraft are $\xi_0 = [0.45, 0.05, 0.45]^\top$ m and $\eta_0 = [-0.02, 0.05, 0]^\top$ rad, and initial velocities are assumed as zeros. Here, in this simulation, it is assumed that the external disturbances on the aerodynamic forces and moments are given as $A_x = 1$ N at $t = 5$ s, $A_y = 1$ N at $t = 15$ s, $A_z = 1$ N at $t = 25$ s, $A_\phi = 1$ Nm at $t = 10$ s, $A_\theta = 1$ Nm at $t = 20$ s, and $A_\psi = 1$ Nm at $t = 30$ s.

In order to evaluate the flight performance with respect to actuator saturation of quadrotor aircrafts, the input limitations are taken into account. For example, in the z direction, the control input ranges from 0 to 21 N, while torque inputs u_2 and u_3 range from -10 to 10 Nm. Based on the above assumed magnitude, the anti-windup performance can be validated in the following experiments. Alternatively, the sign function in robust term can be chosen as sigmoid function, and thus the control inputs avoid the chattering problem so that the obtained input control signals are acceptable and physically realizable.

Simulation results using adaptive control without actuator saturation are shown in Figs. 6.12, 6.13, 6.14, and 6.15, and performance is noticeable in these figures especially at the time when force and moment disturbances occur. Figures 6.12 and 6.14 present the position tracking responses in 3-D space and tracking errors, respectively, from which it can be seen that starting from an initial position far from the reference, the proposed control strategy is able to make the vehicle follow the reference trajectory in a small delay. Besides, the adaptive law helps eliminate the effects caused by sustained disturbances, and three positions successfully track the desired position references with different phases simultaneously.

Figures 6.13 and 6.15 show the results of the attitude tracking and attitude tracking errors, respectively. It can be observed that the inner adaptive controller makes the vehicle track its rotational reference trajectory even when each degree of freedom is affected by the moment disturbances. The estimated values of unknown system dynamics and disturbances are well tuned by the adaptive law. Furthermore, the desired attitude angles ϕ_r and θ_r calculated by the translational controller vary with time, but the adaptive controller achieves null steady-state error. From these

Table 6.2 Parameters of the quadrotor aircraft

Symbol	m	g	l	I_ϕ	I_θ	I_ψ
Value	0.74	9.81	0.2	0.004	0.004	0.0084
Unit	kg	m/s^2	m	$kg \cdot m^2$	$kg \cdot m^2$	$kg \cdot m^2$

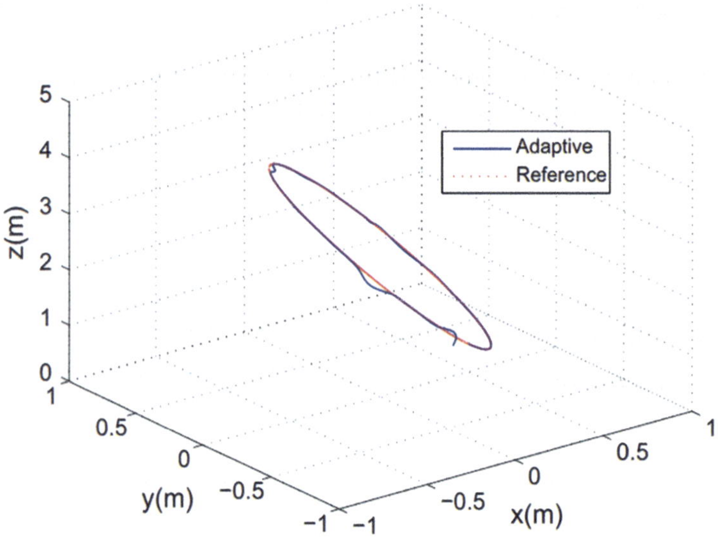

Fig. 6.12 Adaptive path following without saturation

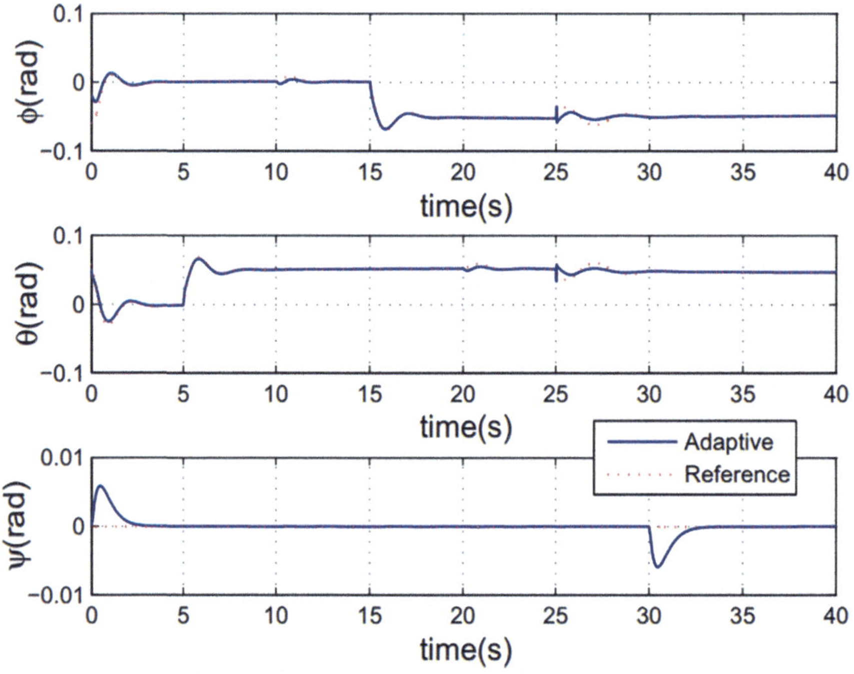

Fig. 6.13 Adaptive attitude tracking without saturation

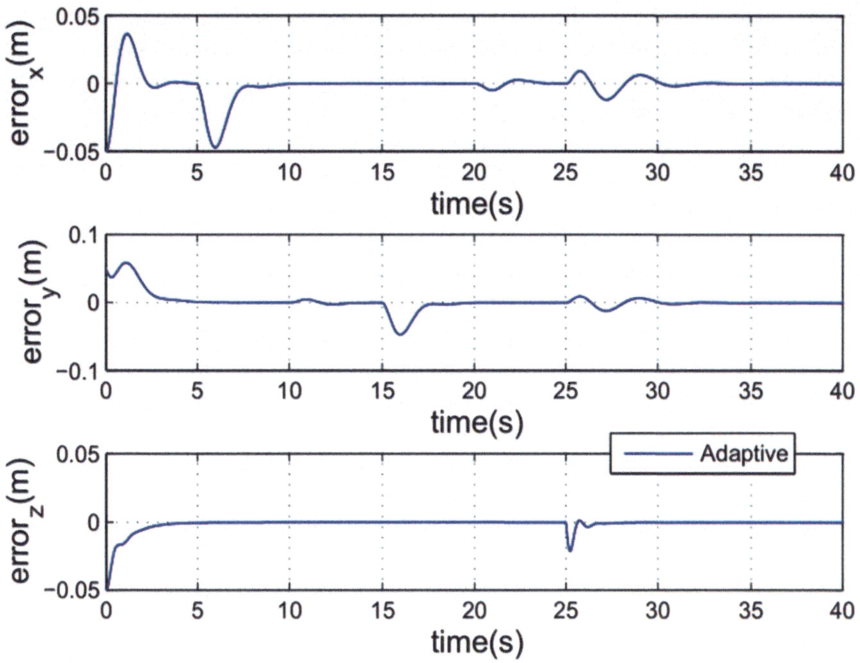

Fig. 6.14 Adaptive position tracking error without saturation

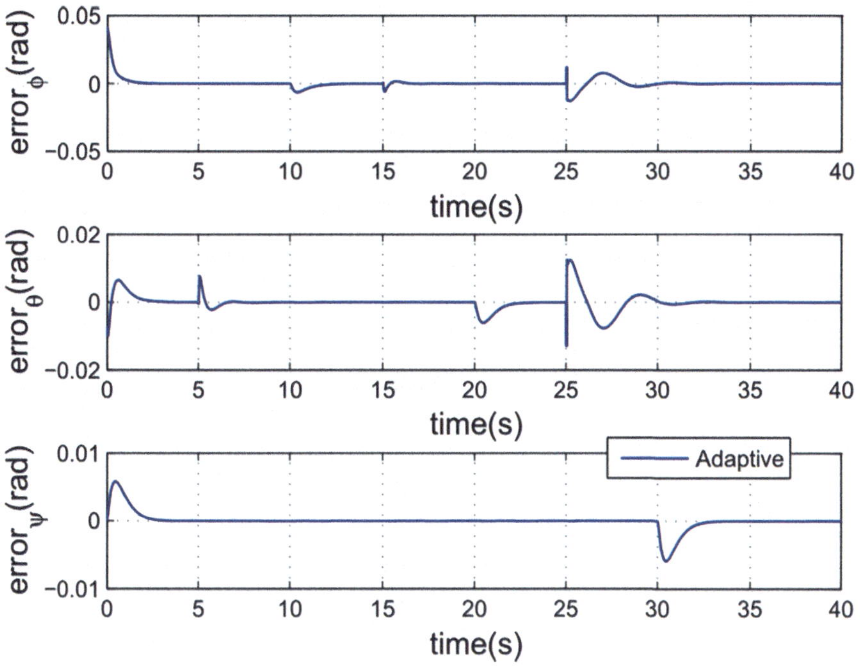

Fig. 6.15 Adaptive attitude tracking error without saturation

figures, the designed controller succeeds in reaching the desired attitude trajectory, while disturbances and model parameter uncertainty are well compensated.

Figures 6.16 and 6.17 are plotted to show the position and attitude tracking errors under the adaptive control laws (6.59) and (6.60) with input saturations. At $t = 25$ s, it is noticeable to see the position tracking error of the z direction in Fig. 6.16 and control signal u_T in Fig. 6.18 suffers from a drastic oscillation caused by the limitations of control input value, and it takes a longer time before convergence. From these figures, one can observe that windup phenomenon caused by actuator saturation leads to a delay in convergence time of the closed-loop system control, which may cause physical accident to the real-time flight of quadrotor aircrafts.

In contrast, simulation results using the proposed adaptive robust control (6.73) and (6.76) under the same saturation are presented in Figs. 6.19 and 6.20, with Fig. 6.20 showing the position tracking errors and Fig. 6.20 showing the tracking errors of attitude angles, respectively. In Figs. 6.19 and 6.20, it is worthy to mention that the proposed approach can obtain the less tracking errors compared with the saturated system under adaptive controller without introducing the anti-windup term, and the proposed controller reduces the convergence time in the presence of actuator saturation and disturbances. Figure 6.21 shows the corresponding control inputs, and it can be observed that the control signals return to normal states quickly after saturation vanishes especially in the z direction, while those under adaptive control law face several oscillations. Moreover, the adaptive gain Γ_i and robust gain β_i

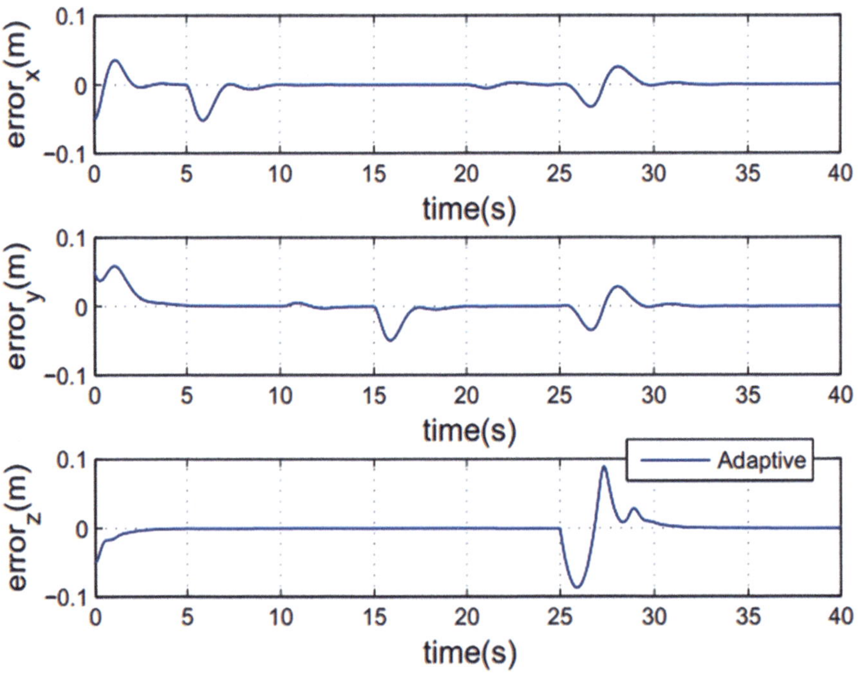

Fig. 6.16 Adaptive position tracking error with saturation

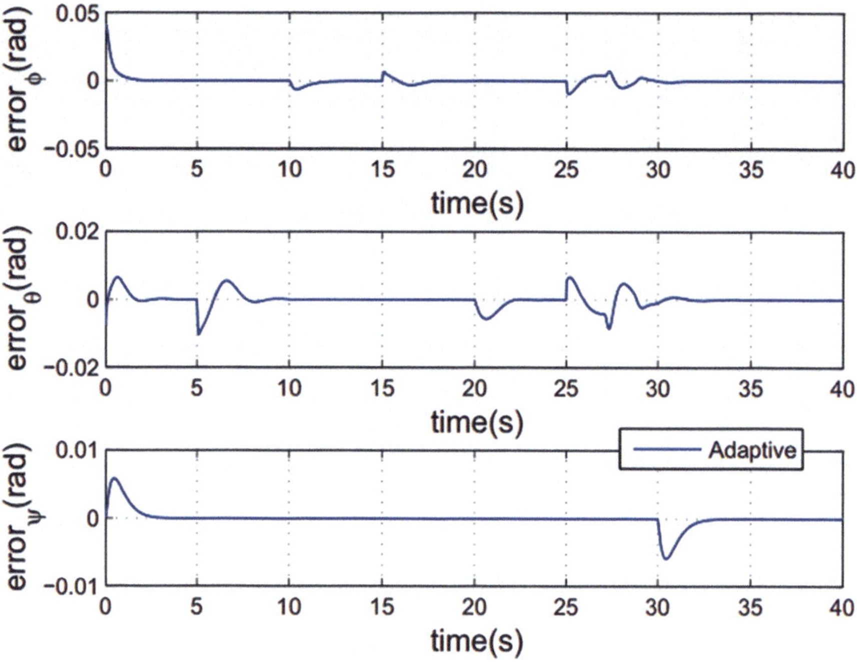

Fig. 6.17 Adaptive attitude tracking error with saturation

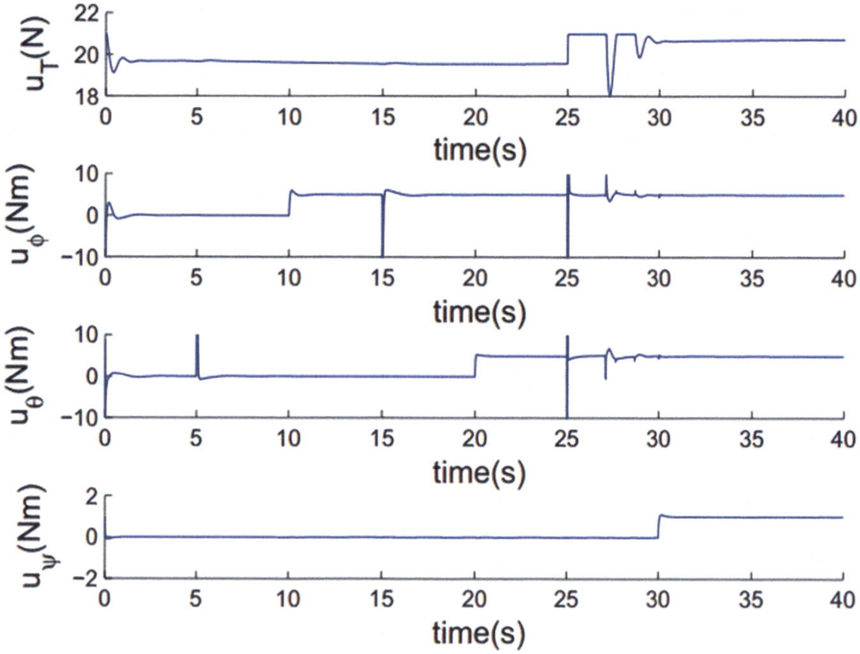

Fig. 6.18 Control inputs of adaptive control with saturation

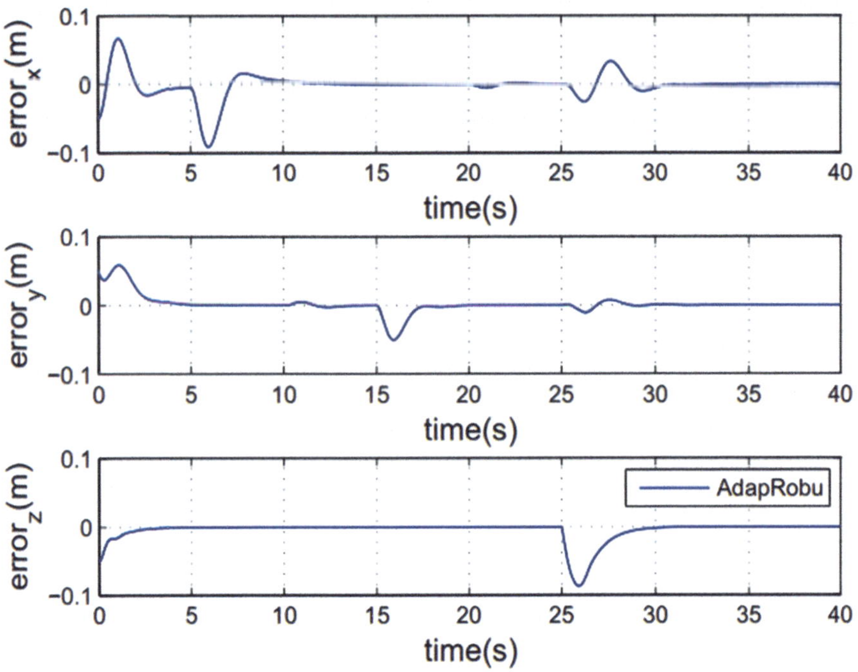

Fig. 6.19 Adaptive robust position tracking error with saturation

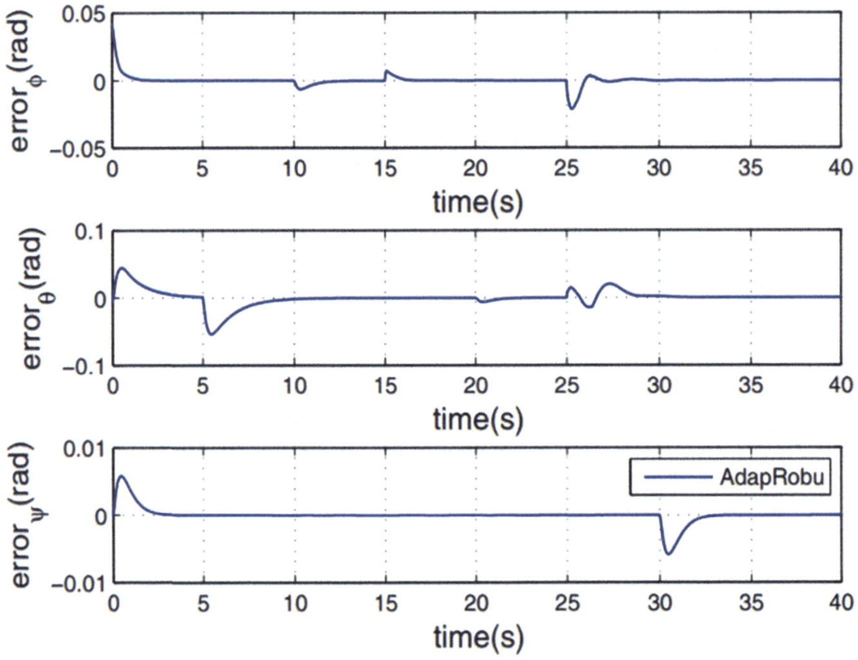

Fig. 6.20 Adaptive robust attitude tracking error with saturation

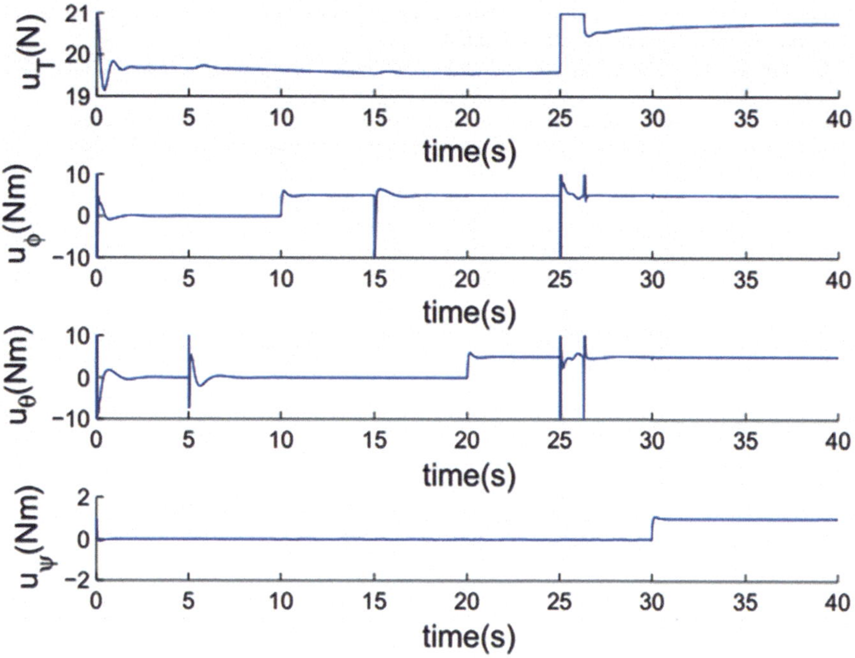

Fig. 6.21 Control input of adaptive robust control with saturation

decide the convergence rate of its robustness performance. Thus, the proposed law is able to cover many kinds of unknown system parameters and six degrees of freedom sustained disturbances. This further confirms the high robustness of the proposed adaptive robust anti-windup control approach.

References

1. Aaltonen I, Arvola A, Heikkila P, et al. Hello Pepper, may I tickle you? Children's and adults' responses to an entertainment robot at a shopping mall. In: Proceedings of the Companion of the 2017 ACM/IEEE International Conference on Human-Robot Interaction. ACM, 2017, 53–54.
2. Dekan M, Duchon F, Jurisica L, et al. iRobot create used in education. Journal of Mechanics Engineering and Automation, 2013, 3(4): 197–202.
3. Cremer S, Mastromoro L, Popa D O. On the performance of the Baxter research robot. In: 2016 IEEE International Symposium on Assembly and Manufacturing (ISAM). IEEE, 2016: 106–111.
4. Li J. Design optimization of amazon robotics. Automation, Control and Intelligent Systems, 2016, 4(2): 48.
5. euRoboticsaisbl. [Online]. Robotics 2020 Strategic research agenda for robotics in Europe. Available: https://ec.europa.eu/research/industrial_technologies/pdf/robotics-ppp-roadmap_en.pdf.

6. Ministry of Economy T, Industry. [Online]. Japan's robot strategy. Available: https://www.meti.go.jp/english/press/2015/0123_01.html.
7. Siegwart R, Nourbakhsh I R, Scaramuzza D. Introduction to autonomous mobile robots. MIT press, 2011: 15–76.
8. Madokoro H, Sato K, Shimoi N. Development of micro air vehicle using aerial photography for safe rowing and coaching. In: 2016 16th International Conference on Control, Automation and Systems (ICCAS). IEEE, 2016, 1–6.
9. Paneque J L, Martínez-de Dios J R, Ollero A. Multi-sensor 6-dof localization for aerial robots in complex gnss-denied environments. In: 2019 IEEE/RSJ International Conference on Intelligent Robots and Systems (IROS). IEEE, 2019: 1978–1984.
10. Didiok Makings. Product: JX-1900 PRO. [Online]. Available: https://didiokmakings.com/page150.
11. Jenssen R, Roverso D. Intelligent monitoring and inspection of power line components powered by UAVs and deep learning. IEEE Power and energy technology systems journal, 2019, 6(1): 11–21.
12. Teledyne Flir. Product: Black Recon. [Online]. Available: https://www.youtube.com/watch?v=vyDWoaNa6Rc.
13. Orsag M, Korpela C, Oh P. Modeling and control of MM-UAV: Mobile manipulating unmanned aerial vehicle. Journal of Intelligent & Robotic Systems, 2013, 69(14): 227–240.
14. Chandler P, Rasmussen S, Pachter M. UAV cooperative path planning. In: AIAA Guidance, Navigation, and Control Conference and Exhibit. 2000, 4370–4379.
15. Chen H, Wang X m, Li Y. A survey of autonomous control for UAV. In: 2009 International Conference on Artificial Intelligence and Computational Intelligence. IEEE, 2009, 2: 267–271.
16. Gupte S, Mohandas P I T, Conrad J M. A survey of quadrotor unmanned aerial vehicles. In: 2012 Proceedings of IEEE Southeastcon. IEEE, 2012, 1–6.
17. Tayebi A, McGilvray S. Attitude stabilization of a VTOL quadrotor aircraft. IEEE Transactions on Control Systems Technology, 2006, 14(3): 562–571.
18. Olfati-Saber R. Nonlinear control of underactuated mechanical systems with application to robotics and aerospace vehicles. Cambridge: Massachusetts Institute of Technology; 2001.
19. Raffo GV, Ortega MG, Rubio FR. An integral predictive/nonlinear H∞ control structure for a quadrotor helicopter. Automatica. 2010;46(1):29–39.
20. Sadati N, Ghadami R. Adaptive multi-model sliding mode control of robotic manipulators using soft computing. Neurocomputing. 2008;71(13–15):2702–10.
21. Cui Y, Zhang H, Wang Y, et al. Adaptive neural dynamic surface control for a class of uncertain nonlinear systems with disturbances. Neurocomputing. 2015;165:152–8.
22. Kemper M, Fatikow S. Impact of center of gravity in quadrotor helicopter controller design. IFAC Proceedings Volumes, 2006, 39(16): 157–162.
23. Xiong JJ, Zheng EH. Position and attitude tracking control for a quadrotor UAV. ISA Trans. 2014;53(3):725–31.
24. Alexis K, Nikolakopoulos G, Tzes A. Model predictive quadrotor control: Attitude, altitude and position experimental studies. IET Control Theory & Applications, 2012, 6(12): 1812–1827.
25. Raffo G V, Ortega M G, Rubio F R. Robust nonlinear control for path tracking of a quadrotor helicopter. Asian Journal of Control, 2015, 17(1): 142–156.

Chapter 7
Unmanned Aerial Robot Manipulation in Transmission Lines and Industrial Pipelines

7.1 Background and Motivation

Robot technology and equipment are currently cutting-edge research fields and are the focus of national defense technology and the national economy. Traditional robots mostly work in the form of passive observation, which greatly limits their application scenarios. Therefore, robots with the ability to interact with environments are the focus of future research and development. On the basis of information observation-type robots, active operation-type mobile robots can be obtained by adding active execution mechanisms (such as manipulator arms, dexterous hands, etc.). Typical cases include unmanned detonation robots (UDRs), unmanned aerial manipulators (UAMs), space manipulators, and unmanned underwater manipulators (UUMs), as shown in Fig. 7.1.

Typically, active operating mechanisms are mounted on aerial robots to implement aerial physical interaction tasks [1, 2]. Such novel combined aerial systems are called UAMs, which perform the advantages of both the wide-range maneuverability of the flying platform and the highly dexterous operation characteristics of the operating mechanism [3]. Therefore, transforming rotary-wing UAVs from passive observation to active operation is essential for expanding their application scope and leading the new research field of rotary-wing UAVs worldwide. Research on UAMs is of great significance in promoting the improvement and application of relevant theoretical systems for mobile robots, specifically summarized as follows:

1. Integrating active operating mechanisms into rotary-wing drones to form a functional hoverable aircraft system has broad application prospects, but it also faces significant technical challenges and theoretical bottlenecks. Drones, especially rotary-wing drones, possess the characteristics of vertical takeoff and landing, hovering in the air, and strong maneuverability. Drones equipped with active operating manipulators can significantly enhance their operational capabilities in various environments, opening up a wide range of operational

Y. Wang et al., *Electric Power Robots*,
https://doi.org/10.1007/978-981-96-2512-3_7

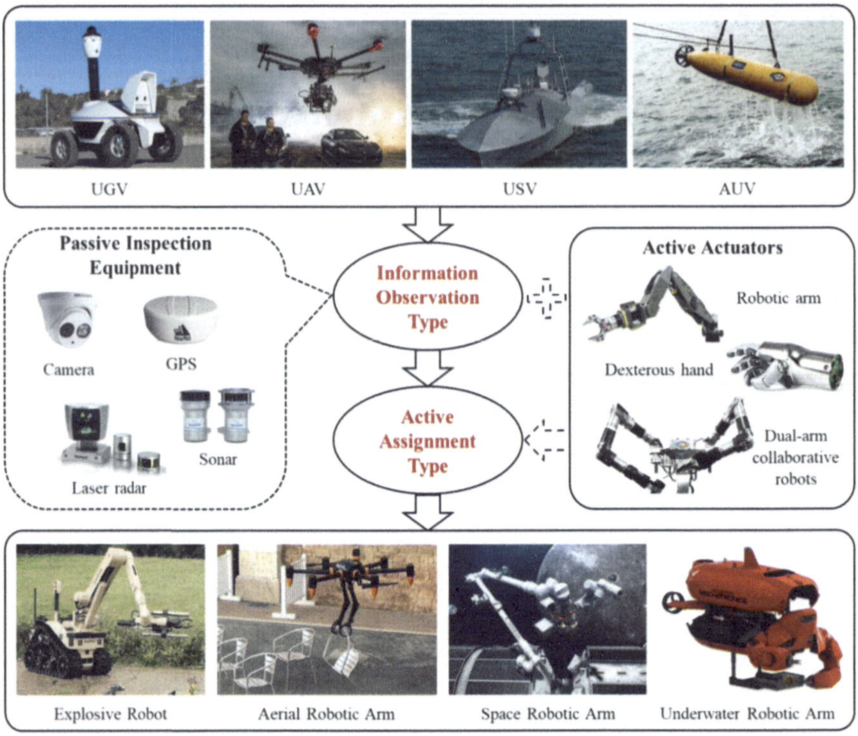

Fig. 7.1 Development diagram of passive observation-type to active operation-type robots

Fig. 7.2 UAVs with active manipulators having broad application prospects. (**a**) Polar scientific sampling. (**b**) Power hazardous environment operations. (**c**) New concept of multi-terrain mobile platform

applications. For instance, during polar scientific expeditions, UAMs can efficiently undertake tasks such as remote sample collection for wide-area unmanned scientific research and environmental monitoring (Fig. 7.2a), which greatly reduce the labor intensity and risks for scientific researchers. In industrial settings, UAMs can replace personnel for transporting goods and performing tasks in hazardous and complex environments (Fig. 7.2b). Moreover, with the movement

Table 7.1 Comparisons of coupling and disturbance intensity for several active work-type mobile robots

Combined systems	Ground robots with active mechanisms	Underwater robots with active mechanisms	Space robots with active mechanisms	Aerial robots with active mechanisms
Coupling with mechanisms	None	Medium	High	**High**
Disturbed by external factors	Weak	Strong	None	**Strong**

capabilities of active operating mechanisms, UAMs can aid in landing in challenging terrain conditions and perform unmanned detection of surface features, expanding the application of drones in other domains. In the long term, the fusion of drone flight capabilities with the flexibility of active operating mechanisms can break through traditional paradigms and give rise to a new concept of a multi-habitat mobile platform (Fig. 7.2c).

2. The study of key scientific issues in operational rotary-wing UAVs has significant implications for the improvement of mobile robot system theory. Due to the underactuated nature [4] and unstable characteristics susceptible to external disturbances of rotary-wing UAVs, the operational rotary-wing UAV system formed by adding active operating mechanisms exhibits stronger coupling and instability compared to ground mobile manipulators, underwater manipulators, space manipulators, and other active operation mobile robot systems. Table 7.1 compares the coupling and disturbance intensity of several active operation mobile robots. Based on the characteristics and difficulties faced by operational rotary-wing UAV systems, it is believed that when studying operational hovering UAV systems, the following three key scientific issues need to be focused on:

   (a) Dynamic modeling and analysis: The presence of active operating mechanisms causes the center of gravity position and disturbances to vary rapidly during its operation. As a result, rotary-wing UAVs need to adjust their flight attitude to cope with this issue, thereby leading to more complex unmodeled dynamics (especially aerodynamics).
   (b) Coordinated control under strong coupling and external disturbances: The coupling between the active operating mechanism and the rotary-wing UAV is one of the most significant features of the operational rotary-wing UAV system, and it is an important challenge in its autonomous control.
   (c) Autonomous control and real-time autonomous planning challenges under underactuated conditions: Due to the underactuated nature of rotary-wing UAVs, installing detection sensors (such as cameras, lasers, etc.) on the body also has underactuated characteristics, making position and attitude control coupled, while also making real-time autonomous planning more challenging.

The resolution of these three scientific issues is essential for the successful development of operational rotary-wing UAV systems and ensuring their motion performance. It also holds significant reference value for the autonomous control research of surface and deep space operational mobile robot systems. In addition, autonomous control under complex conditions has always been a research hotspot and challenge in the field of robotics. Researching these issues will also promote the further improvement of mobile robot system theory, especially the analysis and control theory of complex systems such as maneuverable hovering aircrafts.

7.2 Kinematic and Dynamic Modeling for UAM

7.2.1 Kinematic Modeling

7.2.1.1 Coordinate System Establishment

In the rigid body system, in order to completely describe the position and relationship of a rigid body in space, it is necessary to establish an individual coordinate system for each rigid body (usually with the origin at the center of mass) and a unified reference coordinate system for the entire system. Th$\{\mathcal{F}_I : O_I - x_I y_I z_I\}$ en, the posture of the rigid body can be reflected by describing the posture between the rigid body coordinate system or relative to the reference coordinate system. For unmanned aerial work robots, their coordinate system is shown in Fig. 7.3. represents the world coordinate system, and z_I is vertically upward, which is the opposite

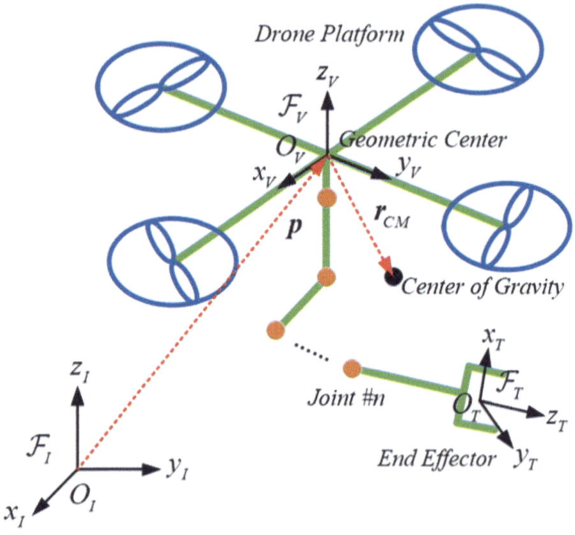

Fig. 7.3 Kinematic model of unmanned aerial work robot

direction of gravity. $\{\mathcal{F}_V : O_V - x_V y_V z_V\}$ represents the body coordinate system of the unmanned aerial vehicle platform, and its coordinate origin is generally at the geometric center of the drone, with the z_V direction pointing upward. The $\{\mathcal{F}_B : O_B - x_B y_B z_B\}$ is the base coordinate system of the manipulator, and its origin is the installation point of the manipulator and the unmanned aerial vehicle platform. The $\{\mathcal{F}_{L_i} : O_{L_i} - x_{L_i} y_{L_i} z_{L_i}\}$ is the coordinate system of the i-th joint ($i = 1, \cdots, n$) of the manipulator, and the O_{L_i} is the origin of the joint coordinate system. $\{\mathcal{F}_E : O_E - x_E y_E z_E\}$ and $\{\mathcal{F}_T : O_T - x_T y_T z_T\}$ are the end-effector and tool coordinate systems, respectively. The axes of the two coordinate systems are in the same direction, differing only by a displacement offset vector. The transformation matrix from the robot base coordinate system $\mathcal{F}_B$ to the body coordinate system $\mathcal{F}_V$ is a fixed homogeneous transformation matrix ${}^V\boldsymbol{T}_B \in \mathbb{P}^{4 \times 4}$, with its rotation matrix being ${}^V\boldsymbol{R}_B \in \mathbb{P}^{3 \times 3}$ and displacement offset vector being ${}^V\boldsymbol{t}_B \in \mathbb{P}^3$.

The coordinate variables of the system use $\boldsymbol{p} = (x, y, z)^\top$ and $\boldsymbol{\Phi} = (\varphi, \vartheta, \psi)^\top$ to represent the position vector of the drone in the world coordinate system $\mathcal{F}_I$ and the Euler angle vector (roll angle, pitch angle, and yaw angle), respectively. $\boldsymbol{\theta} = (\theta_1, \theta_2, \cdots, \theta_n)^\top$ represents the joint space vector of the manipulator. Let $\boldsymbol{q}$ represent the generalized coordinate variables of the entire system, and the variable $\boldsymbol{q}$ is defined as

$$\boldsymbol{q} = (\boldsymbol{p}^\top \ \ \boldsymbol{\Phi}^\top \ \ \boldsymbol{\theta}^\top)^\top \in \mathbb{R}^{6+n} \tag{7.1}$$

7.2.1.2 Kinematic Modeling and Analysis

The kinematic modeling of an unmanned aerial robotic operation mainly establishes the coordinate system change relationship between generalized coordinate variables and the tool coordinate system, as well as the relationship between velocity and static force after differentiation. For the manipulator system, generally speaking, its end pose representation can be expressed using homogeneous transformation matrices, that is, the homogeneous transformation matrix ${}^I\boldsymbol{T}_T \in \mathbb{P}^{4 \times 4}$ from the tool coordinate system $\mathcal{F}_T$ to the world coordinate system $\mathcal{F}_I$ can be represented as

$${}^I\boldsymbol{T}_T = \begin{pmatrix} {}^I\boldsymbol{R}_T & \boldsymbol{p}_T \\ \boldsymbol{0}_{3\times 1} & 1 \end{pmatrix} \tag{7.2}$$

where ${}^1\boldsymbol{R}_T \in SO(3)$ and $\boldsymbol{p}_T \in \mathbb{P}^3$ represent the rotation matrix and displacement vector, respectively. Since the unmanned aerial work robot is composed of the flight platform and manipulator, its positive kinematics can be obtained by matrix transformation:

$$^{I}\boldsymbol{T}_T = {}^{I}\boldsymbol{T}_V {}^{V}\boldsymbol{T}_B \underbrace{{}^{B}\boldsymbol{T}_{L_1}^{L_1}\boldsymbol{T}_{L_2} \cdots {}^{L_{n-1}}\boldsymbol{T}_{L_n} {}^{L_n}\boldsymbol{T}_E {}^{E}\boldsymbol{T}_T}_{{}^{B}\boldsymbol{T}_{L_n}} \tag{7.3}$$

In this case, ${}^{V}\boldsymbol{T}_B \in \mathbb{P}^{4\times 4}$, ${}^{L_n}\boldsymbol{T}_E = \mathbb{I}_{4\times 4}$, ${}^{E}\boldsymbol{T}_T \in \mathbb{P}^{4\times 4}$, and ${}^{B}\boldsymbol{T}_{L_n} \in \mathbb{R}^{4\times 4}$ are known constant matrices. The matrix ${}^{B}\boldsymbol{T}_{L_n}$ can be recursively solved by the D-H (Denavit-Hartenberg) parameters of the robot manipulator [5]. It is worth noting that in formula (7.3), the rotation matrix ${}^{I}\boldsymbol{R}_T$ is used to represent the pose; although it has nine components, each column of it has the constraint of being orthogonal to each other and having a magnitude of one. Therefore, only three degrees of freedom are needed for the rotation matrix. In addition to using the rotation matrix to represent the pose, other methods such as Euler angles, quaternions, exponential scale functions, etc. can also be used for pose representation, and they can be converted with each other under certain conditions. For example, if Euler angles are employed to describe the end pose, then formula (7.3) can be expressed as

$$\boldsymbol{x}_T = \begin{pmatrix} \boldsymbol{p}_T \\ \boldsymbol{\Phi}_T \end{pmatrix} = \begin{pmatrix} f_l(\boldsymbol{q}) \\ f_r(\boldsymbol{q}) \end{pmatrix} = f(\boldsymbol{q}) \tag{7.4}$$

where $\boldsymbol{x}_T$ represents the position and orientation description vector of the end-effector coordinate system $\mathcal{F}_T$ in the world coordinate system $\mathcal{F}_I$. $f_l(\boldsymbol{q})$, $f_r(\boldsymbol{q})$, and $f(\boldsymbol{q})$ can be calculated using formula (7.3).

For robot autonomous control, it generally operates in the velocity space and requires differentiation of the system's forward kinematics to obtain the Jacobian matrix of the velocity space. Firstly, for an unmanned aerial vehicle platform, let $\dot{\boldsymbol{p}}$ and ${}^{V}\dot{\boldsymbol{p}}$ represent the linear velocities of the UAV geometric center point O_V in the world coordinate system $\mathcal{F}_I$ and the body coordinate system $\mathcal{F}_V$, respectively. The relationship between these two can be expressed by the rotation matrix ${}^{I}\boldsymbol{R}_V \in SO(3)$ of the UAV:

$$\dot{\boldsymbol{p}} = {}^{I}\boldsymbol{R}_V {}^{V}\dot{\boldsymbol{p}} \tag{7.5}$$

Similarly, the angular velocity of the drone platform is represented as $\mathcal{F}_I$ and $\mathcal{F}_V$ in the world coordinate system $w = (w_x, w_y, w_z)^{\top}$ and body coordinate system ${}^{V}w = ({}^{V}w_x, {}^{V}w_y, {}^{V}w_z)^{\top}$, respectively. Among them, the angular velocity w in the world coordinate system can be obtained by transforming the derivative of Euler angles $\dot{\boldsymbol{\Phi}}$ of the drone platform. The specific transformation relationship is given as follows:

$$\boldsymbol{w} = \boldsymbol{R}_{\Phi}\dot{\boldsymbol{\Phi}} \tag{7.6}$$

$$R_\Phi = \begin{pmatrix} \cos\vartheta & 0 & -\cos\varphi\sin\vartheta \\ 0 & 1 & \sin\varphi \\ \sin\vartheta & 0 & \cos\varphi\cos\vartheta \end{pmatrix} \tag{7.7}$$

where $\boldsymbol{R}_\Phi$ is a transformation matrix that converts the Euler angle velocity of the drone platform into angular velocity, so the angular velocity $\mathcal{F}_V$ of the drone platform in the body coordinate system ${}^V\boldsymbol{w}$ can be expressed as

$${}^V\boldsymbol{w} = \left({}^I\boldsymbol{R}_V\right)^\top \boldsymbol{R}_\Phi \dot{\boldsymbol{\Phi}} = \boldsymbol{Q}\dot{\boldsymbol{\Phi}} \tag{7.8}$$

where $\boldsymbol{Q} = \left({}^I\boldsymbol{R}_V\right)^\top \boldsymbol{R}_\Phi \in \mathbb{R}^{3\times3}$. Similarly, for the end effector of the robotic arm, $\boldsymbol{p}_T$, $\dot{\boldsymbol{p}}_T$, $\boldsymbol{\Phi}_T$, and $\boldsymbol{w}_T$ are used to represent the position, linear velocity, Euler angles, and angular velocity in the coordinate system $\mathcal{F}_T$ with respect to the world coordinate system $\mathcal{F}_I$, respectively. The following relationships exist:

$$\boldsymbol{p}_T = \boldsymbol{p} + {}^I\boldsymbol{R}_V {}^V\boldsymbol{p}_T \tag{7.9}$$

where ${}^V\boldsymbol{p}_T$ is the position vector of $\mathcal{F}_T$ in the body coordinate system $\mathcal{F}_V$. By differentiating formula (7.9), the expression of $\dot{\boldsymbol{p}}_T$ can be obtained:

$$\dot{\boldsymbol{p}}_T = \dot{\boldsymbol{p}} + [\boldsymbol{w}]_\times {}^I\boldsymbol{R}_V {}^V\boldsymbol{p}_T + {}^I\boldsymbol{R}_V {}^V\dot{\boldsymbol{p}}_T \tag{7.10}$$

$$\boldsymbol{w}_T = \boldsymbol{w} + {}^I\boldsymbol{R}_V {}^V\boldsymbol{w}_T \tag{7.11}$$

In this case, $[\boldsymbol{a}]_\times$ is the skew-symmetric matrix of any vector $\boldsymbol{a}$ [6]. ${}^V\dot{\boldsymbol{p}}_T$ and ${}^V\boldsymbol{w}_T$ represent the linear velocity and angular velocity of the tool coordinate system $\mathcal{F}_T$ and the body coordinate system $\mathcal{F}_V$, respectively, and they can be obtained through the Jacobian matrix of the manipulator's kinematics:

$$\begin{aligned} V\dot{\boldsymbol{p}}_T &= {}^V\boldsymbol{R}_B\left({}^B\dot{\boldsymbol{p}}_E - {}^B\boldsymbol{R}_E(\boldsymbol{\theta})^E\boldsymbol{t}_T{}^B_\times\boldsymbol{w}_E\right) \\ &= {}^V\boldsymbol{R}_B\left(\boldsymbol{B}_t(\boldsymbol{\theta})\dot{\boldsymbol{\theta}} - \left[{}^B\boldsymbol{R}_E(\boldsymbol{\theta})^E\boldsymbol{t}_T\right]_\times \boldsymbol{B}_r(\boldsymbol{\theta})\dot{\boldsymbol{\theta}}\right) \end{aligned} \tag{7.12}$$

$${}^V\boldsymbol{w}_T = {}^V\boldsymbol{R}_B {}^B\boldsymbol{w}_E = {}^V\boldsymbol{R}_B\boldsymbol{B}_r(\boldsymbol{\theta})\dot{\boldsymbol{\theta}} \tag{7.13}$$

where ${}^B\dot{\boldsymbol{p}}_E = \boldsymbol{B}_t(\boldsymbol{\theta})\dot{\boldsymbol{\theta}}$ and ${}^B\boldsymbol{w}_E = \boldsymbol{B}_r(\boldsymbol{\theta})\dot{\boldsymbol{\theta}}$. $\boldsymbol{B}_t \in \mathbb{P}^{3\times n}$ and $\boldsymbol{B}_r \in \mathbb{P}^{3\times n}$ are the linear velocity and angular velocity components of the Jacobian matrix of the manipulator, respectively. To simplify the movement model of unmanned aerial robots, combining formulas (7.6), (7.10), (7.11), (7.12), and (7.13) yields the following equations:

$$\begin{aligned}\dot{\boldsymbol{p}}_T &= \dot{\boldsymbol{p}} + [\boldsymbol{w}]_\times {}^I\boldsymbol{R}_V {}^V\boldsymbol{p}_T + {}^I\boldsymbol{R}_V {}^V\boldsymbol{R}_B\left(\boldsymbol{B}_l(\boldsymbol{\theta})\dot{\boldsymbol{\theta}} - \left[{}^B\boldsymbol{R}_E(\boldsymbol{\theta}){}^E\boldsymbol{t}_T\right]_\times \boldsymbol{B}_r(\boldsymbol{\theta})\dot{\boldsymbol{\theta}}\right) \\ &= \begin{pmatrix}\mathbb{I}_{3\times 3} & -\left[{}^I\boldsymbol{R}_V {}^V\boldsymbol{p}_l\right]_\times \boldsymbol{R}_\Phi & {}^I\boldsymbol{R}_V {}^V\boldsymbol{R}_B\left(\boldsymbol{B}_l(\boldsymbol{\theta}) - \left[{}^B\boldsymbol{R}_E(\boldsymbol{\theta}){}^E\boldsymbol{t}_T\right]_\times \boldsymbol{B}_r(\boldsymbol{\theta})\right)\end{pmatrix} \\ &\quad \begin{pmatrix}\dot{\boldsymbol{p}}^\top & \dot{\boldsymbol{\Phi}}^\top & \dot{\boldsymbol{\theta}}^\top\end{pmatrix}^\top \\ &= \boldsymbol{J}_l\dot{\boldsymbol{q}}\end{aligned} \tag{7.14}$$

$$\begin{aligned}\boldsymbol{w}_T &= \boldsymbol{w} + {}^I\boldsymbol{R}_V {}^V\boldsymbol{R}_B\boldsymbol{B}_r(\boldsymbol{\theta})\dot{\boldsymbol{\theta}} \\ &= \begin{pmatrix}\boldsymbol{0}_{3\times 3} & R_\Phi & {}^IR_V^V R_B B_r(\theta)\end{pmatrix}\begin{pmatrix}p^\top & \dot{\boldsymbol{\Phi}}^\top & \dot{\boldsymbol{\theta}}^\top\end{pmatrix}^\top \\ &= \boldsymbol{J}_r\dot{\boldsymbol{q}}\end{aligned} \tag{7.15}$$

where $\mathbb{I}_{i\times i}$ is a unit matrix of $i \times i$ and $\boldsymbol{0}_{i\times i}$ is a zero matrix of $i \times i$. $\boldsymbol{J}_l \in \mathbb{P}^{3\times(6+n)}$ and $\boldsymbol{J}_r \in \mathbb{P}^{3\times(6+n)}$ represent the linear velocity and angular velocity components of the Jacobian matrix of system kinematics, respectively. By combining formulas (7.14) and (7.15), it can be written in a unified form as

$$\dot{\boldsymbol{x}}_T = \begin{pmatrix}\dot{\boldsymbol{p}}_T \\ \dot{\boldsymbol{\Phi}}_T\end{pmatrix} = \begin{pmatrix}\dot{\boldsymbol{p}}_T \\ \boldsymbol{R}_{\Phi_T\boldsymbol{w}_T}\end{pmatrix} = \begin{pmatrix}\boldsymbol{J}_l \\ \boldsymbol{R}_{\Phi_T\boldsymbol{J}_r}\end{pmatrix}\dot{q} = \boldsymbol{J}_T\dot{\boldsymbol{q}} \tag{7.16}$$

where $\dot{\boldsymbol{x}}_T = \left(\dot{\boldsymbol{p}}_T^\top, \dot{\boldsymbol{\Phi}}_T^\top\right)^\top$ represents the velocity vector. $\boldsymbol{R}_\Phi$ is the rotation matrix, but the angle is the Euler angle of the end tool coordinate system. $\boldsymbol{J}_T = \left(\boldsymbol{J}_l^\top, \left(\boldsymbol{R}_{\Phi_r}\boldsymbol{J}_r\right)^\top\right)^\top \in \mathbb{R}^{6\times(6+n)}$ represents the kinematic Jacobian matrix of the unmanned aerial work robot.

7.2.2 UAM Dynamic Modeling

For establishing the robot dynamic model, there are generally two methods: the recursive Newton-Euler equation method and the Euler-Lagrange equation method. The former uses a recursive method to derive the system model, fully considering the dynamic coupling effect between the flying platform and the operating mechanism; the latter is deduced based on energy conservation, treating the entire composite system as a coupled system. For the dynamic modeling of a single drone platform, the recursive Newton-Euler equation method is relatively mature [7]. In addition, since unmanned aerial work robots usually move slowly or remain stationary in non-contact modes, their essential difference from drones lies in the fact that the center of gravity of the system does not coincide with the geometric center of gravity. Therefore, this section proposes a static modeling method for unmanned aerial work robots in non-contact mode. For contact mode, the Euler-Lagrange equation method is used for modeling. The following sections will provide detailed

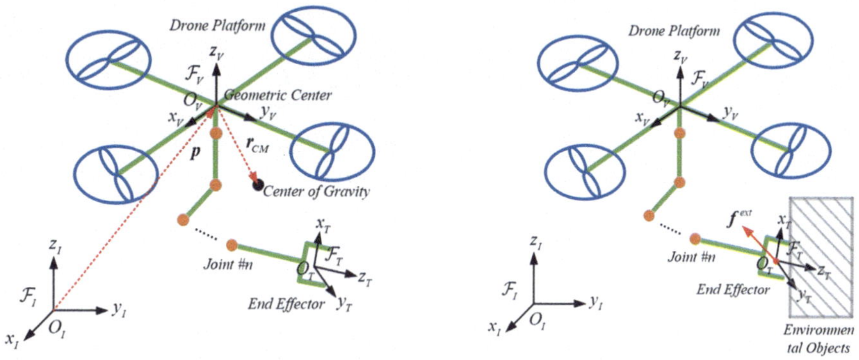

(a) Non-contact modal dynamic model schematic diagram.

(b) Contact modal dynamics model schematic diagram.

Fig. 7.4 Diagram of the dynamic model for unmanned aerial work robots. (**a**) Non-contact modal dynamic model schematic diagram. (**b**) Contact modal dynamic model schematic diagram

introductions to these two modeling methods. Figure 7.4 is a schematic diagram of these two modeling methods.

7.2.2.1 Dynamic Modeling Under Non-contact Mode

For the dynamic modeling under non-contact modality, the key is to regard the unmanned aerial robot as a quasi-static center-of-gravity offset drone platform and use the Newton-Euler equation method for modeling. As shown in Fig. 7.4a, $\boldsymbol{r}_{CM}$ is the position vector of the system's center of mass (CoM) in the body coordinate system $\mathcal{F}_V$. $\boldsymbol{V} = ({}^I\boldsymbol{R}_V)^\top \dot{\boldsymbol{p}}$ and $\boldsymbol{\Omega} = {}^V\boldsymbol{w}$ represent the speed and angular velocity of the drone in its own body coordinate system, respectively. Then, according to the Newton-Euler equation method, the translational motion and rotational motion equations of the system are given as [8]

$$m\ddot{\boldsymbol{p}} = F - m\dot{\boldsymbol{\Omega}} \times \boldsymbol{r}_{CM} - 2m\boldsymbol{\Omega} \times \dot{\boldsymbol{r}}_{CM} - m\boldsymbol{\Omega} \times (\boldsymbol{\Omega} \times \boldsymbol{r}_{CM}) - m\ddot{\boldsymbol{r}}_{CM} \tag{7.17}$$

$$\boldsymbol{I}\dot{\boldsymbol{\Omega}} = \boldsymbol{\Gamma} - \boldsymbol{\Omega} \times \boldsymbol{I}\boldsymbol{\Omega} - \dot{\boldsymbol{r}}_{CM} \times \boldsymbol{P} - \boldsymbol{r}_{CM} \times \dot{\boldsymbol{P}} - (\boldsymbol{\Omega} \times \boldsymbol{r}_{CM}) \times \boldsymbol{P} - \boldsymbol{r}_{CM} \times (\boldsymbol{\Omega} \times \boldsymbol{P}) \tag{7.18}$$

where m represents the total mass of the system (including the unmanned aerial vehicle platform and the operating mechanism) and $\boldsymbol{P} = m\boldsymbol{V}$ is the additional torque caused by the center-of-gravity offset. $\boldsymbol{F} \in \mathcal{F}_I$ and $\boldsymbol{\Gamma} \in \mathcal{F}_v$ represent external forces and moments, respectively, which are synthesized by the lift provided by the drone's rotor. $\boldsymbol{I} \in \mathbb{P}^{3 \times 3}$ represents the inertia tensor matrix of the system at the center-of-gravity point. Since the unmanned aerial robot can be considered as static when its manipulator moves slowly in non-contact mode, there are $\dot{\boldsymbol{r}}_{CM} = \boldsymbol{0}$ and $\ddot{\boldsymbol{r}}_{CM} = \boldsymbol{0}$. The small terms $(\boldsymbol{\Omega} \times \boldsymbol{r}_{CM}) \times \boldsymbol{P}$ and $\boldsymbol{r}_{CM} \times (\boldsymbol{\Omega} \times \boldsymbol{P})$ in Eq. (7.18) are ignored. Then Eqs. (7.17) and (7.18) can be rewritten as

$$m\ddot{p} = \boldsymbol{F} - m\,\dot{\boldsymbol{\Omega}} \times \mathbf{r}_{CM} - m\boldsymbol{\Omega} \times (\boldsymbol{\Omega} \times \mathbf{r}_{CM}) \tag{7.19}$$

$$\boldsymbol{I}\,\dot{\boldsymbol{\Omega}} = \boldsymbol{\Gamma} - \boldsymbol{\Omega} \times \boldsymbol{I\Omega} - r_{CM} \times \dot{\boldsymbol{P}} \tag{7.20}$$

For a quadrotor drone platform, $\boldsymbol{F}$ and $\boldsymbol{\Gamma}$ in Eqs. (7.19) and (7.20) can be calculated from the speed of each rotor. The angular velocity of these four rotors is represented by $c_i(i = 1, \cdots, 4)$, and the lift j_i and reaction force of each rotor g_i are expressed by [9]

$$\begin{cases} j_i = b_j c_i^2 \\ g_i = b_g c_i^2 \end{cases}, \quad i = 1, \cdots, 4 \tag{7.21}$$

where b_j and $b_g$are constants, and their sizes are related to air density, drag coefficient, propeller geometric parameters, etc. Therefore, $\boldsymbol{F}$ and $\boldsymbol{\Gamma}$ are

$$\boldsymbol{F} = {}^{I}\boldsymbol{R}_V T \boldsymbol{e}_z - mg\boldsymbol{e}_z \tag{7.22}$$

$$\boldsymbol{\Gamma} = (\boldsymbol{\tau}_x \quad \boldsymbol{\tau}_y \quad \boldsymbol{\tau}_z)^{\top} \tag{7.23}$$

$$\begin{pmatrix} \boldsymbol{T} \\ \boldsymbol{\tau}_x \\ \boldsymbol{\tau}_y \\ \boldsymbol{\tau}_z \end{pmatrix} = \begin{pmatrix} b_j & b_j & b_j & b_j \\ Db_j & 0 & -Db_j & 0 \\ 0 & -Db_j & 0 & Db_j \\ -b_g & b_g & -b_g & b_g \end{pmatrix} \begin{pmatrix} c_1^2 \\ c_2^2 \\ c_3^2 \\ c_4^2 \end{pmatrix} \tag{7.24}$$

where parameter D is a constant that represents half of the distance between the two rotors of a quadrotor drone. In Eq. (7.22), both the total lift T and the system gravity component are aligned with the coordinate axis z_V through the unit vector $\boldsymbol{e}_z = (0, 0, 1)^{\top}$.

7.2.2.2 Dynamic Modeling Under Contact Mode

In dynamic modeling within the contact mode, the unified system of the unmanned aerial vehicle platform and the manipulator is considered, and the Euler-Lagrange equation method is employed for modeling and analysis. As shown in Fig. 7.4b, it is assumed that the end effector of the manipulator is subjected to environmental reaction forces and moments $f^{ext} = \left(f_l^{ext\top}, f_r^{ext\top}\right)^{\top} \in \mathbb{R}^6$ under the contact mode, where $f_l^{ext} = \left(f_{lx}^{ext}, f_{ly}^{ext}, f_{lz}^{ext}\right)^{\top} \in \mathbb{R}^3$ represents the reaction force and $f_r^{ext} = \left(f_{rx}^{ext}, f_{ry}^{ext}, f_{rz}^{ext}\right)^{\top} \in \mathbb{R}^3$ represents the reaction moment. As the Euler-Lagrange equation method is an energy-based modeling approach, the initial step involves calculating the total energy $\mathcal{L} = \mathcal{K} - \mathcal{U}$ of the entire system, where $\mathcal{K}$ and $\mathcal{U}$ represent the kinetic and potential energy of the system, respectively. The specific expression of the Euler-Lagrange equation is expressed as

$$\frac{\mathrm{d}}{\mathrm{d}t}\frac{\partial \mathcal{L}}{\partial \dot{q}} - \frac{\partial \mathcal{L}}{\partial q} = \tau + \tau^{ext} \tag{7.25}$$

where $\tau^{ext} = \boldsymbol{J}_T^{\top} f^{ext} \in \mathbb{R}^{6+n}$ represents the environmental reaction forces and torques converted into joint space. τ signifies the generalized joint space forces and torques corresponding to the system's generalized coordinate variable q. For a manipulator, its joint torque can be directly obtained from the joint actuators. However, for an unmanned aerial vehicle platform, the situation is quite different. Taking a quadrotor UAV as an example, the corresponding torque calculation is expressed as

$$\begin{pmatrix} \tau_1 \\ \tau_2 \\ \tau_3 \end{pmatrix} = T \begin{pmatrix} \cos\psi & \sin\psi & 0 \\ \sin\psi & -\cos\psi & 0 \\ 0 & 0 & 1 \end{pmatrix} \begin{pmatrix} \varphi_d \\ \vartheta_d \\ \cos\varphi\cos\vartheta \end{pmatrix}, \begin{pmatrix} \tau_4 \\ \tau_5 \\ \tau_6 \end{pmatrix} = \left(\left({}^{I}\boldsymbol{R}_V\right)^{\top} \boldsymbol{R}_{\Phi}\right)^{\top} \begin{pmatrix} \tau_x \\ \tau_y \\ \tau_z \end{pmatrix} \tag{7.26}$$

In the computation of the total energy in formula (7.25), the kinetic energy and potential energy of the system are calculated separately. According to the composition of the system, the total kinetic energy $\mathcal{K}$ is

$$\mathcal{K} = \mathcal{K}_V + \sum_{i=1}^{n} \mathcal{K}_i \tag{7.27}$$

$$\mathcal{K}_V = \frac{1}{2} m_V \dot{\boldsymbol{p}}^{\top} \dot{\boldsymbol{p}} + \frac{1}{2} \dot{\boldsymbol{\Phi}}^{\top} \mathbf{R}_{\Phi}^{\top} {}^{I}\mathbf{R}_V \boldsymbol{I}_V \left({}^{I}\boldsymbol{R}_V\right)^{\top} \boldsymbol{R}_{\Phi} \dot{\boldsymbol{\Phi}} \tag{7.28}$$

$$\mathcal{K}_i = \frac{1}{2} m_i \dot{\boldsymbol{p}}_i^{\top} \dot{\boldsymbol{p}}_i + \frac{1}{2} \boldsymbol{w}_i^{\top} \left({}^{I}\boldsymbol{R}_V {}^{V}\boldsymbol{R}_i\right) \boldsymbol{I}_i \left({}^{I}\boldsymbol{R}_V {}^{V}\boldsymbol{R}_i\right)^{\top} \boldsymbol{w}_i \tag{7.29}$$

where $\boldsymbol{m}_V$ and $\boldsymbol{I}_V \in \mathbb{P}^{3\times 3}$ represent the mass and inertia matrix of the drone platform, respectively. Similarly, $\boldsymbol{m}_i$ and $\boldsymbol{I}_i \in \mathbb{P}^{3\times 3}$ denote the mass and inertia matrix of the i-th link of the manipulator. ${}^{V}\boldsymbol{R}_i$ is the rotation matrix from the i-th joint coordinate system $\mathcal{F}_{L_i}$ to the body coordinate system F_V. $\dot{\boldsymbol{p}}_i$ and w_i are the linear velocity and angular velocity of the i-th link in the world coordinate system F_I, with their specific calculations being

$$\dot{\boldsymbol{p}}_i = \dot{\boldsymbol{p}} + [\boldsymbol{w}]_\times {}^I\boldsymbol{R}_V {}^V\boldsymbol{p}_i + {}^I\boldsymbol{R}_V \boldsymbol{B}_{t,i}\dot{\boldsymbol{\theta}} \tag{7.30}$$

$$\boldsymbol{w}_i = \boldsymbol{w} + {}^I\boldsymbol{R}_V \boldsymbol{B}_{r,i}\dot{\boldsymbol{\theta}} \tag{7.31}$$

where ${}^V\boldsymbol{p}_i$ represents the position vector of the i-th link of the robot manipulator in the body coordinate system, which is a function of the joint angles of the robot manipulator. $\boldsymbol{B}_{t,\ i}$ and $\boldsymbol{B}_{r,\ i}$ represent the Jacobian matrix line velocity and angular velocity components from the i-th joint to the body coordinate system, respectively. By using Eqs. (7.8), (7.28), (7.29), (7.30), and (7.31), the total kinetic energy equation (7.27) can be simplified as

$$\mathcal{K} = \frac{1}{2}\dot{\boldsymbol{q}}^\top \boldsymbol{M}(\boldsymbol{q})\dot{\boldsymbol{q}} \tag{7.32}$$

where $\boldsymbol{M}(\boldsymbol{q})$ is a $(6 + n) \times (6 + n)$ symmetric positive definite inertia matrix, and the calculation of its element blocks is given as follows:

$$\boldsymbol{M}(\boldsymbol{q}) = \begin{pmatrix} \boldsymbol{M}_{11} & \boldsymbol{M}_{12} & \boldsymbol{M}_{13} \\ \boldsymbol{M}_{21} & \boldsymbol{M}_{22} & \boldsymbol{M}_{23} \\ \boldsymbol{M}_{31} & \boldsymbol{M}_{32} & \boldsymbol{M}_{33} \end{pmatrix} \tag{7.33}$$

where $\boldsymbol{M}_{11} = \left(m_V + \sum_{i=1}^{n} m_i\right)\mathbb{I}_{3\times 3} \in \mathbb{R}^{3\times 3}$, $\boldsymbol{M}_{22} = \boldsymbol{Q}^\top I_V \boldsymbol{Q} + \sum_{i=1}^{n} \times \left(m_i \boldsymbol{R}_\Phi^\top [{}^I\boldsymbol{R}_V {}^V\boldsymbol{p}_i]_\times^\top [{}^I\boldsymbol{R}_V {}^V\boldsymbol{p}_i]_\times \boldsymbol{R}_\Phi + \boldsymbol{Q}^{\top V}\boldsymbol{R}_i I_i ({}^V\boldsymbol{R}_i)^\top \boldsymbol{Q}\right) \in \mathbb{R}^{3\times 3}$, $\boldsymbol{M}_{33} = \sum_{i=1}^{n}\left(m_i \boldsymbol{B}_{t,i}^\top \boldsymbol{B}_{t,i} + \boldsymbol{B}_{r,i}^{\top V}\boldsymbol{R}_i I_i ({}^V\boldsymbol{R}_i)^\top \boldsymbol{B}_{r,i})\right) \in \mathbb{R}^{n\times n}$, $\boldsymbol{M}_{12} = \boldsymbol{M}_{21}^\top = -\sum_{i=1}^{n}\left(m_i [{}^I\boldsymbol{R}_V {}^V\boldsymbol{p}_i]_\times \boldsymbol{R}_\Phi\right) \in \mathbb{R}^{3\times 3}$, $\boldsymbol{M}_{13} = \boldsymbol{M}_{31}^\top = \sum_{i=1}^{n}\left(m_i^I \boldsymbol{R}_V \boldsymbol{B}_{t,i}\right) \in \mathbb{R}^{3\times n}$, and $\boldsymbol{M}_{23} = \boldsymbol{M}_{32}^\top = \sum_{i=1}^{n}\left(\boldsymbol{Q}^{\top V}\boldsymbol{R}_i I_i ({}^V\boldsymbol{R}_i)^\top \boldsymbol{B}_{r,i} - m_i \boldsymbol{R}_\Phi^\top [{}^I\boldsymbol{R}_V^V \boldsymbol{p}_i]_\times^\top {}^I\boldsymbol{R}_V \boldsymbol{B}_{t,i}\right) \in \mathbb{R}^{3\times n}$.

For the system potential energy, it is also the sum of the potential energies of each link of the drone platform and the manipulator. The total potential energy $\mathcal{U}$ of the entire system can be expressed as

$$\mathcal{U} = \mathcal{U}_V + \sum_{i=1}^{n} \mathcal{U}_i \tag{7.34}$$

where $\mathcal{U}_V$ represents the potential energy of the drone platform and $\mathcal{U}_i$ represents the potential energy of the i-th link of the manipulator. The potential energy of this system is mainly gravitational potential energy. According to the formula for calculating gravitational potential energy, $\mathcal{U}_V$ and $\mathcal{U}_i$ are given by

$$\mathcal{U}_V = m_V g \boldsymbol{e}_z^\top \boldsymbol{p} \tag{7.35}$$

$$\mathcal{U}_i = m_i g \mathbf{e}_z^\top \left(\mathbf{p} + {}^I\mathbf{R}_V {}^V\mathbf{p}_i\right) \tag{7.36}$$

where $g = 9.8\,\text{m/s}^2$ represents the value of gravitational acceleration and $\boldsymbol{e}_z = (0, 0, 1)^\top$ is an operator that projects the target vector onto the z axis direction.

At this point, after calculating and analyzing the total kinetic energy and potential energy of the system, the dynamic equations of the system can be derived by combining Eqs. (7.25), (7.27), and (7.34):

$$\boldsymbol{M}(\boldsymbol{q})\ddot{\boldsymbol{q}} + \boldsymbol{C}(\boldsymbol{q}, \dot{\boldsymbol{q}})\dot{\boldsymbol{q}} + \boldsymbol{G}(\boldsymbol{q}) = \boldsymbol{\tau} + \boldsymbol{\tau}^{ext} \tag{7.37}$$

where $\boldsymbol{M(q)} \in \mathbb{P}^{(6+n)\times(6+n)}$ is a symmetric positive definite inertia matrix, and its specific calculation is shown in formula (7.33). $\boldsymbol{G}(\boldsymbol{q}) = \partial\mathcal{U}/\partial\boldsymbol{q} \in \mathbb{R}^{6+n}$ is the gravity term vector, $\boldsymbol{C}(\boldsymbol{q}, \dot{\boldsymbol{q}}) \in \mathbb{R}^{(6+n)\times(6+n)}$ is the Coriolis matrix, and the calculation of its elements is as follows:

$$c_{ij} = \sum_{k=1}^{n+6} \frac{1}{2}\left(\frac{\partial m_{ij}}{\partial q_k} + \frac{\partial m_{ik}}{\partial q_j} + \frac{\partial m_{jk}}{\partial q_i}\right)\dot{q}_k \tag{7.38}$$

where m_{ij} is the function of the i-th row and j-th column in matrix $\boldsymbol{M(q)}$.

The derived formula (7.37) is strictly speaking the dynamic equation of a robot's joint space. For the need to control the end effector in Cartesian space and interact with the environment, the robot's dynamic equation is transformed into Cartesian space. By multiplying both sides of formula (7.37) by the transpose inverse Jacobian matrix $\boldsymbol{J}_T^{-1}$ as referenced in formula (7.36), the following expression is obtained:

$$\boldsymbol{J}_T^{-\top}\boldsymbol{M}(\boldsymbol{q})\ddot{\boldsymbol{q}} + \boldsymbol{J}_T^{-\top}\boldsymbol{C}(\boldsymbol{q}, \dot{\boldsymbol{q}})\dot{\boldsymbol{q}} + \boldsymbol{J}_T^{-\top}\boldsymbol{G}(\boldsymbol{q}) = \boldsymbol{J}_T^{-\top}\boldsymbol{\tau} + \boldsymbol{J}_T^{-\top}\boldsymbol{\tau}^{ext} \tag{7.39}$$

Secondly, the system speed space Jacobian can be obtained by differentiation:

$$\ddot{\boldsymbol{x}}_T = \dot{\boldsymbol{J}}_T\dot{\boldsymbol{q}} + \boldsymbol{J}_T\ddot{\boldsymbol{q}} \Rightarrow \ddot{\boldsymbol{q}} = \boldsymbol{J}_T^{-1}\ddot{\boldsymbol{x}}_T - \boldsymbol{J}_T^{-1}\dot{\boldsymbol{J}}_T\dot{\boldsymbol{q}} \tag{7.40}$$

The dynamic equation of the system in Cartesian space can be obtained by substituting Eqs. (7.16) and (7.40) into Eq. (7.39):

$$\boldsymbol{M}_x(\boldsymbol{q})\ddot{\boldsymbol{x}}_T + \boldsymbol{C}_x(\boldsymbol{q}, \dot{\boldsymbol{q}})\dot{\boldsymbol{x}}_T + \boldsymbol{G}_x(\boldsymbol{q}) = \boldsymbol{J}_T^{-\top}\boldsymbol{\tau} + \boldsymbol{f}^{ext} \tag{7.41}$$

where $\boldsymbol{C}_x(\boldsymbol{q}, \dot{\boldsymbol{q}}) = \boldsymbol{J}_T^{-\top}\boldsymbol{C}(\boldsymbol{q}, \dot{\boldsymbol{q}})\boldsymbol{J}_T^{-1} - \boldsymbol{M}(\boldsymbol{q})\dot{\boldsymbol{J}}_T\boldsymbol{J}_T^{-1}$, $\boldsymbol{G}_x(\boldsymbol{q}) = \boldsymbol{J}_T^{-\top}\boldsymbol{G}(\boldsymbol{q})$, and $\boldsymbol{f}^{ext} = \boldsymbol{J}_T^{-\top}\boldsymbol{\tau}^{ext}$.

7.3 Visual Servo Controller for UAM

7.3.1 Practical Visual Servo Control for Aerial Manipulation with a Spherical Projection Model

7.3.1.1 Reference Frames and Kinematic Model

The reference frames of the aerial manipulator are shown in Fig. 7.5. Let $\{\mathcal{I} : O_I - x_I y_I z_I\}$ denote the right-hand inertial coordinate frame and z_I denote the inverse vertical direction. $\{\mathcal{B} : O_b - x_b y_b z_b\}$ represents the body frame of the quadrotor platform. The geometric center of the quadrotor coincides in O_b. $\{\mathcal{D}_i : O_i - x_i y_i z_i\}$ is the body frame of the i-link, where the subscript $i = 1, 2, 3, 4$ denotes the link number of the robot arm. $\{\mathcal{T} : O_t - x_t y_t z_t\}$ is the gripper frame, and the axis direction is the same as $\mathcal{D}_4$. The generalized coordinate variables $\boldsymbol{p} = (x, y, z)^\top$ and $\boldsymbol{\Phi} = (\varphi, \vartheta, \psi)^\top$ are the position and triple Euler roll-pitch-yaw angles of the quadrotor expressed in the inertial frame, respectively. The joint angle vector $\boldsymbol{\theta} = (\theta_1, \theta_2, \theta_3, \theta_4)^\top$ of the manipulator at the zero position is shown in Fig. 7.6. Let vector $\boldsymbol{q}$ represent all the generalized coordinate variables, which can be written as

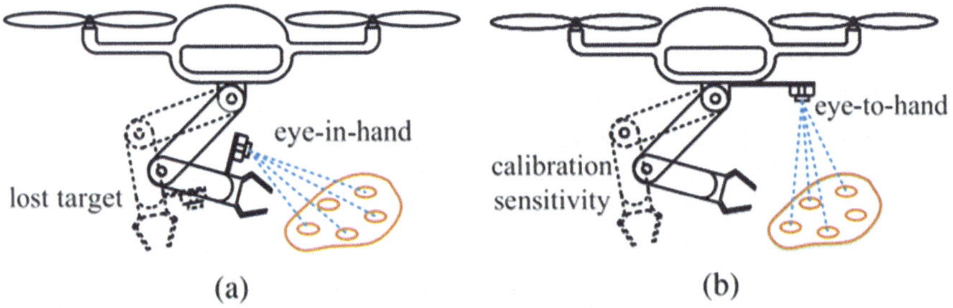

Fig. 7.5 Two configurations of a camera installation in a visual servo control. (**a**) Camera is mounted at the end of the robot arm (eye-in-hand). (**b**) Camera is mounted on the flight platform (eye-to-hand)

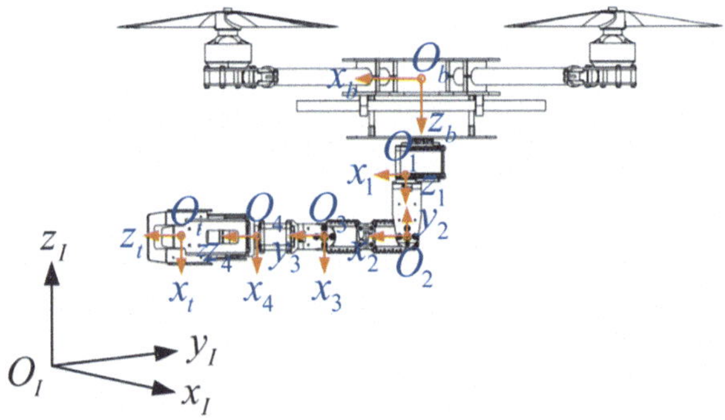

Fig. 7.6 Reference coordinate frames and the kinematic model of the UAM

$$q = \begin{pmatrix} p^\top & \Phi^\top & \theta^\top \end{pmatrix}^\top \in \mathbb{R}^{10\times 1}. \tag{7.42}$$

Let p_t represent the position of the gripper with respect to $\mathcal{I}$. Therefore, $\dot{p}$, ω, $\dot{p}_t$, and ω_t represent the translational velocity and angular velocity of the quadrotor and the translational velocity and angular velocity of the gripper, respectively. Note that all the variables are defined with respect to the inertial coordinate frame $\mathcal{I}$. The kinematic model of the aerial manipulator aims to calculate the expression of p_t and its derivatives with the independent variables q.

Let p_t^b be the position of the gripper described in $\mathcal{B}$. Then, p_t can be calculated by using the chain rule as

$$p_t = p + R_b^I p_t^b \tag{7.43}$$

where $R_b^I \in SO(3)$ is the rotation transformation matrix from frame $\mathcal{B}$ to $\mathcal{I}$ and p_t^b is a function of the joint angle θ. In addition, $\dot{p}_t^b$ and ω_t^b are the translational velocity and angular velocity of the gripper described in $\mathcal{B}$ and can be expressed by the Jacobian matrices of the manipulator, $J_l \in \mathbb{P}^{3\times 4}$ and $J_r \in \mathbb{P}^{3\times 4}$, as

$$\dot{p}_t^b = J_l \dot{\theta} \tag{7.44}$$

$$\omega_t^b = J_r \dot{\theta}. \tag{7.45}$$

Taking the derivative of (7.43) and using (7.44) and (7.45), the translational and angular velocities of the gripper with respect to $\mathcal{I}$ can be written as

$$\dot{p}_t = \dot{p} + sk(\omega) R_b^I p_t^b + R_b^I J_l \dot{\theta} \tag{7.46}$$

$$\omega_t = \omega + R_b^I J_r \dot{\theta} \tag{7.47}$$

where the notation $sk(\omega)$ denotes the skew-symmetric matrix such that $sk(\omega)v = \omega \times v$ for the vector cross product $\times$ and any vector v. Using the generalized coordinate variables q, (7.46) and (7.47) can be rewritten in the matrix form given by

$$\dot{p}_t = \left(\mathbb{I}_{3\times 3} - sk\left(R_b^I p_t^b\right) T R_b^I J_l\right)\dot{q} = B_l \dot{q} \tag{7.48}$$

$$\omega_t = \left(0_{3\times 3} T R_b^I J_r\right)\dot{q} = B_r \dot{q} \tag{7.49}$$

where $B_l \in \mathbb{R}^{3\times 9}$ and $B_r \in \mathbb{P}^{3\times 9}$ are the transformation matrices that convert $\dot{q}$ to the translational and angular velocities of the gripper. $\mathbb{I}_{3\times 3}$ and $0_{3\times 3}$ are the identity and zero matrices, respectively. T is also a transformation matrix that converts Euler angle rates $\dot{\Phi}$ to angular velocity ω and is defined as

$$\boldsymbol{T} = \begin{pmatrix} \cos\vartheta & 0 & -\cos\varphi\sin\vartheta \\ 0 & 1 & \sin\varphi \\ \sin\vartheta & 0 & \cos\varphi\cos\vartheta \end{pmatrix} \tag{7.50}$$

7.3.1.2 Visual Servo Control Design

Choosing visual feedback features that are strongly correlated with the target position and orientation is very important for visual servo control. In some of the previous literature [10], spherical visual features have been shown to be effective for visual servo control of a quadrotor. It is capable of decoupling the position and orientation dynamics of the system and can use the centroid-based design methods to eliminate the dependence on depth information of target features. For an aerial manipulator, spherical visual features can also be used by installing the camera in the end effector of the manipulator.

The geometric relationship of the sphere projection is shown in Fig. 7.7. Let $\{\mathcal{C} : O_c - x_c y_c z_c\}$ be a Cartesian frame attached to the focal point of the camera, and the axis direction is the same as the end-effector frame $\mathcal{T}$. The object to be grabbed is represented by n feature points, and $\boldsymbol{Q}_i = (X_i, Y_i, Z_i)^\top$ denotes the i-th point. Let $\boldsymbol{q}_i \in \mathbb{P}^{3\times 1}$ and $\boldsymbol{s}_i \in \mathbb{P}^{3\times 1}$ be the corresponding position of $\boldsymbol{Q}_i$ in the focal normalized plane and the sphere normalized imaging plane, respectively. Thus, it is obtained:

$$\boldsymbol{s}_i = \frac{\boldsymbol{q}_i}{|\boldsymbol{q}_i|} = \frac{\boldsymbol{Q}_i}{|\boldsymbol{Q}_i|} \tag{7.51}$$

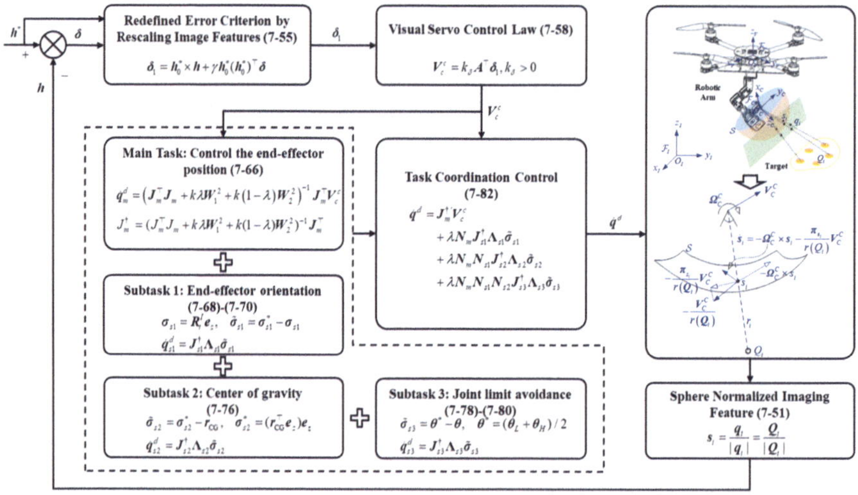

Fig. 7.7 Visual servo control and geometric relationship of the sphere projection

where $|\mathbf{q}|$ represents the Euclidean norm of any vector $\boldsymbol{q} \in \mathbb{P}^{n \times 1}$, $|\ \boldsymbol{q}\ | = \sqrt{\boldsymbol{q}^{\top}\boldsymbol{q}}$. Let (u_i, v_i) be the image coordinate of $\boldsymbol{Q}_i$, and (u_i, v_i) is the intersection of the optical axis z_c and the image plane. Thus, $\boldsymbol{q}_i = (q_{xi}, q_{yi}, 1)^{\top}$ can be rewritten as follows:

$$\begin{cases} q_{xi} = \dfrac{f}{\alpha_x}(u - u_0) = \dfrac{X_i}{Z_i} \\ q_{yi} = \dfrac{f}{\alpha_y}(v - v_0) = \dfrac{Y_i}{Z_i} \end{cases} \tag{7.52}$$

where $\alpha_x > 0$ and $\alpha_y > 0$ are, respectively, the magnification factors of the imaging plane to the object plane in the x and y axes and f is the focal length of the camera. Benefiting from the previous literature [11], the dynamic of an image point $\mathbf{s}_i$ for a spherical camera of the image surface with a radius of unity is

$$\dot{\mathbf{s}}_i = -\boldsymbol{\Omega}_c^c \times \boldsymbol{s}_i - \frac{\boldsymbol{\pi}_{s_i}}{r_i}\boldsymbol{V}_c^c \tag{7.53}$$

where $\mathbf{V}_c^c$ and $\mathbf{\Omega}_c^c$ are the linear and angular velocities of the camera with respect to the camera frame $\mathcal{C}$. $\boldsymbol{r}_i = |\ \boldsymbol{Q}_i|$ is the relative depth of the target point $\boldsymbol{Q}_i$, and $\boldsymbol{\pi}_{s_i} = \left(\mathbb{I}_{3\times 3} - \boldsymbol{s}_i\boldsymbol{s}_i^{\top}\right)$ is the projection $\boldsymbol{\pi}_s : \mathbb{R}^3 \rightarrow T_s\mathcal{K}^2$ onto the tangent space of sphere $\mathcal{K}^2$ at the point $\mathbf{s}_i \in \mathcal{K}^2$. It can be seen that though the third component of $\mathbf{s}_i$ is nonlinearly related to the depth of the observed target, $\mathbf{s}_i$ can be calculated using only the image coordinates of the feature. In the following, a visual servo control algorithm will be developed that does not require depth information for image features.

For n target points, the use of the centroid technology has many advantages. The image centroid is highly robust to pixel noise and can be easily computed in real time. Thus, the unnormalized spherical centroid of $\mathbf{s}_i$ is defined as $\boldsymbol{h} = \sum_{i=1}^{n}\mathbf{s}_i$. Let $\boldsymbol{h}^*$ be the desired target image feature vector expressed in the camera-fixed frame. The feedback visual error is defined as

$$\boldsymbol{\delta} = \boldsymbol{h} - \boldsymbol{h}^* \tag{7.54}$$

where $\boldsymbol{h}^*$ is set to $\boldsymbol{R}_I^c\boldsymbol{b}$, and $\boldsymbol{b} \in \mathbb{P}^{3 \times 1}$ is a known desired direction of the visual feature in the inertial frame $\mathcal{I}$. The feature h is measured relative to the camera frame and not in the inertial frame, and it is necessary to map the desired set point into the camera frame before an image-based error can be defined. It is highlighted that the feedback visual error $\boldsymbol{\delta}$ given by (7.54) yields an ill-conditioned feedback matrix. To address this issue, studies [12] and [13] proposed a redefined error criterion by rescaling the image features

$$\boldsymbol{\delta}_1 = \boldsymbol{h}_0^* \times \boldsymbol{h} + \gamma\boldsymbol{h}_0^*\left(\boldsymbol{h}_0^*\right)^{\top}\boldsymbol{\delta} \tag{7.55}$$

where $\boldsymbol{h}_0^* = \boldsymbol{h}^*/|\mathbf{h}^*| \in \mathbb{R}^{3\times 1}$ corresponds to a normalized version of h^*, and $\gamma \in \mathbb{P}$ is a positive constant to be suitably tuned. By using the redefined visual error criterion $\boldsymbol{\delta}_1$, the kinematics of $\boldsymbol{\delta}_1$ can be derived by (7.51) as follows:

$$\dot{\boldsymbol{\delta}}_1 = -\boldsymbol{\Omega}_c^c \times \boldsymbol{\delta}_1 - \boldsymbol{AHV}_c^c \tag{7.56}$$

where $\boldsymbol{H} = \sum_{i=1}^{n} (\boldsymbol{\pi}_{s_i}/r_i) \in \mathbb{R}^{3\times 3}$ and $\boldsymbol{A} = -sk(\boldsymbol{h}_0^*) + \gamma \boldsymbol{h}_0^* (\boldsymbol{h}_0^*)^\top \in \mathbb{R}^{3\times 3}$. Note that (7.56) satisfies the passivity-like structure properties, and an error storage function S is defined as

$$S = \frac{1}{2}|\boldsymbol{\delta}_1|^2. \tag{7.57}$$

Taking the time derivative of the error function S and substituting into (4.139) yields

$$\dot{S} = -\boldsymbol{\delta}_1^\top \boldsymbol{AHV}_c^c. \tag{7.58}$$

If the control law is chosen as $\boldsymbol{V}_c^c = k_\delta \boldsymbol{A}^\top \boldsymbol{\delta}_1, \quad k_\delta > 0$, the derivative of the storage function S becomes $\dot{S} = -k_\delta \boldsymbol{\delta}_1^\top \boldsymbol{AHA}^\top \boldsymbol{\delta}_1$. Since $\boldsymbol{H}$ is a positive definite matrix and $\boldsymbol{A}$ is nonsingular, matrix $\boldsymbol{AHA}^\top$ is symmetric positive definite. Thus, the Lyapunov theory guarantees that $\boldsymbol{\delta}_1$ converges exponentially to zero. Note that the designed control rate $\boldsymbol{V}_c^c$ and angular velocity $\boldsymbol{\Omega}_c^c$ are independent of each other.

The aerial manipulator described herein is a redundant system with 10-DOFs, of which the quadrotor platform has 6-DOFs (four controllable and two uncontrollable) and the manipulator has 4-DOFs. The inverse kinematics of the UAM are not unique for a given end-effector position and orientation. This allows us to perform some subtasks, such as camera posture, system center of gravity (CoG), and joint limit, while ensuring that the main task is completed. In this subsection, several low-priority subtasks useful for the main task of the aerial manipulator are proposed:

1. Main Task: Control the end-effector position.

   Since the camera is fixed on the end effector, let us take the camera velocity $\boldsymbol{V}_c^c$ as the end-effector velocity with respect to time. To move the gripper to the desired position, it is necessary to distribute $\boldsymbol{V}_c^c$ into the coordinate variables $\dot{\mathbf{q}}$, which are constrained as

$$\mathbf{V}_c^c = \boldsymbol{J}_{\text{main}} \dot{\boldsymbol{q}} \tag{7.59}$$

   where $\boldsymbol{J}_{\text{main}}$ is the Jacobian matrix linking $\dot{\boldsymbol{q}}$ to the end-effector velocity $\boldsymbol{V}_c^c$. Consider $\boldsymbol{V}_c^c$ as a sum of the velocity added by the arm kinematics and the quadrotor movement in camera frame $\mathcal{C}$. According to (7.38) and (7.44), it is obtained:

$$J_{\text{main}} = R_I^c \dot{p}_t = \left(R_I^c, \ -R_I^c sk(p_t)T, R_b^c J_l\right) \tag{7.60}$$

where R_I^c and R_b^c are the rotation matrices from frame $\mathcal{I}$ to C and $\mathcal{B}$ to C, respectively. The general solution for the $\dot{\mathbf{q}}$ is

$$\dot{q} = J_{\min}^{\dagger} V_c^c \tag{7.61}$$

where the $J_{\text{main}}^{\dagger} \in \mathbb{R}^{10\times 3}$ is the pseudoinverse matrix of J_{main} and $J_{\text{main}}^{\dagger} = J_{\text{main}}^{\top}\left(J_{\text{main}} J_{\text{main}}^{\top}\right)^{-1}$ if J_{main} has full row rank.

The pseudoinverse solution (7.61) provides a meaningful solution in the sense of least squares, regardless of whether (7.59) is underspecified, square, or overspecified. This ability makes the pseudoinverse the most attractive technique in redundancy resolution. However, a major drawback exists associated with this solution. The solution given by (7.61) may generate configurations of joints, where $J_{\text{main}}^{\dagger}$ is no longer full rank. Near singular configurations, the norm of the solution obtained by (7.61) becomes very large. Therefore, to solve this issue, a damped least squares scheme should be used. To explain, when the UAM uses visual guidance to grasp a target object, two representative cases can be considered. One case is that when the end effector is far away from the target, it is better to move the quadrotor platform than the robot arm, which can avoid the manipulator stretching fully to reach the singular point. Another case is that when the end effector is close to the target, it is better to move the robot arm than the quadrotor platform, since the movement of the quadrotor in the x and y directions is associated with ϕ and ϑ motion and can lead to the undesirable motion of the end effector. For the UAM, the motions of the flight platform and the manipulator are quite different. Thus, a dynamic penalty parameter λ and matrices W_1 and W_2 are designed to separate the variables of the flight platform and the robot arm. The cost function is designed as follows:

$$F = \frac{1}{2}\left|J_{\text{main}}\dot{q} - V_c^c\right|^2 + \frac{1}{2}k\lambda\left|W_1\dot{q}\right|^2 + \frac{1}{2}k(1-\lambda)\left|W_2\dot{q}\right|^2 \tag{7.62}$$

where V_c^c is the desired end-effector velocity, $k > 0$ is a positive real number parameter, and W_1 and W_2 are the matrices that separate the quadrotor platform variables and robot arm joint variables of $\dot{q}$, as shown below:

$$W_1 = \begin{pmatrix} \mathbb{I}_{6\times 6} & 0_{6\times 4} \\ 0_{4\times 6} & 0_{4\times 4} \end{pmatrix}; \quad W_2 = \begin{pmatrix} 0_{6\times 6} & 0_{6\times 4} \\ 0_{4\times 6} & \mathbb{I}_{4\times 4} \end{pmatrix}. \tag{7.63}$$

The dynamic penalty parameter λ is associated with the Euclidean norm of visual error $\boldsymbol{\delta}$ and is defined as follows:

$$\lambda = \begin{cases} 1, & \text{if} \quad |\boldsymbol{\delta}| \leq \delta_L \\ \dfrac{\delta_H - |\boldsymbol{\delta}|}{\delta_H - \delta_L}, & \text{if} \quad \delta_L < |\boldsymbol{\delta}| < \delta_H \\ 0, & \text{if} \quad |\boldsymbol{\delta}| \geq \delta_H \end{cases} \tag{7.64}$$

where δ_L and δ_H are the critical distances such that the movement of the end effector is entirely from the robot arm or quadrotor platform. Taking the derivative of the cost function F, it is obtained:

$$\frac{\partial F}{\partial \dot{\boldsymbol{q}}^\top} = \boldsymbol{J}_m^\top \boldsymbol{J}_m \dot{\boldsymbol{q}} - \boldsymbol{J}_m^\top \boldsymbol{V}_c^c + k\lambda \mathbf{W}_1^2 \dot{\boldsymbol{q}} + k(1-\lambda)\mathbf{W}_2^2 \dot{\boldsymbol{q}}. \tag{7.65}$$

From (7.65), solving the partial derivative of the cost function F for the $\dot{\boldsymbol{q}}$ results in

$$\dot{\boldsymbol{q}}_m^d = \left(\boldsymbol{J}_m^\top \boldsymbol{J}_m + k\lambda \mathbf{W}_1^2 + k(1-\lambda)\mathbf{W}_2^2\right)^{-1} \boldsymbol{J}_m^\top \boldsymbol{V}_c^c. \tag{7.66}$$

Now, let $\boldsymbol{J}_m^{\dagger\,\prime} = \left(\boldsymbol{J}_m^\top \boldsymbol{J}_m + k\lambda \mathbf{W}_1^2 + k(1-\lambda)\mathbf{W}_2^2\right)^{-1} \boldsymbol{J}_m^\top$ represent the modified pseudoinverse of the matrix of $\boldsymbol{J}_m$, and the null space of the Jacobian matrix $\boldsymbol{J}_m$ becomes

$$\boldsymbol{N}_m = \left(\mathbb{I}_{10\times 10} - \boldsymbol{J}_m^{\dagger\,\prime} \boldsymbol{J}_m\right). \tag{7.67}$$

2. Subtask 1: End-effector orientation.

   In this subtask, a spherical coordinate system is used to describe the gripper orientation control task. With this constraint task, the gripper can grasp the target from an arbitrary direction. Let $\boldsymbol{R}_t^I = \boldsymbol{R}_b^I \boldsymbol{R}_t^b$ represent the rotation transformation matrix from $\mathcal{T}$ to $\mathcal{I}$, and the inverse of the gravity direction is $\boldsymbol{e}_z = (0, 0, 1)^\top$. The subtask can be described as

$$\sigma_{s1} = \mathbf{R}_t^I \mathbf{e}_z. \tag{7.68}$$

where σ_{s1} is the subtask that means the projection of z_t in the unit direction axes of the inertial frame. The derivative of (7.68) is

$$\dot{\sigma}_{s1} = \frac{\partial(\boldsymbol{R}_t^I \boldsymbol{e}_z)}{\partial \boldsymbol{q}} \dot{\boldsymbol{q}} = \boldsymbol{J}_{s1} \dot{\boldsymbol{q}} \tag{7.69}$$

where $\boldsymbol{J}_{s1}$ is the Jacobian matrix of the subtask. Let $\sigma_{s1}^* = \left(\sin\varphi_{s1}^* \cos\theta_{s1}^*, \sin\varphi_{s1}^* \sin\theta_{s1}^*, \cos\varphi_{s1}^*\right)^\top$ be the desired value of σ_{s1} that represents the desired end-effector orientation expressed in the spherical coordinate system. $\widetilde{\sigma}_{s1} = \sigma_{s1}^* - \sigma_{s1}$ is the task error, and the desired $\dot{\boldsymbol{q}}$ of the subtask is

$$\dot{q}_{s1}^{d} = J_{s1}^{\dagger} \Lambda_{s1} \widetilde{\sigma}_{s1} \tag{7.70}$$

where $J_{s1}^{\dagger}$ is the pseudoinverse matrix of J_{s1} and $\Lambda_{s1} \in \mathbb{P}^{3\times 3}$ is a positive definite diagonal matrix of gains. The null space of the Jacobian matrix J_{s1} is

$$N_{s1} = \left(\mathbb{I}_{10\times 10} - J_{s1}^{\dagger} J_{s1}\right). \tag{7.71}$$

3. Subtask 2: Center of gravity.
   If the arm and quadrotor CoG are not vertically aligned, the motion of the arm produces an undesired torque on the quadrotor base that perturbs the system attitude and position. Thus, minimizing the CoG offset vector r_{CG} is very significant for stability control of the quadrotor. The offset vector r_{CG} is a function of the joint angle configuration defined as

$$r_{\mathrm{CG}} = \frac{1}{m} \sum_{i=1}^{4} m_i p_{gi}^{b} \tag{7.72}$$

where m_i and p_{gi}^{b} are the mass and the position of the CoG of the i-th arm link, respectively. $m = m_0 + \sum_{i=1}^{4} m_i$ is the total mass of the system, and m_0 is the quadrotor mass. The differential relationship between r_{CG} and $\dot{q}$ can be solved by taking the derivative of (7.72) and yields

$$\dot{r}_{\mathrm{CG}} = J_{s2}\dot{q} = \left(0_{3\times 6} J_{g}^{b}\right) \begin{pmatrix} \dot{p}^{\top} \\ \dot{\Phi}^{\top} \\ \dot{\theta}^{\top} \end{pmatrix} \tag{7.73}$$

where J_{s2} is the subtask Jacobian matrix and $J_{g}^{b} \in \mathbb{R}^{3\times 4}$ is the CoG Jacobian matrix expressed in $\mathscr{B}$ and defined as follows:

$$J_{g}^{b} = \frac{\partial r_{\mathrm{CG}}}{\partial \theta} = \left(\frac{\partial r_{\mathrm{CG}}}{\partial \theta_1} \; \frac{\partial r_{\mathrm{CG}}}{\partial \theta_2} \; \frac{\partial r_{\mathrm{CG}}}{\partial \theta_3} \; \frac{\partial r_{\mathrm{CG}}}{\partial \theta_4}\right) \tag{7.74}$$

$$\frac{\partial r_{\mathrm{CG}}}{\partial \theta_i} = \frac{\sum_{j=1}^{4} m_j}{m} sk\left(R_i^b e_z\right) r_{gi}^{b}; r_{gi}^{b} = \frac{1}{m} R_i^b \sum_{j=i}^{4} m_j p_{gj}^{b}. \tag{7.75}$$

Let $\sigma_{s2}^{*} = \left(r_{\mathrm{CG}}^{\top} e_z\right) e_z$ be the desired task value, and $\widetilde{\sigma}_{s2} = \sigma_{s2}^{*} - r_{\mathrm{CG}}$ is the task error. The desired $\dot{q}$ of the subtask is

$$\dot{q}_{s2}^{d} = J_{s2}^{\dagger}\Lambda_{s2}\widetilde{\sigma}_{s2} \tag{7.76}$$

where $J_{s2}^{\dagger}$ is the pseudoinverse matrix of J_{s2} and $\Lambda_{s2} \in \mathbb{P}^{3 \times 3}$ is a positive definite diagonal matrix of gains. The null space of the Jacobian matrix J_{s2} is

$$N_{s2} = \left(\mathbb{I}_{10 \times 10} - J_{s2}^{\dagger}J_{s2}\right). \tag{7.77}$$

4. Subtask 3: Joint limit avoidance.

   For an aerial manipulator system, the goal of joint limit avoidance is to keep the robot arm joint far from its limits. One possible solution to avoid mechanical joint limits is to bring the joints to a central position

$$\theta^{*} = \frac{\theta_L + \theta_H}{2} \tag{7.78}$$

where θ_L and θ_H are the vectors of the low and high joint limits, respectively. The differential relationship between the robot arm joint angles θ and q can be described as

$$\dot{\theta} = J_{s3}\dot{q} \tag{7.79}$$

where $J_{s3} = (\mathbb{O}_{4 \times 6}, \mathbb{I}_{4 \times 4})$ is the subtask Jacobian matrix. Let $\widetilde{\sigma}_{s3} = \theta^{*} - \theta$ be the task error, and the desired $\dot{\mathbf{q}}$ of the subtask is

$$\dot{q}_{s3}^{d} = J_{s3}^{\dagger}\Lambda_{s3}\widetilde{\sigma}_{s3} \tag{7.80}$$

where $J_{s3}^{\dagger}$ is the pseudoinverse matrix of J_{s3} and $\Lambda_{s3} \in \mathbb{P}^{3 \times 3}$ is a positive definite diagonal matrix of gains. The null space of the Jacobian matrix J_{s3} is

$$N_{s3} = \left(\mathbb{I}_{10 \times 10} - J_{s3}^{\dagger}J_{s3}\right). \tag{7.81}$$

For the redundancy features of the aerial manipulator system, as mentioned above, multiple tasks (one main task and three subtasks) can be prioritized to try to accomplish simultaneously most of the tasks. Based on the previous research literature [14], a possible generalization of the task priority for aerial manipulation is

$$\dot{q}^{d} = J_{m}^{\dagger'}V_{c}^{c} + \lambda N_m J_{s1}^{\dagger}\Lambda_{s1}\widetilde{\sigma}_{s1} + \lambda N_m N_{s1} J_{s2}^{\dagger}\Lambda_{s2}\widetilde{\sigma}_{s2} + \lambda N_m N_{s1} N_{s2} J_{s3}^{\dagger}\Lambda_{s3}\widetilde{\sigma}_{s3} \tag{7.82}$$

where $\dot{q}^{d}$ is the desired derivation of the system configuration for the whole task and λ is the dynamic penalty parameter, as mentioned earlier. This algorithm has a clear geometrical interpretation, and the task error associated with the lower-priority task is further projected in the null space of the sole higher-priority task.

7.3.1.3 Simulation Results

The proposed approach has been tested in a simulator developed in the MATLAB/Simulink environment. The simulation of the camera imaging model is based on the Machine Vision Toolbox [15]. A quadrotor with a weight of 2.7 kg and a propeller wheelbase of 55 cm endowed with a 4-DOF robot arm plus an end effector has been employed. A camera is attached to the end effector and has the same direction of z_t that is used for the eye-in-hand configuration. A simulation scenario, including grasping an object using visual guidance of the redundant system, is designed to show the performance of the aerial manipulation with the proposed method. The visual markers are characterized by four feature points forming a square of 10 cm of segment length lying on the ground with the center offset (0.6, 0.3). The quadrotor platform's initial position with respect to the fixed inertial frame is (−0.4, −0.4, 2), and (−π/3, 3π/4, 0, 0) is taken as the robot arm initial joint value for suitable imaging of the markers. The desired position of the end effector is $\boldsymbol{p}_t^* = (0.6, 0.3, 0.3)^\top$. The involved tuning gains are empirically set as follows: $\gamma = 2$, $\delta_L = 0.1$, $\delta_H = 0.8$, $k = 0.3$, $\boldsymbol{\Lambda}_{s1} = \text{diag}(1, 1, 1)$, $\boldsymbol{\Lambda}_{s2} = \text{diag}(0.1, 0.1, 0.1)$, and $\boldsymbol{\Lambda}_{s3} = \text{diag}(1, 1, 1)$. The total simulation time is 30 s and divided into two phases. In the first stage, from 0 to 15 s, the desired gripper orientation is set to be $\varphi_{s1}^* = \pi$ and $\theta_{s1}^* = 0$. In the second stage, from 15 to 30 s, the desired gripper orientation is set to be $\varphi_{s1}^* = 2\pi/3$ and $\theta_{s1}^* = -2\pi/3$. The same parameters are used to compare the two cases with and without λ. The corresponding results are shown in Figs. 7.8, 7.9, and 7.10.

As seen from Figs. 7.8a and 7.9a, the image features $\boldsymbol{\delta}_1$ specified in (7.55) can be achieved for both cases. The UAM can reach the grasping point under the proposed visual servo control. However, the trajectory of the quadrotor and end effector is quite different with and without λ parameters. In Fig. 7.8a, due to the adjustment of

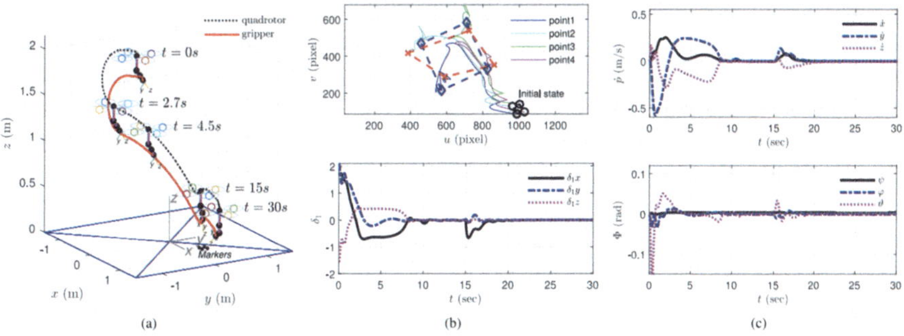

Fig. 7.8 Simulation results of the proposed control scheme with λ. (**a**) Corresponding path of UAM. (**b**) Evolution of the visual features (u_i, v_i) and the redefined visual error $\boldsymbol{\delta}_1$. (**c**) Quadrotor velocity $\dot{\boldsymbol{p}}$ and the triple Euler roll-pitch-yaw angles $\boldsymbol{\Phi}$

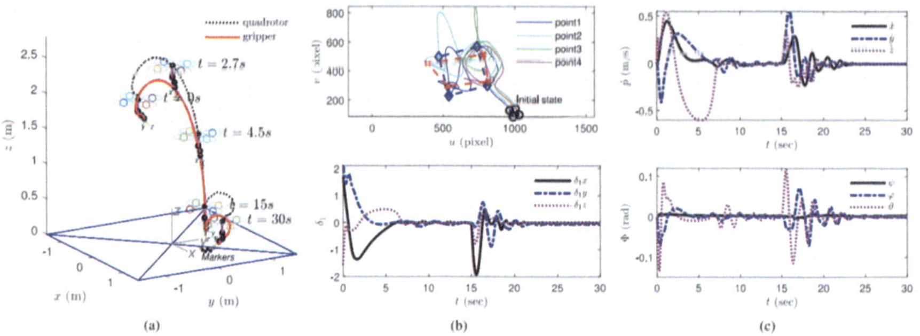

Fig. 7.9 Simulation results of the proposed control scheme without λ. (**a**) Corresponding path of UAM. (**b**) Evolution of the visual features (u_i, v_i) and the redefined visual error $\boldsymbol{\delta}_1$. (**c**) Quadrotor velocity $\dot{\boldsymbol{p}}$ and the triple Euler roll-pitch-yaw angles $\boldsymbol{\Phi}$

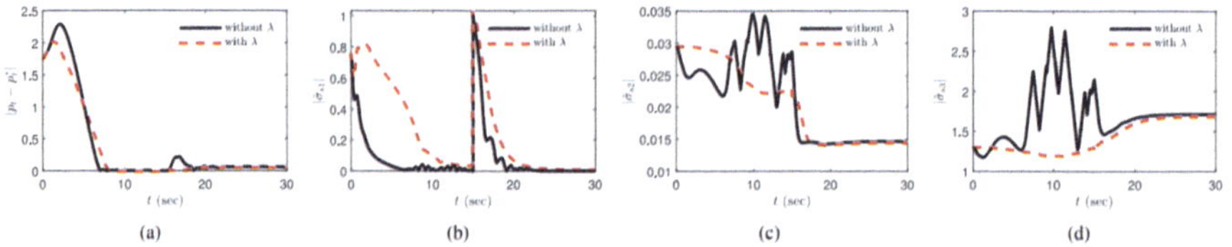

Fig. 7.10 Comparison effect of λ parameter on each task. (**a**) Comparison of the norm of the end-effector position error $|\boldsymbol{p}_t - \boldsymbol{p}_t^*|$. (**b**) Comparison of the norm of the gripper orientation error $|\widetilde{\sigma}_{s1}|$. (**c**) Comparison of the norm of the CoG offset error $|\widetilde{\sigma}_{s2}|$. (**d**) Comparison of the norm of the joint limit avoidance error $|\widetilde{\sigma}_{s3}|$

λ, the wide range of motion of the UAM is mainly due to the quadrotor, and the robotic arm only starts to move when it is close to the target. In contrast, in Fig. 7.9a, the quadrotor and the robotic arm have the same motion priority due to the absence of adjustment of λ. Therefore, in the initial stage, the robotic arm moves rapidly to reduce the modified image features $\boldsymbol{\delta}_1$, which makes the convergence speed of δ_{1x}, δ_{1y}, and δ_{1z} in Fig. 7.9b faster than that in Fig. 7.8b. To show the details, the evolution of the visual features (u_i, v_i) and the redefined visual error $\boldsymbol{\delta}_1$ are shown in Figs. 7.8b and 7.9b. The initial locations of the visual features are shown with four circles in black and move smoothly to a square with a diamond at 15 s. This means the end effector is looking down vertically to the markers as previously set. From 15 to 30 s, the visual features become an irregular quadrilateral with a cross in red achieved by setting the target gripper orientation to be nonvertical. The visual error vectors in Fig. 7.9b exhibit a faster convergence speed compared to those in Fig. 7.8b, especially in the components in the x and y directions. However, for the visual feature evolution, Fig. 7.9b is more complex compared with Fig. 7.8b because of the movement of the robot arm at the beginning. This indicates that λ can increase the system stability and robustness. The quadrotor velocity $\dot{\boldsymbol{p}}$ and triple Euler roll-pitch-yaw angles $\boldsymbol{\Phi}$ shown in Figs. 7.8c and 7.9c, respectively, also reveal the

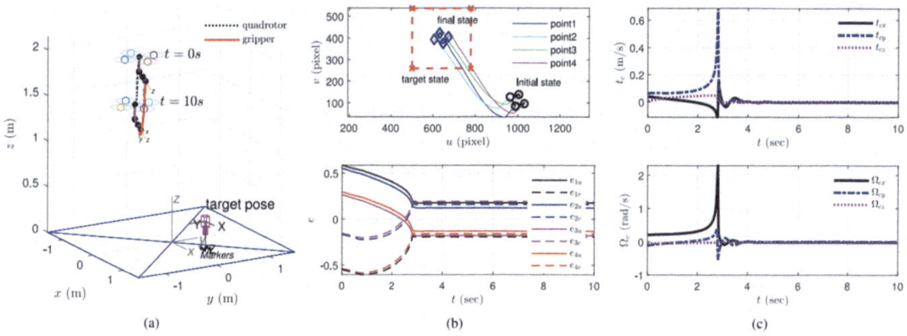

Fig. 7.11 Simulation results of the uncalibrated visual servo control. (**a**) Corresponding path of UAM. (**b**) Evolution of the visual features (u_i, v_i) and the visual error **e**. (**c**) Evolution of the six components of the camera velocity $\boldsymbol{v}_c = \left(\boldsymbol{t}_c^{\mathrm{T}}, \boldsymbol{\Omega}_c^{\mathrm{T}}\right)^{\mathrm{T}}$

influence of λ. In Fig. 7.9c, both the velocity $\dot{\boldsymbol{p}}$ and the Euler angles $\boldsymbol{\Phi}$ of the quadrotor are greater than the variation in Fig. 7.8c, especially on the z axis velocity component $\dot{z}$. This is because the movement of the robot arm also affects the motion of the quadrotor when the visual error $\boldsymbol{\delta}_1$ is large. The quadrotor is maneuvered at a high speed to accommodate the rapid movement of the robotic arm.

In Fig. 7.10, the effect of λ on each task is explained in detail. Overall, the task error curve with λ is smoother, which is more apparent in Fig. 7.10c and d. For the end-effector position error $\left|\boldsymbol{p}_t - \boldsymbol{p}_t^*\right|$ in Fig. 7.10a, since it is the main task in the task-priority control, the effect of the λ is not obvious. However, for Fig. 7.10b, the gripper orientation error $\widetilde{\sigma}_{s1}$ converges faster to the steady state in the absence of λ than when it is included. This phenomenon can be explained from the whole process. Comparing the gripper trajectories in Figs. 7.8a and 7.9a, the motion of the UAM in Fig. 7.8a is primarily derived from the quadrotor in the initial stage (from 0 to 10 s). The quadrotor is an underactuated system, and its pitch and roll attitude angles are uncontrollable variables, so the convergence speed of the end-effector orientation is not obvious. In Fig. 7.9a, since the motion of the quadrotor and the robotic arm are the same during the entire process, the movement of the robotic arm can adjust the orientation error of the gripper at the beginning stage, so the convergence speed is faster in Fig. 7.10b.

For the purpose of comparison and a better evaluation, the performance of our proposed visual servo control is compared with the uncalibrated visual servo control [16]. The initial state of the UAM is consistent with the one implemented above, and the results are shown in Fig. 7.11. Figure 7.11a shows the corresponding path of the UAM under the uncalibrated visual servo control, where the UAM reaches a local minimum and is far away from the desired configuration. As seen in Fig. 7.11b, each component of the visual error e has a fast exponential decrease with the same convergence speed, causing straight-line trajectories in the image plane, but the steady-state error is not exactly zero. It is also clear from Fig. 7.11c that the evolution of the six components of the camera velocity $\boldsymbol{v}_c$ tends to zero. In fact, the image

Jacobian matrix in [16] is essentially the same as that in [17]. The difference lies in the online estimation of the camera focal length and depth information of target features. According to [17], when the number of DOFs of the target features is larger than the degrees of freedom of the camera (6-DOFs), there will be a local minimum problem, which is more obvious at a long distance. For the visual servo control proposed in this article, the spherical projection method is adopted to decouple the linear velocity and angular velocity of the image Jacobian matrix, which avoids the disadvantage of the local minimum. At the same time, it is still effective when the target features are far away. It is worth pointing out that the uncalibrated visual servo control proposed in [16] is also effective when the target feature distance is relatively short.

7.3.1.4 Experimental Results

The overall system architecture used to perform the experiments for the validation of the proposed visual servo control for the UAM system in this section is shown in Fig. 7.12. The aerial platform is controlled by a custom-made electronic flight controller board that consists of an inertial measurement unit MPU9250 to provide information about the attitude and direction of the vehicle. The high-level proposed

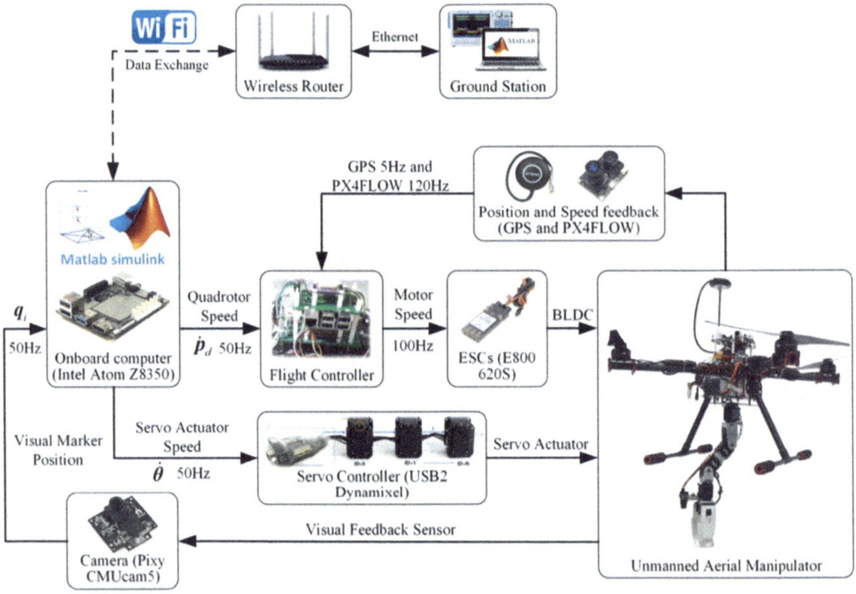

Fig. 7.12 Overall system architecture used to perform the experiments for the validation of the proposed visual servo control

visual servo control scheme runs on an online ATOM computer of 1.8 GHz with 4 GB of RAM, and it can be tuned to obtain sufficiently fast closed-loop visual servo dynamics of the system by using MATLAB with the speed of 50 Hz. The image sensor named Pixy CMUcam5 that is preprogrammed with the identity and tracking algorithm is adapted to provide visual information for the onboard computer via universal asynchronous receiver/transmitter (UART). The robot manipulator is designed as a series of four joints by using the ROBOTIS Dynamixel MX-28T servo steering gear. The actuator is easily controlled with the USB2Dynamixel converter.

For efficient implementation, a real-time and high-accuracy positioning system is necessary. Unlike the other articles described previously that use a high-precision positioning system (such as VICON) for experiments in an indoor laboratory, all experiments in this section are carried out outdoors by fusing the GPS/INS and the optical flow sensor to obtain accurate position and speed data. The output frequency of the optical flow module named PX4FLOW is 120 Hz, which can be used to provide a highly dynamic response for the position system. The PX4FLOW is also integrated with an ultrasonic sensor to provide height data. The communication between the ground station and the onboard computer is performed using MATLAB over a Wi-Fi data link.

The corresponding results are shown in Figs. 7.13, 7.14, 7.15, and 7.16. In Fig. 7.13, some snapshots of the experiments are provided. First, the quadrotor is manually operated to move above the target, while the joints of the robot arm move to the initial angle to ensure that the target is within the field of view of the camera. Figure 7.13a and b shows this process. Figure 7.13c is the starting point of the proposed visual servo algorithm that takes effect at the time of 23 s, and the camera can better detect the marker points. The entire optimal grasping process lasts for 34 s, as shown in Fig. 7.13d–h. It can be seen from the figure that the robot can reach the grasping position under the visual servo control. Some detailed visual servo process data are shown in Figs. 7.14, 7.15, and 7.16. Figure 7.14 shows the evolution of the visual features (u_i, v_i). Figure 7.15 shows the variation in the visual error features $\boldsymbol{\delta}_1$ in the visual servo control. The closed-loop performance of the system maintains an error of approximately 50 pixels around the desired position (see Fig. 7.14). The authors believe that the most significant source of the error is due to aerodynamic disturbances and measurement noise. Figure 7.16 shows the variation in the quadrotor velocity $\dot{\boldsymbol{p}}$ and the robot arm joint angle velocity $\dot{\boldsymbol{\theta}}$. As seen from the figure, benefiting from λ, only the quadrotor is in motion 26 s before. From then on, the robotic arm begins to move to take full advantage of its dexterity. From the perspective of the proposed framework, a coordinated optimization grasp has been achieved for a wide range of scenarios.

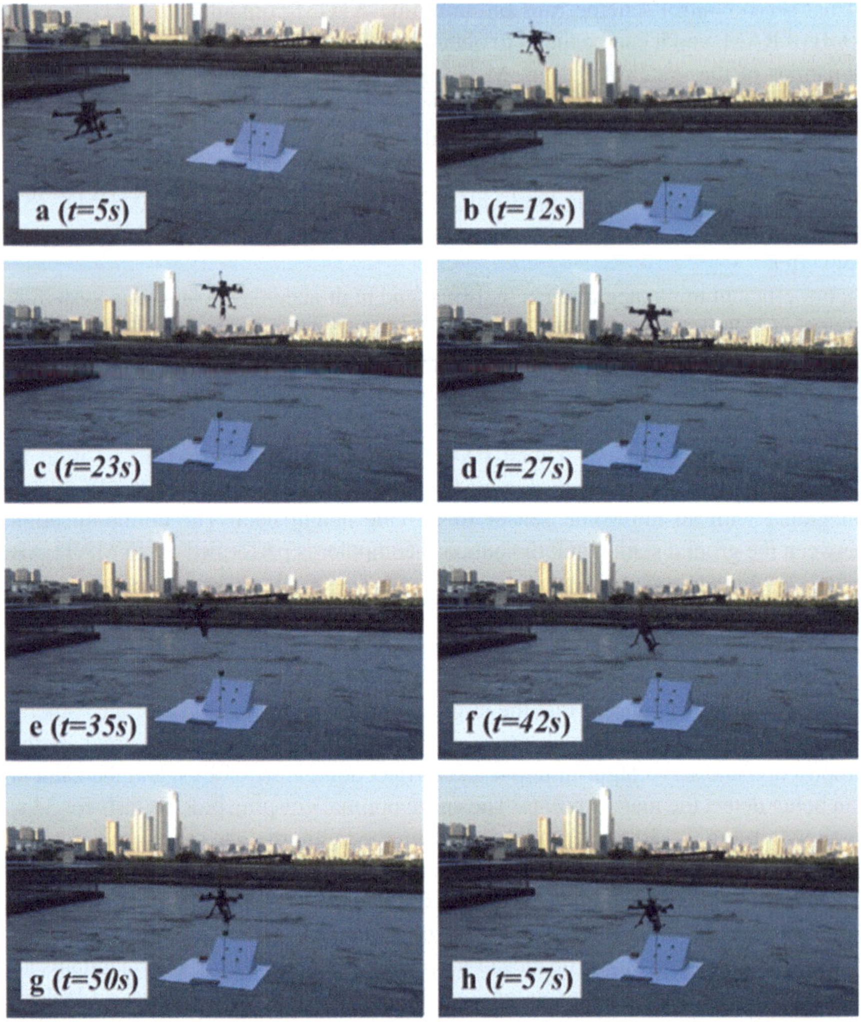

Fig. 7.13 Some snapshots of the experiments

7.3.2 *Image Dynamic-Based Visual Servo Control for Unmanned Aerial Manipulators with a Virtual Camera*

7.3.2.1 System Model

As shown in Fig. 7.17, let W and B represent the inertial reference frame and body reference frame of UAM, respectively. The origin of the body reference frame B is located at $\xi = (x, y, z)^\top$ with respect to the inertial frame W. The triple $\boldsymbol{\eta} = (\phi, \theta, \varphi)^\top$

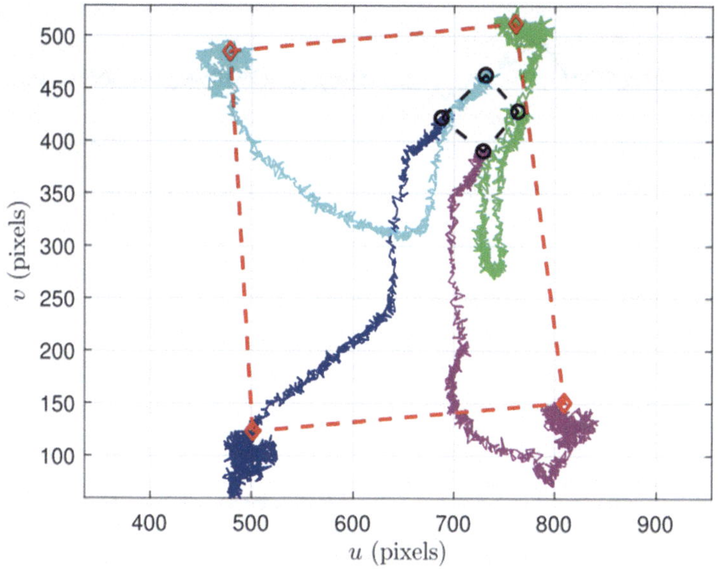

Fig. 7.14 Evolution of the visual features (u_i, v_i) in the image space

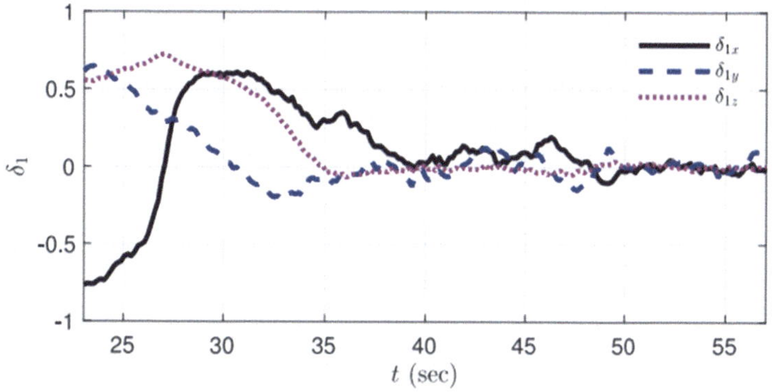

Fig. 7.15 Visual error $\boldsymbol{\delta}_1$ during visual servoing

of Euler roll-pitch-yaw angles are used to describe the attitude of UAM in the inertial frame W. The orientation of the body reference frame B is given by a rotation matrix $\boldsymbol{R}$, where $\boldsymbol{R} = \text{Rot}(Z, \varphi)\text{Rot}(Y, \theta)\text{Rot}(X, \phi)$. Let $\boldsymbol{v} = (v_x, v_y, v_z)^\top$ denote the linear velocities of the UAM with respect to the inertial frame W and $\boldsymbol{w} = (w_x, w_y, w_z)^\top$ denote the angular velocities with respect to the body reference frame B. The force $\boldsymbol{F} = (0, 0, f)^\top$ and the torque $\boldsymbol{\Gamma} = (\tau_x, \tau_y, \tau_z)^\top$ are generated by four rotors in the body reference frame B, respectively. Let $\boldsymbol{q} = (\theta_1, \theta_2, \theta_3)^\top$ be the onboard manipulator joint vector. Let E be the reference frame fixed with the onboard manipulator end effector (e.g., gripper). Let $\boldsymbol{r}_G = (x_G, y_G, z_G)^\top$ be the barycenter position of the UAM system in the body reference frame B. According to the multibody centroid solution

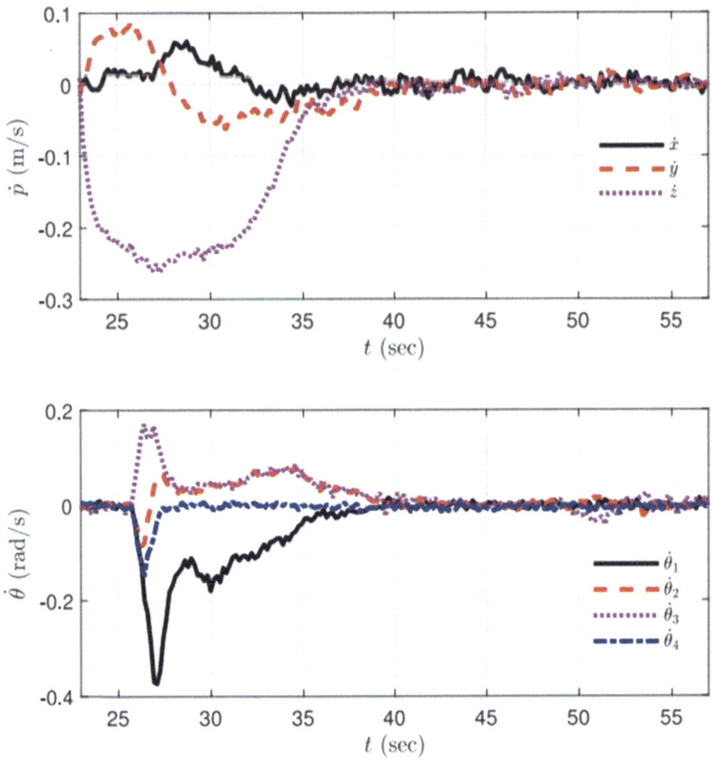

Fig. 7.16 Quadrotor velocity $\dot{\boldsymbol{p}}$ and robot arm joint angle velocity $\dot{\boldsymbol{\theta}}$

method, the offset vector $\boldsymbol{r}_G$ is a function of the onboard manipulator joint angle configuration as follows:

$$\boldsymbol{r}_G = \frac{\sum_i^N m_i r_{Gi}}{m}, \quad N = 1, 2, 3 \tag{7.83}$$

where m is the mass of the UAM. m_i and r_{Gi} represent the mass and position of the center of gravity of the i-th onboard manipulator link, respectively.

Generally, the position $\boldsymbol{\xi}$, yaw angle φ, and onboard manipulator joint angles $\boldsymbol{q}$ are adopted as the visual servo variables for security consideration. Therefore, the generalized coordinate variables of the UAM system are defined as follows:

$$\boldsymbol{Q} = (\boldsymbol{\xi}^\top, \varphi, \boldsymbol{q}^\top)^\top \in \mathbb{R}^{7\times 1} \tag{7.84}$$

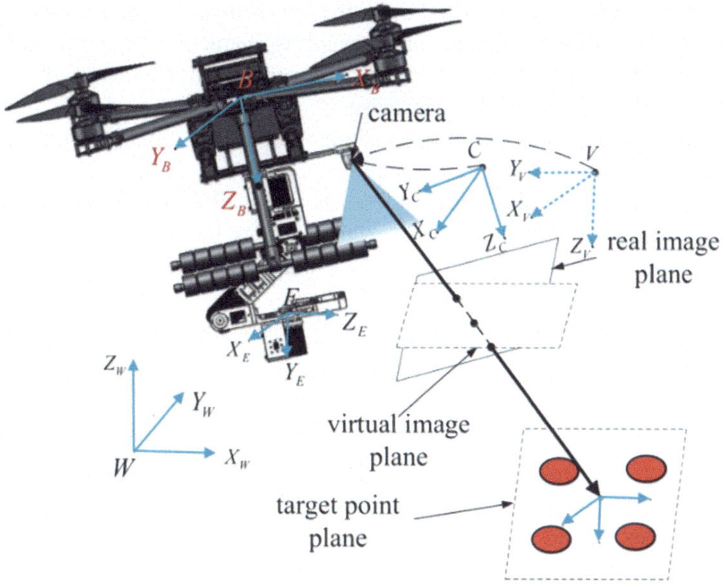

Fig. 7.17 Relationship among the UAM, target point plane, real image plane, and virtual image plane

According to the Newton-Euler method [18], the UAM flying platform and onboard manipulator are considered as two different subsystems. Considering the interaction of the two subsystems, the dynamic equations of the UAM are given by

$$\begin{cases} m\dot{\mathbf{v}} = mg\mathbf{E}_3 - \mathbf{RF} \\ \underbrace{+2m\mathbf{w}\times\mathbf{v} + m\dot{\mathbf{w}}\times\mathbf{r}_G + m\mathbf{w}\times(\mathbf{w}\times\mathbf{r}_G)}_{\mathbf{a}} \\ \mathbf{J}\dot{\mathbf{w}} = -\mathbf{w}\times\mathbf{J}\mathbf{w} + \boldsymbol{\Gamma} \\ -\underbrace{[(\mathbf{w}\times\mathbf{r}_G)\times m\mathbf{v} + \mathbf{r}_G\times(\mathbf{w}\times m\mathbf{v}) + \mathbf{r}_G\times m\dot{\mathbf{v}}]}_{\mathbf{b}} \end{cases} \tag{7.85}$$

where $\mathbf{J}$ is the body frame inertia, $\times$ denotes the cross product, and $\boldsymbol{E}_3 = (0, 0, 1)^\top$ is a unit vector in the inertial frame W. $\mathbf{a} = (a_x, a_y, a_z)^\top$ and $\mathbf{b} = (b_x, b_y, b_z)^\top$ are extra items generated because the UAM center of gravity does not coincide with the origin of the body reference frame B.

When the UAM moves, the target plane is not parallel to the image plane. As a result, the analytical form of the interaction matrix between image motion and UAM motion becomes complicated and loses the decoupled structure [19]. Besides, since the underactuation of the UAM flying platform, the platform is required to tilt to produce translational motion. The tilt of the UAM flying platform causes the image error to increase.

However, as shown in Fig. 7.17, the defined virtual image plane is always parallel to the target plane, which maintains the decoupling characteristics of the interaction matrix sustainedly. When the UAM flying platform is tilted, the image error defined in the virtual image plane does not increase.

Then, the image points are projected onto the virtual image plane, and the corresponding image features are defined. Let C and V represent the camera reference frame and the virtual camera reference frame of UAM, respectively. The reference frames V and C have the same origin. Besides, the virtual camera reference frame V has the same yaw angle as reference frame C, and its roll angle and pitch angle are always zero.

The rotation matrix ${}^V\boldsymbol{R}_C$ between reference frame C and V is given by

$$ {}^V\boldsymbol{R}_C = \mathrm{Rot}(Y,\theta)\mathrm{Rot}(X,\phi) = \begin{bmatrix} \cos\theta & \sin\theta\sin\phi & \sin\theta\cos\phi \\ 0 & \cos\phi & -\sin\phi \\ -\sin\theta & \cos\theta\sin\phi & \cos\theta\cos\phi \end{bmatrix} \tag{7.86} $$

Let $\boldsymbol{p} = (u, v)^\top$ be the image pixel coordinate of the target point. For any target point, $\boldsymbol{P}_C = (X_C, Y_C, Z_C)^\top$ represents the coordinate in the reference frame C. The normalized coordinate $\overline{\boldsymbol{P}}_C = (u, n, 1)^\top$ is represented as follows:

$$ \overline{\boldsymbol{P}}_C = \frac{\boldsymbol{P}_C}{Z_C} = \boldsymbol{K}^{-1}(\boldsymbol{p}^\top, 1)^\top \tag{7.87} $$

where $\boldsymbol{K}$ is the known camera intrinsic matrix and Z_C can be obtained by a depth camera.

Then, the relationship between the normalized coordinates $\overline{\boldsymbol{P}}_C$ in the reference frame C and $\overline{\boldsymbol{P}}_V = ({}^v u, {}^v n, 1)^\top$ in the reference frame V can be obtained as follows:

$$ \overline{\boldsymbol{P}}_V = \alpha \boldsymbol{R}_C \overline{\boldsymbol{P}}_C \tag{7.88} $$

where $\alpha = 1/\left((0,0,1)^V\boldsymbol{R}_C\overline{\boldsymbol{P}}_C\right)$.

Considering that there are $M(M > 1)$ target points, let ${}^v u_g = I_{10}/I_{00}$ and ${}^v n_g = I_{01}/I_{00}$ be the center of the target points in the virtual image plane, where $I_{ij} = \sum_{k=1}^{M} {}^v u_k^i \, {}^v n_k^j$, $({}^v u_k, {}^v n_k)$ is the coordinate of the k-th target point in the reference frame V.

Then, the image features are defined as follows [20]:

$$ \begin{cases} h_1 = h_3^v u_g, h_2 = h_3^v v_g, h_3 = \sqrt{\dfrac{c^*}{c}} \\ h_4 = \dfrac{1}{2}\arctan\left(\dfrac{2^v\mu_{11}}{\mu_{20} - {}^v\mu_{02}}\right) \end{cases} \tag{7.89} $$

where ${}^{v}\mu_{ij} = \sum_{k=1}^{N} \left({}^{v}u_k - {}^{v}u_g\right)^i \left({}^{v}n_k - {}^{v}n_g\right)^j$ and $c={}^{v}\mu_{20}+{}^{v}\mu_{02}$. c^* is the desired value of c. c^* can be obtained when the UAM is in the desired position. At that moment, the depth value z of the target points is also the desired value z^* in the reference frame C.

Define $\boldsymbol{h} = (h_1, h_2, h_3)^\top$ and $\boldsymbol{h}_o = (h_1, h_2, h_3, h_4)^\top$. The dynamics of the image features in the virtual image plane are

$$\begin{cases} \dot{\boldsymbol{h}} = -(\dot{\varphi}\boldsymbol{E}_3) \times \boldsymbol{h} - \frac{1}{z^*}{}^{v}\boldsymbol{v} \\ \dot{h}_4 = -\dot{\varphi} \end{cases} \tag{7.90}$$

where ${}^{v}\boldsymbol{v} = ({}^{v}v_x, {}^{v}v_y, {}^{v}v_z)^\top$ is the linear velocity of the body in the reference frame V, ${}^{v}\boldsymbol{v}={}^{V}\boldsymbol{R}_C\mathbf{v}$.

Therefore, the translational dynamics of the UAM can be obtained in the reference frame V as follows:

$$\begin{cases} {}^{v}\dot{\boldsymbol{v}} = -(\dot{\varphi}\boldsymbol{E}_3) \times {}^{v}\boldsymbol{v} + {}^{v}\boldsymbol{f} \\ {}^{v}\boldsymbol{f} = -\dfrac{{}^{V}\boldsymbol{R}_C f \boldsymbol{E}_3}{m} + g\boldsymbol{E}_3 + \boldsymbol{A} \end{cases} \tag{7.91}$$

where $\boldsymbol{A} = \mathrm{Rot}(Z, \varphi)a/m = (A_x, A_y, A_z)^\top$.

From (7.85), (7.90), and (7.91), the complete image dynamic model of the UAM system is written, including UAM dynamics and image feature dynamics, as follows:

$$\begin{cases} \dot{\boldsymbol{h}} = -(\dot{\varphi}\boldsymbol{E}_3) \times \boldsymbol{h} - \dfrac{1}{z^*}\boldsymbol{v}, \dot{h}_4 = -\dot{\varphi} \\ {}^{v}\dot{\boldsymbol{v}} = -(\dot{\varphi}\boldsymbol{E}_3) \times {}^{v}\boldsymbol{v} + {}^{v}\boldsymbol{f} \\ {}^{v}\boldsymbol{f} = -\dfrac{{}^{V}\boldsymbol{R}_C f \boldsymbol{E}_3}{m} + g\boldsymbol{E}_3 + \boldsymbol{A} \\ \boldsymbol{J}\dot{\boldsymbol{w}} = -\boldsymbol{w} \times \boldsymbol{J}\boldsymbol{w} + \boldsymbol{\Gamma} - \boldsymbol{b} \end{cases} \tag{7.92}$$

7.3.2.2 Visual Servo Control Design

Figure 7.18 shows the flowchart of the proposed control method. First, the position and attitude controller with center-of-gravity compensation is designed. Then, the image feature kinematics is popularized and applied to the end-effector positioning of the onboard manipulator. In addition, considering the sequence of movements of the UAM flying platform and the onboard manipulator, the UAM motion distribution strategy is designed to control the sequence of the UAM movements.

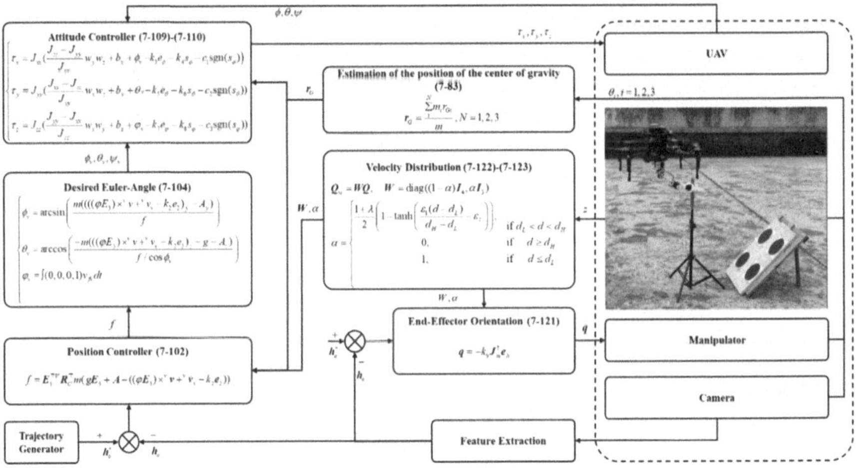

Fig. 7.18 Flowchart of the proposed control method

1. UAM flying platform image-based visual servo position controller
   The desired value of image features $\boldsymbol{h}_o^*$ is defined as follows:

$$\boldsymbol{h}_o^* = \left(\boldsymbol{h}^*, h_4^*\right)^\top = \left(h_1^*, h_2^*, h_3^*, h_4^*\right)^\top. \tag{7.93}$$

The target points are fixed relative to the earth, so $\dot{h}_4^* = \ddot{h}_4^* = 0$. The image feature error e_1 is defined as follows:

$$e_1 = \boldsymbol{h}_o - \boldsymbol{h}_o^*. \tag{7.94}$$

The first and second derivatives of image feature error e_1 are as follows:

$$\begin{cases} \dot{e}_1 = \dot{\boldsymbol{h}}_o = \boldsymbol{L}\boldsymbol{v}_{fv} \\ \ddot{e}_1 = \ddot{\boldsymbol{h}}_o = \dot{\boldsymbol{L}}\boldsymbol{v}_{fv} + \boldsymbol{L}\dot{\boldsymbol{v}}_{fv} \end{cases} \tag{7.95}$$

where $\boldsymbol{L}$ is the image Jacobian matrix defined as

$$\boldsymbol{L} = \begin{bmatrix} -\dfrac{1}{z^*} & 0 & 0 & h_2 \\ 0 & -\dfrac{1}{z^*} & 0 & -h_1 \\ 0 & 0 & -\dfrac{1}{z^*} & 0 \\ 0 & 0 & 0 & -1 \end{bmatrix}, \boldsymbol{v}_{fv} = \left({}^{v}\boldsymbol{v}_v^\top, \dot{\varphi}_v\right)^\top.$$

A sliding surface $\boldsymbol{s}$ is defined as follows:

$$s = \gamma_1 e_1 + \dot{e}_1 \quad (7.96)$$

where γ_1 is a positive constant.

Define the Lyapunov function V_1 as follows:

$$V_1 = \frac{1}{2} s^\top s \quad (7.97)$$

Taking the time derivative of V_1 and using (7.95) and (7.96), we obtain the following expression:

$$\dot{V}_1 = s^\top \dot{s} = s^\top (\gamma_1 \dot{e}_1 + \ddot{e}_1) = s^\top (\gamma_1 L v_{fv} + \dot{L} v_{fv} + L \dot{v}_{fv}). \quad (7.98)$$

Thus, we design the following virtual controller:

$$\dot{v}_{fv} = -(\gamma_1 I_4 + L^{-1} \dot{L}) v_{fv} - k_1 L^{-1} \mathbf{s} \quad (7\text{-}99)$$

where $k_1 > 0$ can guarantee $\dot{V}_1 = s^\top \dot{s} = -k_1 s^\top s \leq -2k_1 V_1$ and I_4 is a 4×4 identity matrix. Therefore, the error $\mathbf{e}_1$ has asymptotic stability and converges to zero.

Then, the v_{fv} as the virtual input signal of the next controller is obtained by solving the differential equation (7-99), where $v_{fv} = \left({}^V v_v^\top, \dot{\varphi}_v\right)^\top$ and ${}^V v_v = \left({}^V v_{vx}, {}^V v_{vy}, {}^V v_{vz}\right)^\top$. Define the velocity error e_2 and the Lyapunov function V_2 as follows:

$$e_2 = {}^V v - {}^V v_v, \; V_2 = \frac{1}{2} e_2^\top e_2. \quad (7.100)$$

Taking the time derivative of V_2 and using (7.94) yields the following:

$$\begin{aligned} \dot{V}_2 &= e_2^\top \dot{e}_2 = e_2^\top ({}^V \dot{v} - {}^V \dot{v}_v) = e_2^\top (-(\dot{\varphi} E_3) \times {}^V v + {}^V f - {}^V \dot{v}_v) \\ &= e_2^\top \left(-(\dot{\varphi} E_3) \times {}^V v - \frac{{}^V R_C f E_3}{m} + g E_3 + A - {}^V \dot{v}_v \right). \end{aligned} \quad (7.101)$$

Therefore, the position controller is designed as follows:

$$f = E_3^\top {}^V R_C^\top m (g E_3 + A - ((\dot{\varphi} E_3) \times {}^V v + {}^V \dot{v}_v - k_2 e_2)) \quad (7.102)$$

where k_2 is a positive constant.

Then, the following inequality can be obtained:

$$\dot{V}_2 = -k_2 e_2^\top e_2 \leq -2k_2 V_2 \quad (7.103)$$

where $\dot{V}_2$ is definite negative. Therefore, the e_2 has asymptotic stability and converges to zero.

Combining (7.86), (7.91), and (7.102), the desired roll angle ϕ_v, desired pitch angle θ_v [21, 22], and desired yaw angle φ_v can be obtained by the following:

$$\begin{cases} \phi_v = \arcsin\left(\dfrac{m\left(((\dot{\varphi}\boldsymbol{E}_3)\times{}^v\boldsymbol{v}+{}^v\dot{\boldsymbol{v}}_v - k_2e_2)_y - \boldsymbol{A}_y\right)}{f}\right) \\ \theta_v = \arccos\left(\dfrac{-m(((\dot{\varphi}\boldsymbol{E}_3)\times{}^v\boldsymbol{v}+{}^v\dot{\boldsymbol{v}}_v - k_2e_2)_z - g - \boldsymbol{A}_z)}{f/\cos\phi_v}\right) \\ \varphi_v = \displaystyle\int(0,0,0,1)\boldsymbol{v}_{fv}\mathrm{d}t. \end{cases} \tag{7.104}$$

2. UAM flying platform image-based visual servo attitude controller

From the dynamic equation (7.85), the attitude dynamics can be written as three independent equations, respectively, as follows:

$$\begin{cases} \dot{w}_x = \dfrac{\tau_x}{J_{xx}} - \dfrac{J_z z - J_{yy}}{J_{xx}} w_y w_z - b_x \\ \dot{w}_y = \dfrac{\tau_y}{J_{yy}} - \dfrac{J_{xx} - J_{zz}}{J_{yy}} w_x w_z - b_y \\ \dot{w}_z = \dfrac{\tau_z}{J_{zz}} - \dfrac{J_{yy} - J_{xx}}{J_{zz}} w_x w_y - b_z. \end{cases} \tag{7.105}$$

Therefore, the pitch angle, roll angle, and yaw angle can be controlled separately. Taking the roll angle φ as an example, there is the following:

$$\begin{cases} \dot{\phi} = w_x \\ \dot{w}_x = \dfrac{\tau_x}{J_{xx}} - \dfrac{J_z - J_{yy}}{J_{xx}} w_y w_z - b_x. \end{cases} \tag{7.106}$$

The angle error e_ϕ, sliding surface s_ϕ, and Lyapunov function V_ϕ are defined as follows:

$$e_\phi = \phi - \phi_v, s_\phi = \dot{e}_\phi + k_3 e_\phi, V_\phi = \frac{1}{2} s_\phi^\top s_\phi \tag{7.107}$$

where k_3 is a positive constant.

Taking the time derivative of V_ϕ and applying (7.106), we derive the following expression:

$$\begin{aligned} \dot{V}_\phi &= s_\phi^\top \dot{s}_\phi = s_\phi^\top(\ddot{e}_\phi + k_3\dot{e}_\phi) = s_\phi^\top(\ddot{\phi} - \ddot{\phi}_v + k_3\dot{e}_\phi) \\ &= s_\phi^\top\left(\frac{\tau_x}{J_{xx}} - \frac{J_{zz} - J_{yy}}{J_{xx}} w_y w_z - b_x - \ddot{\phi}_v + k_3\dot{e}_\phi\right) \end{aligned} \tag{7.108}$$

By choosing the following parameters:

$$\tau_x = J_{xx}\left(\frac{J_{zz} - J_{yy}}{J_{xx}} w_y w_z + b_x + \ddot{\phi}_v - k_3\dot{e}_\phi - k_4 s_\phi - c_1 \operatorname{sgn}(s_\phi)\right) \tag{7.109}$$

where k_4 and c_1 are positive constants.

Then, there is $\dot{V}_\phi = -k_4 s_\phi^\top s_\phi - c_1 \mid s_\phi \mid \leq -2k_4 V_\phi$, which is definite negative. The error e_ϕ has asymptotic stability and converges to zero.

Similarly, sliding-mode controllers for pitch motion and yaw motion can be derived, as shown in (7.110), respectively:

$$\begin{cases} \tau_y = J_{yy}\left(\dfrac{J_{xx} - J_{zz}}{J_{yy}} w_x w_z + b_y + \ddot{\theta}_v - k_5\dot{e}_\theta - k_6 s_\theta - c_2 \operatorname{sgn}(s_\theta)\right) \\ \tau_z = J_{zz}\left(\dfrac{J_{yy} - J_{xx}}{J_{zz}} w_x w_y + b_z + \ddot{\varphi}_v - k_7\dot{e}_\varphi - k_8 s_\varphi - c_3 \operatorname{sgn}(s_\varphi)\right) \end{cases} \tag{7.110}$$

where $\begin{cases} k_5, k_6, k_7, k_8, c_2, c_3 > 0 \\ e_\theta = \theta - \theta_v, s_\theta = \dot{e}_\theta + k_5 e_\theta \\ e_\varphi = \varphi - \varphi_v, s_\varphi = \dot{e}_\varphi + k_7 e_\varphi \end{cases}$.

The UAM flying platform visual servo controllers based on image dynamics have been designed, including position controllers (7.102) and attitude controllers (7.109) and (7.110). In the process of designing the controller, the influence of the shift of the center of gravity of the system caused by the movement of the manipulator is taken into consideration. In addition, the extra action item brought about by the shift of the center of gravity is regarded as a kind of external disturbance.

3. Onboard manipulator end-effector orientation

Different from other methods [23], the image features of the UAM flying platform are applied to the control of the onboard manipulator simultaneously. In addition, the positioning error of the end effector of the onboard manipulator converges to zero globally. The proposed method eliminates the possibility that the positioning error tends to a local minimum.

Let $\boldsymbol{P}_e = (X_e, Y_e, Z_e)^\top$ be the target point coordinate in the reference frame E. Then, $\boldsymbol{P}_e$ can be calculated by using the chain rule

$$\boldsymbol{P}_e = {}_C^E\boldsymbol{T} + {}_C^E\boldsymbol{R}\boldsymbol{P}_C \tag{7.111}$$

where ${}_C^E\boldsymbol{R}$ and ${}_C^E\boldsymbol{T}$ are the rotation and translation transformation matrix from frame C to E, and they can be calculated by position measurement and the forward kinematics of the robot manipulator. Projecting $\boldsymbol{P}_e$ to the normalized plane is

$$\boldsymbol{p}_e = \left(\frac{X_e}{Z_e}, \frac{Y_e}{Z_e}, 1\right)^\top = (\chi_1, \chi_2, 1)^\top \tag{7.112}$$

Similarly, let $\chi_{1g} = \Pi_{10}/\Pi_{00}$ and $\chi_{1g} = \Pi_{10}/\Pi_{00}$ be the center of the target points, where $\Pi_{ij} = \sum_{k=1}^{N} \chi_{1k}^i \chi_{2k}^j$, (χ_{1k}, χ_{2k}) is the normalized plane coordinate of the k-th $\boldsymbol{P}_e$ point. Define the image features of the visual servo task of the onboard manipulator as follows:

$$\boldsymbol{h}_e = \left(h_x, h_y, h_z\right)^\top = \left(h_z\chi_{1g}, h_z\chi_{2g}, h_z\right)^\top \tag{7.113}$$

where $h_z = \sqrt{\frac{d^*}{d}}$, $d = \delta_{20} + \delta_{02}$, and $\delta_{ij} = \sum_{k=1}^{N} \left(\chi_{1k} - \chi_{1g}\right)^i \left(\chi_{2k} - \chi_{2g}\right)^j$. d^* is the desired value of d.

The dynamic of the image features in the reference frame E is

$$\dot{\boldsymbol{h}}_e = -\frac{1}{z_e^*} v_e \tag{7.114}$$

where z_e^* is the depth desired value of z_e. v_e is the velocity of the end effector in the reference frame W. v_e can be expressed as a function of the end-effector motion as follows [12]:

$$\boldsymbol{R} v_e = \left(\mathbf{I}_{3\times 3}, -\mathrm{sk}\left({}_E^B\boldsymbol{R}\boldsymbol{P}_e\right)\right) \begin{pmatrix} {}^B\boldsymbol{t}_e \\ {}^B\boldsymbol{w}_e \end{pmatrix} = \boldsymbol{J}_v {}^B\boldsymbol{v}_e \tag{7.115}$$

where ${}_E^B\boldsymbol{R}$ represents the rotation transformation matrix from the reference frame E to the reference frame B. ${}^B\boldsymbol{v}_e = \left({}^B\boldsymbol{t}_e, {}^B\boldsymbol{w}_e\right)^\top \in \mathbb{P}^{6\times 1}$ represents the velocity vector at the end effector. The notation $\mathrm{sk}\left({}_E^B\boldsymbol{R}\boldsymbol{P}_e\right)$ denotes the skew-symmetric matrix. ${}^B\boldsymbol{v}_e$ can be expressed as a function of the manipulator joint velocity $\dot{\boldsymbol{q}}$ as follows:

$${}^B\boldsymbol{v}_e = \boldsymbol{J}(\boldsymbol{q})\dot{\boldsymbol{q}} \tag{7.116}$$

where $\boldsymbol{J}(\boldsymbol{q})$ is the velocity Jacobian matrix of the onboard manipulator. By combining (7.114)–(7.116), the dynamic of the image features is

$$\dot{\boldsymbol{h}}_e = -\frac{1}{z_e^*}\boldsymbol{R}^\top \boldsymbol{J}_v \boldsymbol{J}(\boldsymbol{q})\dot{\boldsymbol{q}} = \boldsymbol{J}_m \dot{\boldsymbol{q}}. \tag{7.117}$$

The desired value of image features $\boldsymbol{h}_e^*$ is defined as follows:

$$\boldsymbol{h}_e^* = \left(h_x^*, h_y^*, h_z^*\right) \tag{7.118}$$

Since the target points are fixed, then $\dot{\boldsymbol{h}}_e^* = \boldsymbol{0}$. The image feature error $\mathbf{e}_h$ and Lyapunov function V_h are defined as follows:

$$e_h = \boldsymbol{h}_e - \boldsymbol{h}_e^*, V_h = \frac{1}{2} e_h^\top e_h. \tag{7.119}$$

Taking the time derivative of V_h and using (7.117), we obtain the following:

$$\dot{V}_h = e_h^\top \dot{e}_h = e_h^\top \boldsymbol{J}_m \dot{\boldsymbol{q}}. \tag{7.120}$$

By choosing the following parameters:

$$\dot{\boldsymbol{q}} = -k_9 \boldsymbol{J}_m^\dagger e_h \tag{7.121}$$

where $\boldsymbol{J}_m^\dagger$ is the pseudoinverse matrix of $\boldsymbol{J}_m$ and $\boldsymbol{J}_m^\dagger = \boldsymbol{J}_m^\top \left(\boldsymbol{J}_m \boldsymbol{J}_m^\top\right)^{-1}$ if $\boldsymbol{J}_m$ has full row rank. k_9 is a positive constant.

From (7.120) and (7.121), $\dot{V}_h = -k_9 e_h^\top e_h \leq -2k_9 V_h$ can be obtained, which is negative definite. Therefore, the variable $\mathbf{e}_h$ has asymptotic stability and converges to zero.

4. UAM motion distribution strategy

To handle the motion of the UAM flying platform and the onboard manipulator, respectively, by accounting for different moving capabilities, a motion distribution strategy is proposed for the UAM system. In particular, a generalized velocity vector with the weighted norm is defined to coordinate the UAM flying platform and onboard manipulator. It is shown as follows:

$$\dot{\boldsymbol{Q}}_w = \boldsymbol{W} \dot{\boldsymbol{Q}}, \boldsymbol{W} \in \mathbb{R}^{7 \times 7} \tag{7.122}$$

where $\boldsymbol{W} =$ diag $((1 - \alpha)\boldsymbol{I}_4, \alpha \boldsymbol{I}_3)$. The variable α is related to the relative distance between the camera and the target, and its definition expression is as follows:

$$\alpha = \begin{cases} \frac{1+\lambda}{2}\left(1 - \tanh\left(\frac{\varepsilon_1 (d - d_L)}{d_H - d_L} - \varepsilon_2\right)\right), & \text{if } d_L < d < d_H \\ 0, & \text{if } \quad d \geq d_H \\ 1, & \text{if } \quad d \leq d_L \end{cases} \tag{7.123}$$

where d represents the relative distance between the camera and the target and the variable α will decrease as the distance increases. d_H and d_L are the critical distances from the camera to the target. λ, ε_1, and ε_2 are nonnegative constants.

It can be seen from the weight matrix **W** that the motion of the UAM flying platform plays a leading role when $d \geq d_H$, and the priority level of the onboard manipulator is low, while for $d_L < d < d_H$ or $d \leq d_L$, the UAM flying platform is slowed down and the onboard manipulator is commanded to execute the precise movements. As a result, large-scale movements are implemented first by the UAM flying platform, whereas the precise movements are performed by the onboard manipulator due to its dexterity. Especially when the UAM system is approaching the target, the proposed motion distribution strategy can guarantee an efficient motion of the UAM system.

7.3.2.3 Simulation Results

In this section, simulations based on MATLAB/Simulink are carried out to verify the performances of the proposed controller [24]. First, the simulation results of the proposed control algorithm are given. Then, the comparative performances with the visual servo control based on the spherical projection model proposed in [1] are provided.

In these simulations, the gains of the designed controller in (7.99), (7.102), (7.109), (7.110), and (7.121) are selected as $\gamma_1 = c_1 = c_2 = c_3 = 0.2$, $k_1 = k_2 = 2$, $k_3 = k_4 = k_5 = k_6 = k_7 = k_8 = 5$, and $k_9 = 1$. The UAM dynamic parameters used in the simulation are $m = 4\text{kg}$, $g = 9.81\text{m/s}^2$, and $\boldsymbol{J} = \text{diag}\ \{0.082, 0.082, 0.149\}$ $\text{kg} \cdot \text{m}^2/\text{rad}^2$. The camera intrinsic matrix is $\boldsymbol{K} = [800, 0, 512; 0, 800, 512; 0, 0, 1]$. Besides, the camera's configuration is eye-to-hand. To reduce the amount of numerical calculation in the simulation, the reference frame C is made to coincide with the reference frame B.

The Denavit-Hartenberg parameters of the onboard manipulator are given in Table 7.2. The visual goals are composed of four points. The coordinates of these points in the inertial reference frame $\boldsymbol{W}$ are (0.25, 0.25, 0), (0.25, −0.25, 0), (−0.25, −0.25, 0), and (−0.25, 0.25, 0) m, respectively. The initial position of UAM flying platform is $(-1, -1.2, -5)$ m and its initial Euler angle is $(0, 0, \pi/25)$ rad. The initial joint angle of the onboard manipulator is $(\pi/3, -3\pi/4, \pi/4)$ rad. The desired image feature values are $c^* = 0.5$, $d^* = 0.75$, $\boldsymbol{h}_o^* = (0, 0, 1, 0)^\top$, and $\boldsymbol{h}_e^* = (0, 0, 1)^\top$, respectively. The UAM motion distribution parameters are $\lambda = 0$, $d_L = 1.01$ m, $d_H = 1.1$ m, $\varepsilon_1 = 4\pi$, and $\varepsilon_2 = 2\pi$. There is noise in the UAM system center-of-gravity estimation $\mathbf{r}_G$ and calculating the compensation terms **a** and **b**, so it is easy to cause system instability if directly input into the controller. Therefore, referring to the anti-noise method in the disturbance observer [25], a second-order low-pass filter is designed in series after the compensation terms **a** and **b**, as follows:

Table 7.2 Denavit-Hartenberg parameters of the onboard manipulator

Joint	α_i/(deg)	a_i/(mm)	d_i/(mm)	θ_i
1	−90	0.3	83	θ_1
2	0	148.2	0	θ_2
3	90	220.2	0	θ_3

$$Q(s) = \frac{\Lambda_1 s + \Lambda_2}{s^2 + \Lambda_1 s + \Lambda_2} \tag{7.124}$$

where Λ_1 and Λ_2 are determined by the measurement noise of the system.

Simulation results are shown in Fig. 7.19. Figure 7.19a–g shows the simulation results without sensor noises. Figure 7.19a shows the trajectory of the UAM flying platform in Cartesian space. The trajectory shows that the UAM flying platform movement is smooth and effective. Due to the underactuation of the UAM flying platform, the platform is required to pitch to generate forward/backward motion or roll to generate left/right motion. Therefore, as shown in Fig. 7.19b, at the beginning of the UAM flying platform movement, a certain pitch and roll angle are generated. The image feature error $\mathbf{e}_1$ converges to zero smoothly, as shown in Fig. 7.19c. Then, the UAM movement is divided into three stages. In the first stage from 0 to 19 s, $d \geq d_H$, and the UAM flying platform moves first. In the second stage from 19 to 23.45 s, $d_L < d < d_H$. At this time, the UAM flying platform is already very close to the desired position. With the influence of the motion distribution strategy, the onboard manipulator starts to move slowly. In the third stage from 23.45 to 35 s, $d \leq d_L$, and the movement of the onboard manipulator plays a leading role. As shown in Fig. 7.19d and f, the velocities of the UAM flying platform and onboard manipulator change over time, respectively, and the dotted line represents the dividing line between the first stage and the second stage. When the end effector of the onboard manipulator reaches the desired position, its position in the reference frame B is recorded as $\boldsymbol{P}_B^*$. The positioning error $\boldsymbol{\delta} = \left|\boldsymbol{P}_B - \boldsymbol{P}_B^*\right|$ is shown in Fig. 7.19e. Figure 7.19g shows the trajectory of the feature points on the real image plane. The star-shaped points are the starting points of the movement, and the dot-shaped points are the final desired points. It is observed that the trajectories of four feature points form arc shape in the initial stage. After a short time adjustment, the trajectories of the feature points become linear and concise.

Considering the noise of the onboard sensor, the robustness of the proposed control method is tested against sensor noises. Specifically, a white Gaussian noise (WGN) with a standard deviation (SD) of five pixels is introduced as sensor noises in the image pixel coordinate system. Furthermore, to simulate the sensor noises of the UAM flying platform, the translational velocity $\boldsymbol{v}$ is altered by adding WGN of 3 cm/s SD, and the angular velocity $\boldsymbol{w}$ is modified by WGN of 3°/s SD.

Velocity $\mathbf{v}$ is changed by adding WGN of 3 cm/s SD, and the angular velocity w is changed by WGN of 3°/s SD. Similarly, for the noise of the onboard manipulator joint sensor, the measurement noise of the joint angular velocity $\dot{\boldsymbol{q}}$ is simulated by adding a WGN of 3°/s SD. The gains and parameters of the controller are consistent with the simulation without noise. The simulation results are shown in Fig. 7.19h–n. The proposed controller can realize favorable performances with the sensor noises.

Figure 7.20 shows the simulation results of the visual servo control method using the spherical projection model proposed in [1]. Figure 7.20a illustrates the trajectory formed by the feature points on the real image plane. The trajectories of the image

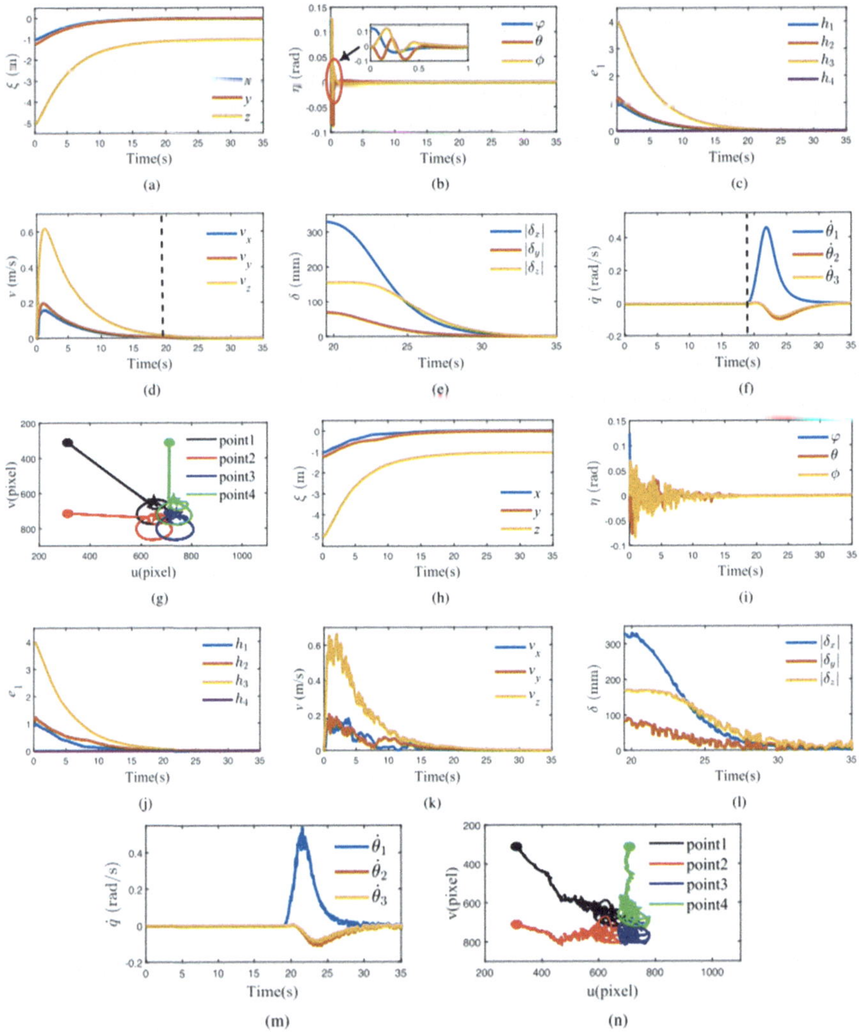

Fig. 7.19 Simulation results of the proposed control method. (**a**–**g**) Simulation results without sensor noises. (**h**–**n**) Simulation results with sensor noises. The star shapes represent the initial position and the dot shapes represent the final position, respectively. (**a**) Position trajectories of the UAM flying platform. (**b**) Attitude of the UAM flying platform. (**c**) Image feature error performances. (**d**) Velocity trajectories of the UAM flying platform. (**e**) Orientational error of the end effector. (**f**) Angular velocity of the onboard manipulator. (**g**) Feature point trajectories in the real image plane. (**h**) Position trajectories of the UAM flying platform. (**i**) Attitude of the UAM flying platform. (**j**) Image feature error performances. (**k**) Velocity trajectories of the UAM flying platform. (**l**) Orientational error of the end effector. (**m**) Angular velocity of the onboard manipulator. (**n**) Feature point trajectories in the real image plane

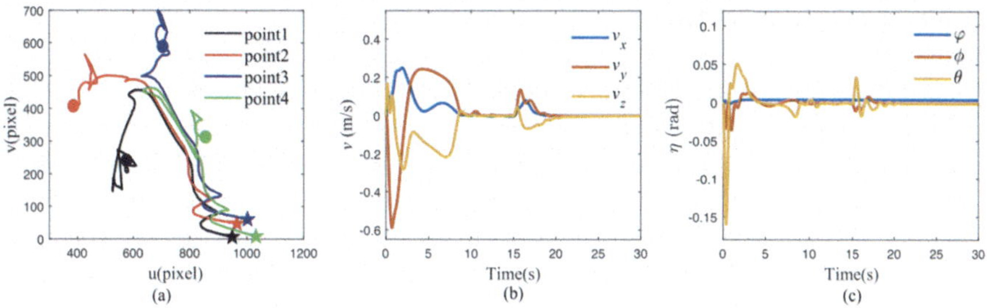

Fig. 7.20 Simulation results using the spherical projection model in [1]. (**a**) Feature point trajectories in the real image plane, where the star shapes represent the initial position and dot shapes represent the final position, respectively. (**b**) Velocity trajectories of the UAM flying platform. (**c**) Attitude performances of the UAM flying platform

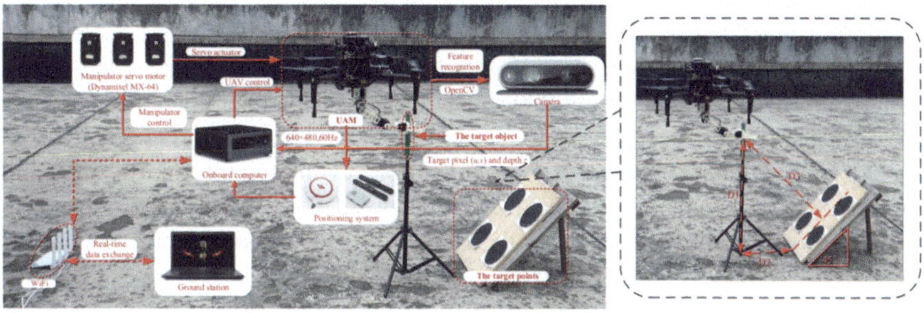

Fig. 7.21 Experimental framework for outdoor visual servo aerial grasping verification

feature are twists and curved. On the contrary, the motion trajectories formed by the feature points on the real image plane are almost straight lines using the proposed control method, as shown in Fig. 7.19g. In addition, it can be observed from Fig. 7.20b and c that the velocity and attitude of the UAM begin to change when it is close to the 15th second. The original stability and balance are disturbed because the movement of the onboard manipulator causes a minor change in the system's center of gravity. According to [1], by designing the center-of-gravity error task function, the error gradually tends to zero. It is a control method on the kinematic level. When the center of gravity of the UAM system changes, an adjustment process is required to make the center of gravity return to the equilibrium state. Meanwhile, the transient instability of the UAM system is not conducive to the precise positioning of the onboard manipulator. For the proposed control method, the extra torque caused by the center-of-gravity deviation is compensated on the dynamic level, avoiding the process of the center-of-gravity adjustment. The stability of the UAM system and the positioning accuracy of the end effector are improved simultaneously.

7.3.2.4 Experimental Results

Outdoor experiments are performed on the UAM system as shown in Fig. 7.21 to verify the proposed method in this section.

The UAM system adopted in the experiment consists of an UAM flying platform, a self-developed 3-D printed 3-DOF manipulator, and a depth camera (Intel RealSense D435i). The UAM flying platform adopts the DJI Matrice 100 weighing 4.39 kg and integrates an inertial measurement unit (IMU), GPS, and Guidance. The IMU and GPS provide real-time UAM flying platform attitude, position, and velocity information at a frequency of 50 Hz. Then, the Guidance is a visual sensor navigation system that can provide height information from the ground. To identify the target points and calculate the distance between the target points and the UAM in real time, a camera is installed on the UAM flying platform with a 45° angle to the horizontal plane. The four black feature points are selected as the visual targets, and the plane where the feature points are located makes a 45° angle with the horizontal plane. The OpenCV-based center coordinate detection method is used for the target point recognition. The onboard manipulator is installed directly under the UAM flying platform and weighs about 900 g. The Denavit-Hartenberg parameters of the onboard manipulator are the same as in Table 7.2. The ROBOTIS Dynamixel MX-64 servo motors are adopted for the three revolute joints of the onboard manipulator and controlled via the USB2Dynamixel converter. The velocity and position of each joint are independently controlled by the motor driver. Besides, the entire UAM system is controlled by an onboard computer and exchanges real-time data with the ground station via Wi-Fi.

The relative position between the target points and the target object is prior known. As shown in Fig. 7.21, the distance $D1 = 75$ cm and $D2 = 30$ cm are set in prior. The position of the target object is the desired position of the end effector of the onboard manipulator. Then, the relative positional relationship between the target points and the UAM can be obtained through the camera. Therefore, the relative position relationship between the target points and the end effector of the onboard manipulator can be obtained favorably by using the forward kinematics of the manipulator. The advantage of this setup is to ensure the onboard manipulator can complete the task of grasping the target object successfully when the target object is not within the camera's view. Figure 7.22 shows snapshots during the experiment. First, the UAM is manually controlled to take off to the initial position of the visual servo procedure and ensure that the target points are within the view of the camera, as shown in Fig. 7.22a1 and b1. Then, the proposed method is going to control the UAM to approach the target object, as shown in Fig. 7.22a2 and b2. When the UAM flying platform approaches the target, the end effector is moved to the target position for grabbing, as shown in Fig. 7.22a3 and b3. Finally, UAM returns to the initial position and landing.

The detailed data of the experimental process are shown in Figs. 7.23 and 7.24. Figure 7.23a shows the trajectories of visual feature points in the real image plane. The circle shapes indicate that the UAM system maintains an error of about ten

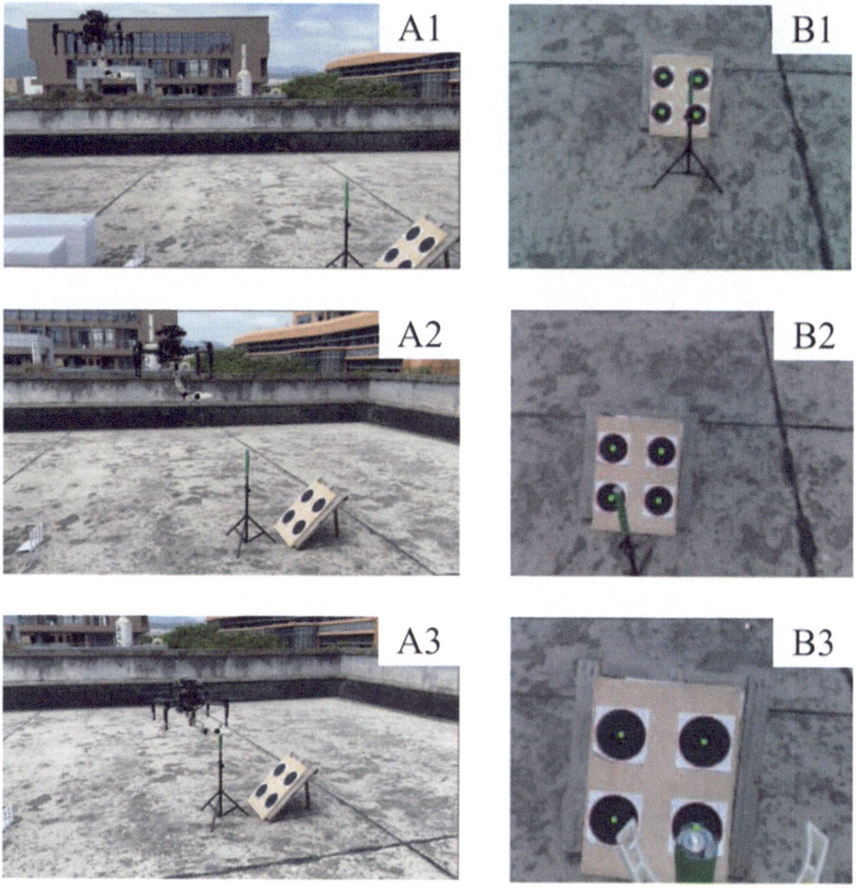

Fig. 7.22 Snapshots of the experiments. The left three images (**a1**, **a2**, and **a3**) come from the third-person perspective, and the right three images (**b1**, **b2**, and **b3**) are the views from onboard camera

pixels near the desired position. The main source of this error is the interference of the external environment and measurement noise. Figure 7.23b shows the linear velocity and angular velocity of the UAM flying platform. It is observed that the velocities are finally stabilized near zero. Figure 7.24a shows the image feature error e_1 and the positioning error $\boldsymbol{\delta}$ of the onboard manipulator end effector. Since the camera is installed obliquely in the experiment, the Z_C coordinate of the camera is inconsistent with the direction of the Z_B coordinate of the UAM flying platform, and the control variable $h_4 = 0$ of the UAM flying platform yaw angle is required. Figure 7.24b shows the movement data of the onboard manipulator. The onboard manipulator does not start to move until about the 15th second, benefiting from the designed motion distribution strategy. The movement of the UAM flying platform plays a leading role before the 15th second. From then on, the onboard manipulator starts to move, and the end effector moves to the position of the target object for

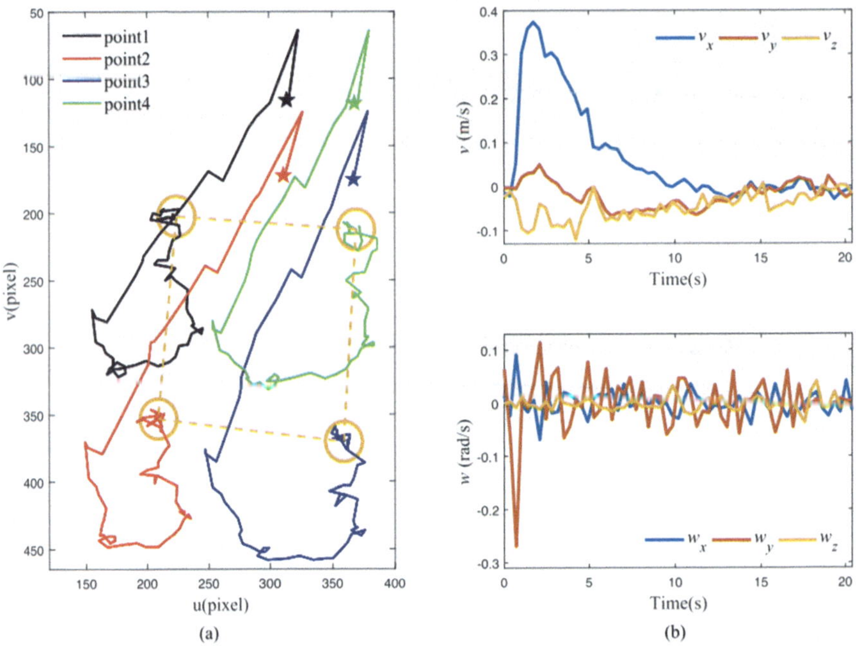

Fig. 7.23 (**a**) Feature point trajectories in the real image plane, where the star shapes represent the initial position and the circle shapes represent the range within ten pixels from the desired position, respectively. (**b**) Linear velocity and angular velocity of the UAM flying platform

manual grasping operations. Finally, the UAM flying platform can approach the target first, and then the end effector of the onboard manipulator can accurately position the target.

7.4 Application: Aerial Manipulation in Transmission Lines and Industrial Pipelines

7.4.1 Autonomous Removing Foreign Objects for Power Transmission Line by Using a Vision-Guided Unmanned Aerial Manipulator

7.4.1.1 Reference Frames and Kinematic Model

As shown in Fig. 7.25, let $\{\mathcal{F}_I : O_I - x_I y_I z_I\}$ represent the right-hand inertial frame and z_I represent the opposite of the direction of gravity. Let $\{\mathcal{F}_V : O_V - x_V y_V z_V\}$ represent the body frame of the drone, and its origin O_V coincides with the geometric center of the drone. Let $\{\mathcal{F}_B : O_B - x_B y_B z_B\}$ represent the base frame of the manipulator. Meanwhile, the base is fixed on the bottom of the drone. Axis

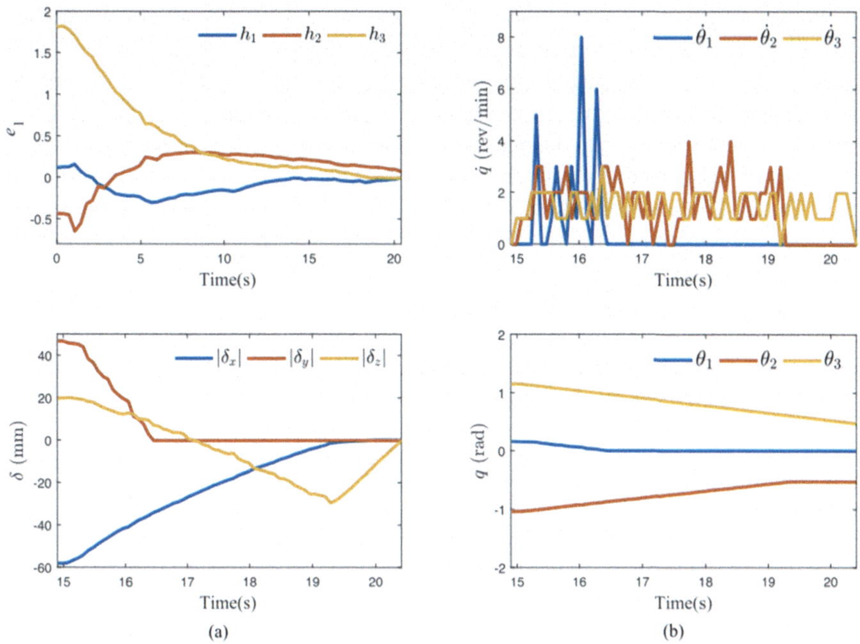

Fig. 7.24 (**a**) Image feature error performances and orientation error performances of the end effector. (**b**) Angular velocity and angle of the onboard manipulator

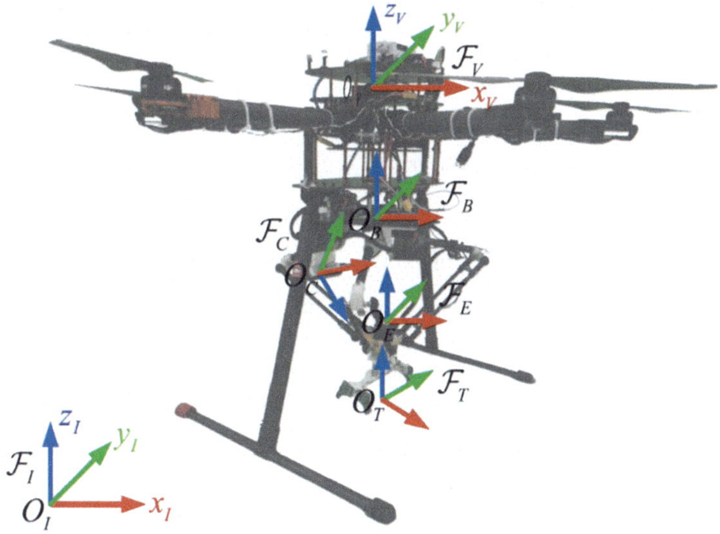

Fig. 7.25 Reference frames of the aerial manipulator

directions of $\mathcal{F}_B$ and $\mathcal{F}_V$ are the same. $\{\mathcal{F}_E : O_E - x_E y_E z_E\}$ is the end-effector frame. $\{\mathcal{F}_T : O_T - x_T y_T z_T\}$ is the tool frame. The axis directions of $\mathcal{F}_E$ and $\mathcal{F}_T$ are the same. Let $\{\mathcal{F}_C : O_C - x_C y_C z_C\}$ be the camera frame, and the direction of z_C is the same as the direction of the camera's optical axis. The transformation matrix ${}^B\mathrm{T}_C \in \mathbb{R}^{4\times 4}$ from the camera frame $\mathcal{F}_C$ to the manipulator base frame $\mathcal{F}_B$ is a constant matrix and can be obtained by hand-eye calibration. Let $\{\mathcal{F}_G : O_G - x_G y_G z_G\}$ denote the foreign object frame. The transformation matrix ${}^B\mathrm{T}_G \in \mathbb{R}^{4\times 4}$ from the foreign object frame $\mathcal{F}_G$ to the camera frame $\mathcal{F}_C$ can be obtained by the vision localization. Let $\mathrm{p} = (x, y, z)^\top$ represent the position state of the drone and $\boldsymbol{\Phi} = (\varphi, \vartheta, \psi)^\top$ represent the Euler angle state of the drone. $\boldsymbol{\Phi}$ represent in the order of roll-pitch-yaw. Meanwhile, p and $\boldsymbol{\Phi}$ are represented in the inertial frame $\mathcal{F}_I$. Let $\boldsymbol{\theta} = (\theta_1, \theta_2, \theta_3, \theta_4)^\top$ represent the joint angle of the manipulator. Let vector q represent the state of the aerial manipulator, which can be described as

$$\mathrm{q} = (\mathrm{p}^\top, \boldsymbol{\Phi}^\top, \boldsymbol{\theta}^\top)^\top \in \mathbb{R}^{10\times 1} \tag{7.125}$$

Firstly, the drone kinematic model can be calculated as follows. Based on the above, p represents the position of the drone in the $\mathcal{F}_I$. Let $\dot{\mathrm{p}}$ and ${}^V\dot{\mathrm{p}}$ represent the linear velocity of the drone expressed in the $\mathcal{F}_I$ and $\mathcal{F}_V$, respectively. Therefore, $\dot{\mathrm{p}}$ can be calculated as

$$\dot{\mathrm{p}} = {}^I\mathrm{R}_V {}^V\dot{\mathrm{p}} \tag{7.126}$$

where ${}^I\mathrm{R}_V \in SO(3)$ describes the rotation relationship from $\mathcal{F}_V$ to $\mathcal{F}_I$. Similarly, let $\mathrm{w} = (w_x, w_y, w_z)^\top$ and ${}^V\mathrm{w} = ({}^Vw_x, {}^Vw_y, {}^Vw_z)^\top$ represent the angular velocity of the drone with respect to frame $\mathcal{F}_I$ and $\mathcal{F}_V$, respectively. The angular velocity can be calculated as

$$\begin{aligned}
&\mathrm{w} = \mathrm{R}_{\boldsymbol{\Phi}}\dot{\boldsymbol{\Phi}} \\
&\mathrm{R}_{\boldsymbol{\Phi}} = \begin{pmatrix} \cos\vartheta & 0 & -\cos\varphi\sin\vartheta \\ 0 & 1 & \sin\varphi \\ \sin\vartheta & 0 & \cos\varphi\cos\vartheta \end{pmatrix} \\
&{}^V\mathrm{w} = ({}^I\mathrm{R}_V)^\top \mathrm{R}_{\boldsymbol{\Phi}}\dot{\boldsymbol{\Phi}} = \mathrm{Q}\dot{\boldsymbol{\Phi}}
\end{aligned} \tag{7.127}$$

where $\mathrm{R}_{\boldsymbol{\Phi}}$ is a transform matrix that can change $\dot{\boldsymbol{\Phi}}$ into w, and $\mathrm{Q} = ({}^I\mathrm{R}_V)^\top \mathrm{R}_{\boldsymbol{\Phi}}$.

Then, let p_T represent the position of the tool frame $\mathcal{F}_T$ in $\mathcal{F}_I$. Therefore, the kinematic model of the manipulator can be calculated as

$$\mathrm{p}_T = \mathrm{p} + {}^I\mathrm{R}_V {}^V\mathrm{p}_T \tag{7.128}$$

where ${}^V\mathrm{p}_T$ is the position of $\mathcal{F}_T$ in $\mathcal{F}_V$. Taking the derivative of (7.128), the linear and angular velocities of the tool frame $\mathcal{F}_T$ in $\mathcal{F}_I$ can be calculated as

$$\begin{aligned} \dot{\mathrm{p}}_T &= \dot{\boldsymbol{p}} + [\boldsymbol{w}]_\times {}^I\boldsymbol{R}_V {}^V\boldsymbol{p}_T + {}^I\boldsymbol{R}_V {}^V\dot{\boldsymbol{p}}_T \\ \mathbf{w}_T &= \mathbf{w} + {}^I\mathbf{R}_V {}^V\mathbf{w}_T \end{aligned} \tag{7.129}$$

where $\dot{\boldsymbol{p}}_T$ and $\boldsymbol{w}_T$ represent the linear and angular velocity of the tool frame $\mathcal{F}_T$ in $\mathcal{F}_I$. $[\boldsymbol{w}]_\times$ represents an antisymmetric matrix that means $[\boldsymbol{w}]_\times \boldsymbol{v} = \boldsymbol{w} \times \boldsymbol{v}$ for every vector $\boldsymbol{v}$. ${}^V\dot{\boldsymbol{p}}_T$represents the linear velocity of the tool frame $\mathcal{F}_T$ in $\mathcal{F}_V$, and ${}^V\boldsymbol{w}_T$ represents the angular velocity of the tool frame $\mathcal{F}_T$ in $\mathcal{F}_V$. They can be expressed as

$$\begin{aligned} V\dot{\boldsymbol{p}}_T &= {}^V\boldsymbol{R}_B\left({}^B\dot{\boldsymbol{p}}_E - \left[{}^B\boldsymbol{R}_E(\boldsymbol{\theta}){}^E\boldsymbol{t}_T\right]_\times {}^B\boldsymbol{w}_E\right) \\ &= {}^V\boldsymbol{R}_B\left(\boldsymbol{B}_t(\boldsymbol{\theta})\dot{\boldsymbol{\theta}} - \left[{}^B\boldsymbol{R}_E(\boldsymbol{\theta}){}^E\boldsymbol{t}_T\right]_\times \boldsymbol{B}_r(\boldsymbol{\theta})\dot{\boldsymbol{\theta}}\right) \end{aligned} \tag{7.130}$$

$${}^V\boldsymbol{w}_T = {}^V\boldsymbol{R}_B {}^B\boldsymbol{w}_E = {}^V\boldsymbol{R}_B\boldsymbol{B}_r(\boldsymbol{\theta})\dot{\boldsymbol{\theta}} \tag{7.131}$$

where $\boldsymbol{B}_r \in \mathbb{P}^{3\times 4}$ and $\boldsymbol{B}_t \in \mathbb{P}^{3\times 4}$ are the Jacobian matrices that change $\dot{\boldsymbol{\theta}}$ into the linear and angular velocities of the frame $\mathcal{F}_T$ in $\mathcal{F}_V$. Using the state of the aerial manipulator $\boldsymbol{q}$, (7.129) can be expressed as

$$\begin{aligned} \dot{\boldsymbol{p}}_T &= \dot{\boldsymbol{p}} + [\boldsymbol{w}]_\times {}^I\boldsymbol{R}_V {}^V\boldsymbol{p}_T + {}^I\boldsymbol{R}_V^V\boldsymbol{R}_B\left(\boldsymbol{B}_l(\boldsymbol{\theta})\dot{\boldsymbol{\theta}} - \left[{}^B\boldsymbol{R}_E(\boldsymbol{\theta}){}^E\boldsymbol{t}_T\right]_\times \boldsymbol{B}_r(\boldsymbol{\theta})\dot{\boldsymbol{\theta}}\right) \\ &= \begin{pmatrix} \mathbb{I}_{3\times 3} \\ -\left[{}^I\boldsymbol{R}_V^V\boldsymbol{p}_T\right]_\times \boldsymbol{R}_\Phi \\ {}^I\boldsymbol{R}_V^V\boldsymbol{R}_B\left(\boldsymbol{B}_l(\boldsymbol{\theta}) - \left[\boldsymbol{B}\boldsymbol{R}_E(\boldsymbol{\theta}){}^E\boldsymbol{t}_T\right]_\times \boldsymbol{B}_r(\boldsymbol{\theta})\right) \end{pmatrix}^\top \dot{\boldsymbol{p}} \\ &= \boldsymbol{J}_l\dot{\boldsymbol{q}} \end{aligned} \tag{7.132}$$

$$\boldsymbol{w}_T = \boldsymbol{w} + {}^I\boldsymbol{R}_V^V\boldsymbol{R}_B\boldsymbol{B}_r(\boldsymbol{\theta})\dot{\boldsymbol{\theta}} = \left(\mathbb{O}_{3\times 3}\boldsymbol{R}_\Phi {}^I\boldsymbol{R}_V^V\boldsymbol{R}_B\boldsymbol{B}_r(\boldsymbol{\theta})\dot{\boldsymbol{q}} = \boldsymbol{J}_r\dot{\boldsymbol{q}}\right. \tag{7.133}$$

where $\mathbb{I}_{3\times 3}$ is the identity matrix and $\mathbb{O}_{3\times 3}$ is the zero matrix. $\dot{\boldsymbol{q}}$ can be converted to $\boldsymbol{w}_T$ and $\dot{\boldsymbol{p}}_T$ by the Jacobian matrices $\boldsymbol{J}_r \in \mathbb{P}^{3\times 10}$ and $\boldsymbol{J}_l \in \mathbb{P}^{3\times 10}$ in $\mathcal{F}_I$.

7.4.1.2 Foreign Object Detection and Location

A foreign object detection and location method that is used to obtain the grasp point is proposed. With the development of RGB-D cameras, point cloud technology has made great technological breakthroughs and research progress in the past years and has been widely used in object detection, location, and recognition. In previous studies [26] and [27], the visual servoing application of point cloud processing technology on UAM proved to be effective. It can not only easily give depth information of objects but also be real time in the feedback. Therefore, as shown in Fig. 7.26, the foreign object detection algorithm in this section uses the point cloud obtained from the Intel RealSense D435 device, and the algorithm is performed in three stages.

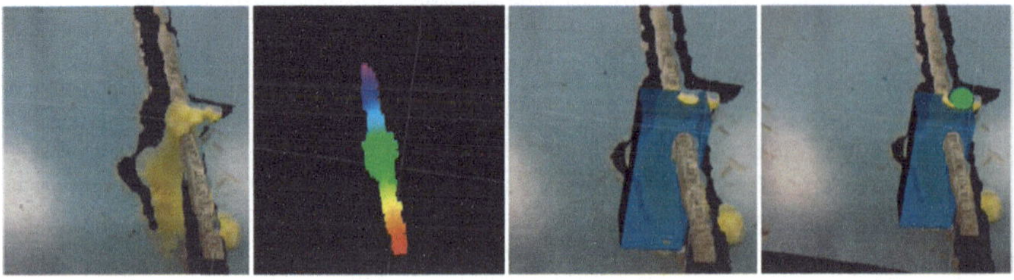

Fig. 7.26 The results of the foreign object detection and location algorithm. (**a**) Input point cloud. (**b**) The result of the line fitting. (**c**) The result of the foreign object detection. (**d**) The results of the foreign object location

First, the point cloud is used to identify and segment transmission lines to obtain outer points. Consider a static 3-D point $\boldsymbol{P}_0 = (X_0, Y_0, Z_0)^\top$ expressed in $\mathcal{F}_C$. Therefore, a line in 3-D can be represented as

$$A(X - X_0) + B(Y - Y_0) + C(Z - Z_0) = 0 \tag{7.134}$$

where $\boldsymbol{v} = (A, B, C)^\top$ is the unit direction vector of the line and $\boldsymbol{P} = (X, Y, Z)^\top$ is any point on the line. They meet the constraints as

$$\boldsymbol{v} \times (\boldsymbol{P} - \boldsymbol{P}_0) = 0 \tag{7.135}$$

Therefore, any line in 3-D can be constrained by the unit direction vector $\boldsymbol{v}$ and a point $\boldsymbol{P}_0$ on the line. We get them by Random Sample Consensus (RANSAC) algorithm to ensure the detection speed and the robustness, which can fit the model from data with noise.

Furthermore, since the foreign objects are outer points of the 3-D line, foreign objects can be obtained by Euler clustering of outer points. Let $\boldsymbol{P}_{outer_i} = (X_{outer_i}, Y_{outer_i}, Z_{outer_i})^\top$, $i = 1, \cdots, N$, represents the point cloud data of the detected target. Therefore, the 3-D bounding box of foreign objects $\boldsymbol{b} = (Len, Wid, Hei)^\top$ can be calculated as

$$\begin{aligned} Len &= \max(X_{outer_i}) - \min(X_{outer_i}) \\ Wid &= \max(Y_{outer_i}) - \min(Y_{outer_i}) \\ Hei &= \max(Z_{outer_i}) - \min(Z_{outer_i}) \end{aligned} \tag{7.136}$$

where *Len*, *Wid*, and *Hei* are the length, width, and height value of the object 3-D bounding box, respectively.

Finally, let $\boldsymbol{P}_G = (X_G, Y_G, Z_G)^\top$ and P_ψ represent the position and the yaw angle of the grasp point expressed in $\mathcal{F}_C$, respectively. Based on the above, we have

$$
\begin{aligned}
\boldsymbol{P}_G &= \frac{1}{N}\sum_{i=1}^{N}\boldsymbol{P}_{outer_i} \\
P_\psi &= \arctan\left(\frac{Y_G}{X_G}\right)
\end{aligned}
\tag{7.137}
$$

Note that the grasp point position $\boldsymbol{P}_G$ will be used to build the vision servo control and the yaw angle P_ψ will be used to control the UAM yaw angle individually.

Figure 7.26 shows the results of the foreign object detection and location algorithm. The input point cloud from the Intel RealSense D435 device is shown in Fig. 7.26a, and the foreign object is tied to the power transmission line. Figure 7.26b shows the results of power transmission line model fitting and detection, which is highlighted in red. The Euler clustering results of the outer points are shown in Fig. 7.26c, and all foreign objects are represented by 3-D bounding box. Figure 7.26d shows the position and yaw angle of the grasp point, which is obtained by the formula (7.137).

7.4.1.3 PBVS for Aerial Manipulation

The position-based visual error can be defined by using the pose of a frame with respect to the desired frame. The pose of a frame can be estimated from the picture using the intrinsic parameters of the camera. Compared to image-based visual servo (IBVS), PBVS can keep the UAM lose the target difficultly, and eye-to-hand configuration can image the whole world. Therefore, in this section, a PBVS control method for aerial manipulator and a new visual error kinematics based on the onboard eye-to-hand camera configuration is proposed.

To separate linear velocity and angular velocity, which can be used to control the translations independently of the rotations, a virtual camera plane is constructed, as shown in Fig. 7.27. Since the pitch and roll angles of the quadrotor are uncontrollable in a stable state, and the yaw angle is an independent control variable, the virtual camera plane is designed in a state where the pitch and roll angles are zero, and the yaw angle is a fixed value, as shown in Fig. 7.27. Let $\{\mathcal{F}_{C^*} : O_{C^*} - x_{C^*}y_{C^*}z_{C^*}\}$ express the virtual camera frame.

For convenience of expression, let $\boldsymbol{h}(t)$ and $\boldsymbol{h}^*(t)$ denote the position vector from the grab point and the UAM end-effector origin point to the virtual camera coordinate system $\mathcal{F}_{C^*}$, respectively. Note that $\boldsymbol{h}^*(t) = \boldsymbol{P}_G$ can be calculated from $\boldsymbol{P}_G$ and the transformation from $\mathcal{F}_C$ to $\mathcal{F}_{C^*}$. $\boldsymbol{h}(t)$ can be calculated from the joint angles $\boldsymbol{\theta}$vv. The goal of this section is that UAM can accurately remove foreign objects, which means $\boldsymbol{h}(t)$ tends to $\boldsymbol{h}^*(t)$. Therefore, the error vector $\mathbf{e}$ can be defined as

$$
e(t) = \boldsymbol{h}^*(t) - \boldsymbol{h}(t) \tag{7.138}
$$

To calculate the visual error in the velocity space, $\boldsymbol{h}^*(t)$ can be represented by another form as

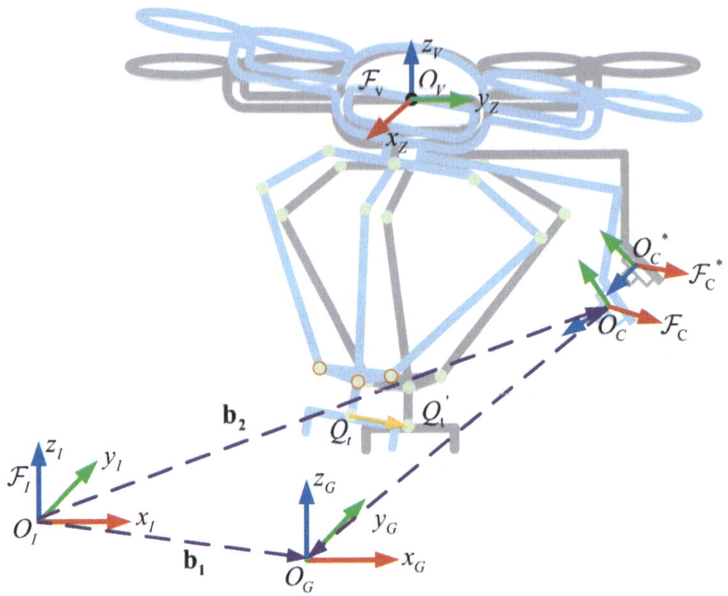

Fig. 7.27 Reference coordinate system and visual motion relationship under the eye-to-hand configuration

$$\boldsymbol{h}^*(t) = {}^{C}\boldsymbol{R}_I(\boldsymbol{b}_1 - \boldsymbol{b}_2) \tag{7.139}$$

where $\boldsymbol{b}_1 \in \mathbb{P}^{3 \times 1}$ is the constant vector that is the foreign object's position in $\mathcal{F}_I$. $\boldsymbol{b}_2 \in \mathbb{P}^{3 \times 1}$ is the vector that represents the position of the virtual camera plane frame $\mathcal{F}_C$ origin point in $\mathcal{F}_I$, as shown in Fig. 7.27. Take the derivative of Eq. (7.139), and we get

$$\dot{\boldsymbol{h}}^* = -\boldsymbol{\Omega}_C^C \times \boldsymbol{h}^* - \mathbf{V}_C^C \tag{7.140}$$

where $\mathbf{V}_C^C$ and $\boldsymbol{\Omega}_C^C$ are the linear and angular velocity of $\mathcal{F}_C$, respectively.

Furthermore, let $\boldsymbol{Q}_t$ represent the origin of the frame $\mathcal{F}_T$. Based on the above results, $\dot{\boldsymbol{Q}}_t$ can be calculated as

$$\dot{\boldsymbol{Q}}_t = {}^{C}\boldsymbol{R}_B\left(\boldsymbol{I}_{3\times 3} - \left[{}^{B}\boldsymbol{R}_E{}^{E}\boldsymbol{t}_T\right]_{\times}\right)\begin{pmatrix} {}^{B}\dot{\boldsymbol{p}}_E \\ {}^{B}\dot{\boldsymbol{w}}_E \end{pmatrix} = \boldsymbol{J}_t{}^{B}\dot{\boldsymbol{x}}_E \tag{7.141}$$

where $\boldsymbol{J}_t = {}^{C}\boldsymbol{R}_B(\boldsymbol{I}_{3\times 3}, [{}^{B}\boldsymbol{R}_E{}^{E}\boldsymbol{t}_T]_{\times})$. ${}^{B}\dot{\boldsymbol{x}}_E = ({}^{B}\dot{\boldsymbol{p}}_E, {}^{B}\dot{\boldsymbol{w}}_E)^{\top}$, ${}^{B}\boldsymbol{R}_E$, and ${}^{E}\boldsymbol{t}_T$ are velocity, rotation transform matrix, and offset vector from $\mathcal{F}_E$ to $\mathcal{F}_T$, respectively. Otherwise, ${}^{B}\dot{\boldsymbol{x}}_E$ can be rewritten as a function of the space velocity $\dot{\boldsymbol{q}}$ of the manipulator joint as

$$
{}^{B}\dot{\boldsymbol{x}}_{E} = \begin{pmatrix} \boldsymbol{B}_{t}(\boldsymbol{\theta}) \\ \boldsymbol{B}_{r}(\boldsymbol{\theta}) \end{pmatrix} \dot{\boldsymbol{\theta}} = \boldsymbol{B}_{e}(\boldsymbol{\theta})\dot{\boldsymbol{\theta}} \tag{7.142}
$$

where $\boldsymbol{B}_e(\boldsymbol{\theta})$ is the Jacobian matrix of the manipulator.

Let $\boldsymbol{h}(t) = \boldsymbol{Q}_t'$ represent the point corresponding to $\boldsymbol{Q}_t$ on the virtual camera plane, and $\boldsymbol{Q}_t'$ can be calculated as

$$
\begin{aligned}
\boldsymbol{h}(t) = \boldsymbol{Q}_t' &= \boldsymbol{Q}_t + {}^{C^*}\boldsymbol{P}_\delta \\
&= \boldsymbol{Q}_t + {}^{C}\boldsymbol{R}_I\left(\boldsymbol{P}_{Q'_t} - {}^{I}\boldsymbol{P}_{Qt}\right) \\
&= \boldsymbol{Q}_t + {}^{C}\boldsymbol{R}_I\left(\left({}^{I}\boldsymbol{P}_{O_{C^*}} + \boldsymbol{Q}_t\right) - \left({}^{I}\boldsymbol{P}_{O_C} + {}^{I}\boldsymbol{R}_{C^*}\boldsymbol{Q}_t\right)\right) \\
&= {}^{I}\boldsymbol{R}_C^{\top}\boldsymbol{Q}_t
\end{aligned} \tag{7.143}
$$

where $\boldsymbol{P}_\delta$ represent the offset vector between $\boldsymbol{Q}_t'$ and $\boldsymbol{Q}_t$ in the frame $\mathcal{F}_C$. Taking the derivative of (7.143), $\dot{\boldsymbol{h}}(t)$ can be written as

$$
\begin{aligned}
\dot{\boldsymbol{h}}(t) &= {}^{I}\boldsymbol{R}_C^{\top}\dot{\boldsymbol{Q}}_t - \boldsymbol{\Omega}_C^C \times {}^{I}\boldsymbol{R}_C^{\top}\boldsymbol{Q}_t \\
&= {}^{I}\boldsymbol{R}_C^{\top}\dot{\boldsymbol{Q}}_t - \boldsymbol{\Omega}_C^C \times \boldsymbol{Q}_t' \\
&= {}^{I}\boldsymbol{R}_C^{\top}\boldsymbol{J}_t^B\boldsymbol{B}_e(\boldsymbol{\theta})\dot{\boldsymbol{\theta}} - \boldsymbol{\Omega}_C^C \times \boldsymbol{h}(t)
\end{aligned} \tag{7.144}
$$

Taking the derivative of $\boldsymbol{e}(t)$ using (7.140), (7.142), (7.143), and (7.144), the derivative of error can be written as

$$
\begin{aligned}
\dot{\boldsymbol{e}} &= -\boldsymbol{\Omega}_C^C \times \boldsymbol{e} - \boldsymbol{V}_C^C - {}^{I}\boldsymbol{R}_C^{\top}\boldsymbol{J}_t\boldsymbol{B}_e\dot{\boldsymbol{\theta}} \\
&= -\boldsymbol{\Omega}_C^C \times \boldsymbol{e} - \boldsymbol{J}_c\dot{\boldsymbol{q}}_c
\end{aligned} \tag{7.145}
$$

where $\boldsymbol{J}_c = (\mathbb{I}_{3\times 3}, {}^{I}\boldsymbol{R}_C\boldsymbol{J}_t\boldsymbol{B}_e) \in \mathbb{R}^{3\times(3+4)}$ and $\dot{\boldsymbol{q}}_c = \left(\boldsymbol{V}_C^C, \dot{\boldsymbol{\theta}}\right)^{\top} \in \mathbb{R}^{3+4}$ is the robot joint space variable.

What needs to be reaffirmed is that if the visual error **e** tends to zero and the UAM is stable, the virtual camera plane will overlap the real physical plane. That means the roll angle and pitch angle will be zero and the yaw angle will tend to the desired value, which meets the previous hypothesis. At the same time, $\boldsymbol{Q}_t'$ will tend to $\boldsymbol{Q}_t$, and $\boldsymbol{P}_\delta$ will tend to zero, when error **e** tends to zero. Therefore, the difference between $\boldsymbol{Q}_t'$ and $\boldsymbol{Q}_t$ will disappear, which finally enable the end effector to grasp the target accurately. Meanwhile, by constructing a virtual camera plane, the linear velocity term and the angular velocity term are separated, which makes it possible to control the linear and angular velocities, respectively.

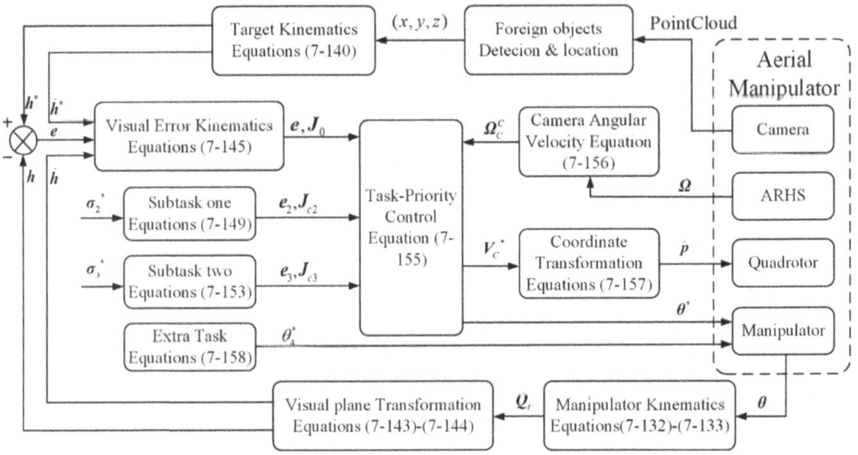

Fig. 7.28 Control architecture of the proposed visual servoing method

7.4.1.4 The Multitask-Priority Control for Aerial Manipulator

A control law derived by the task-priority control method will be designed in the task space, as shown in Fig. 7.28. The proposed control strategy mainly includes three parts. The first is the visual error kinematics. The second is the proposed task-priority control scheme, which includes two subtasks. The last one is the proposed extra-task, which controls the orientation directly. The system inputs of the proposed method are the foreign object position $\boldsymbol{P}_G$, the drone's angular velocity $\boldsymbol{\Omega}$, and the manipulator's joint angles $\boldsymbol{\theta}$. The system outputs are the desired linear velocity of the drone $\mathbf{V}^*$ and the desired joint angular velocity $\dot{\boldsymbol{\theta}}^*$ of the manipulator.

Furthermore, the aerial manipulator we design is ten DOFs, which is redundant relative to the task space of six degrees of freedom. The drone has six DOFs; the rest belong to the manipulator. That means the result of the inverse kinematics is multiple solutions for the desired end-effector pose. Therefore, several low-priority subtasks ensuring the system safety to achieve the optimal state can be designed.

1. Main Task: The end-effector position

   As defined above, when the visual error is zero, the end effector will reach the given position. According to formula (7.145), the following control rates can be designed based on the Lyapunov stability theory:

$$\dot{q}_c = \boldsymbol{J}_c^{\dagger} \boldsymbol{\Lambda}_1 \boldsymbol{e} \tag{7.146}$$

where $\boldsymbol{J}_c^{\dagger} = \boldsymbol{J}_c^{\top}\left(\boldsymbol{J}_c \boldsymbol{J}_c^{\top}\right)^{-1}$ is the pseudoinverse matrix of $\boldsymbol{J}_c$ and $\boldsymbol{\Lambda}_1$ is the positive definite diagonal matrix. Meanwhile, the null space of $\boldsymbol{J}_c$ can be expressed as $\boldsymbol{N}_1$. The result can be calculated as

$$N_1 = \left(\mathbb{I}_{7\times 7} - J_c^{\dagger} J_c\right) \tag{7.147}$$

2. Subtask 1: Camera posture

   In this subtask, the controller will prevent the camera lose the target and keep the foreign object in the center of the view of the camera. Let $\boldsymbol{\sigma}_2 = (X_o, Y_o, Z_o)^{\top}$ be the image feature vector of the projection of the grasp points $\boldsymbol{P}_G$ onto the normalized plane. The purpose of this subtask is to make the target feature point imaging in the center of the camera's field of view; thus, $\boldsymbol{\sigma}_2^* = (0, 0, 0)^{\top}$. The differential relationship of $\boldsymbol{\sigma}_2$ can be written as

$$\begin{aligned}\dot{\boldsymbol{\sigma}}_2 = \begin{pmatrix}\dot{X}_o \\ \dot{Y}_o \\ \dot{Z}_o\end{pmatrix} &= \mathbf{L}_x \begin{pmatrix}\mathbf{V}_C^C \\ \boldsymbol{\Omega}_C^C\end{pmatrix} \\ &= \boldsymbol{L}_{xp} \boldsymbol{V}_C^C + \boldsymbol{L}_{xo} \boldsymbol{\Omega}_C^C \\ &= \boldsymbol{J}_{c2} \dot{\boldsymbol{q}}_c + \bar{\boldsymbol{J}}_{c2} \boldsymbol{\Omega}_C^C\end{aligned} \tag{7.148}$$

   where $\boldsymbol{J}_{c2} = \left(\boldsymbol{L}_{xp}, \mathbb{O}_{3\times 4}\right) \in \mathbb{R}^{3\times 7}$, $\bar{\boldsymbol{J}}_{c2} = \boldsymbol{L}_{xo} \in \mathbb{R}^{3\times 3}$, and $\boldsymbol{L}_x = (\boldsymbol{L}_{xp}, \boldsymbol{L}_{xo})$ is the interaction matrix of the IBVS in [28]. Note that only the control of the aerial platform is needed to accomplish this task. Let us define the error vector as $\boldsymbol{e}_2 = \boldsymbol{\sigma}_2^* - \boldsymbol{\sigma}_2$, and $\dot{\boldsymbol{e}}_2$ can be written as

$$\dot{\boldsymbol{e}}_2 = -\boldsymbol{J}_{c2} \dot{\boldsymbol{q}}_c - \bar{\boldsymbol{J}}_{c2} \boldsymbol{\Omega}_C^C \tag{7.149}$$

   The null space of $\boldsymbol{J}_{c2}$ can be written as

$$\boldsymbol{N}_2 = \left(\mathbb{I}_{7\times 7} - \boldsymbol{J}_{c2}^{\dagger} \boldsymbol{J}_{c2}\right) \tag{7.150}$$

   where $\boldsymbol{J}_{c2}^{\dagger} = \boldsymbol{J}_{c2}^{\top} \left(\boldsymbol{J}_{c2} \boldsymbol{J}_{c2}^{\top}\right)^{-1}$ is the pseudoinverse matrix of $\boldsymbol{J}_{c2}$.

3. Subtask 2: Joint limit avoidance

   In this subtask, the controller will keep the manipulator joint far from its limit. Let $\boldsymbol{\theta}^*$ express the central position of joints, and it can be calculated as

$$\boldsymbol{\theta}^* = \frac{\boldsymbol{\theta}_L + \boldsymbol{\theta}_H}{2} \tag{7.151}$$

   where $\boldsymbol{\theta}_L$ expresses the low limits of each joint of the manipulator and $\boldsymbol{\theta}_H$ expresses the high limits of each joint of the manipulator. Based on the previous section, we have

$$\dot{\boldsymbol{\theta}} = \boldsymbol{J}_{c3} \dot{\boldsymbol{q}}_c \tag{7.152}$$

   where $\boldsymbol{J}_{c3} = \left(\mathbb{O}_{4\times 3}, \mathbb{I}_{4\times 4}\right) \in \mathbb{R}^{4\times 7}$ is the Jacobian matrix of *subtask 2*. Therefore, the error vector can be described as $\boldsymbol{e}_3 = \boldsymbol{\theta}^* - \boldsymbol{\theta}$, and $\dot{\boldsymbol{e}}_3$ can be calculated as

$$\dot{e}_3 = -J_{c3}\dot{q}_c \tag{7.153}$$

The null space of J_{c3} can be written as

$$N_3 = \left(\mathbb{I}_{7\times 7} - J_{c3}^{\dagger} J_{c3}\right) \tag{7.154}$$

where $J_{c3}^{\dagger} = J_{c3}^{\top}\left(J_{c3}J_{c3}^{\top}\right)^{-1}$ is the pseudoinverse matrix of J_{c3}.

In summary, this research is built upon the foundation provided by literature [15]. For the several tasks (one main task and two subtasks) of the aerial manipulator, a feasible solution is

$$\dot{q}_c^d = J_c \Lambda_1 e + N_1\left(J_{c2}^{\dagger}\Lambda_2 e_2 + N_2 J_{c3}^{\dagger}\Lambda_3 e_3\right) \tag{7.155}$$

where $\dot{q}_c^d$ is the desired derivation of q_c and Λ_i $(i = 1, 2, 3)$ is the positive definite diagonal matrix.

The camera velocity V_C^C in the generated corresponding variable $\dot{q}_c$ is expressed in the camera frame $\mathcal{F}_C$. We need to convert it into the body frame $\mathcal{F}_V$ and to the quadrotor control variables. Based on rigid body kinematics, we have

$$\begin{pmatrix} V_C^C \\ \Omega_C^C \end{pmatrix} = \begin{pmatrix} -{}^V R_C^{\top} & -{}^V R_C^{\top}[{}^V t_C]_{\times} \\ 0_{3\times 3} & -{}^V R_C^{\top} \end{pmatrix} \begin{pmatrix} {}^V \dot{p} \\ {}^V w \end{pmatrix} \tag{7.156}$$

where ${}^V\dot{p}$ is the linear velocity of the frame $\mathcal{F}_V$ and ${}^V w$ is the angular velocity of the frame $\mathcal{F}_V$. ${}^V R_C$ and ${}^V t_C$ are rotation matrix and translational vector, respectively. Thence, the desired value of the drone can be calculated as

$$\dot{p} = {}^V\dot{p} = {}^I R_C V_C^C + {}^I R_V \left[{}^V t_C\right]_{\times} R_{\Phi}\dot{\Phi} \tag{7.157}$$

Another independent task is to control the orientation of the end effector. Based on the previous discussions, the final joint of the manipulator is independent of the position of the end effector, the angular velocity of the camera, and the linear velocity of the camera. Therefore, the orientation controller of the end effector and the drone will be designed separately from task-priority control. Let $e_{\psi} = \theta_4 - P_{\psi}$ represent the error between the yaw angle of visual estimation and the joint state of manipulator. Thus, we can design the orientation control rate as

$$\dot{\theta}_4 = -k_{\psi} e_{\psi} \tag{7.158}$$

7.4.1.5 Application: Visual Servo Control of Autonomous Removing Foreign Objects for Power Transmission Line

To better adapt to the task of removing foreign objects, a new gripper with passive adaptability is designed in this section. Compared with most UAMs using serial multi-joint manipulators in the past, this section uses delta parallel robot as the manipulator and designs a new gripper with passive adaptability for the task of power line foreign object removal, which has the following advantages:

1. The parallel mechanism is superior to the serial mechanism in terms of rigidity and accuracy, and the delta manipulator actuators are concentrated on the top mounting plate, which can effectively decrease the weight of the end effector and improve the control accuracy.
2. The gripper with passive adaptability has the capability of adapting to the power transmission line of any diameter and absorbing the impact of the power transmission line while removing foreign objects, ensuring system security.
3. The heating wire has the capability of cutting off the connection point between the floating object and the power transmission line; therefore, there is no residue left.

Figure 7.29 shows the UAM mechanism design overview. The proposed UAM is mainly composed of a hexacopter drone, a delta manipulator, and a foreign object removal mechanism. The system overall structure is shown in Fig. 7.29a. As shown in Fig. 7.29, we chose to remove foreign objects from above the power transmission line, which has the following advantages: first, the foreign objects hang down under the effect of gravity, which makes it easier to remove from the top. Second, the open environment above the power lines can make UAMs safe from collision accidents. Third, by controlling the temperature of the heating wire, foreign objects can be quickly and effectively removed without igniting. The schematic diagram of the UAM foreign object removal process is shown in Fig. 7.29b. The design of the delta manipulator and foreign object removal mechanism is shown in Fig. 7.29c. The delta manipulator has four joints, of which three joints that control the spatial coordinate position of the end effector are mounted on the top and one joint that controls the rotation angle of the end effector is mounted on the end. Carbon fiber is used for the arms and forearms of the delta manipulator to reduce weight. The foreign object removal mechanism is installed at the end of the manipulator and rotates with its fourth joint. It mainly consists of a fixed seat, an elastic compliance mechanism, an adaptive clamping mechanism, an electric heating wire, and a limit switch. The elastic compliance mechanism can ensure the safety of the UAM in contact with the power line and can provide active compliance to adaptively measure and control errors. In addition, the adaptive clamping mechanism can promote full contact between the electric heating wire and the power line to ensure the effective removal of foreign objects.

In this section, indoor experiments are performed on the UAM system as shown in Fig. 7.30 to verify the proposed method in this section. The experiment of removing foreign objects for the power transmission line is introduced, which can

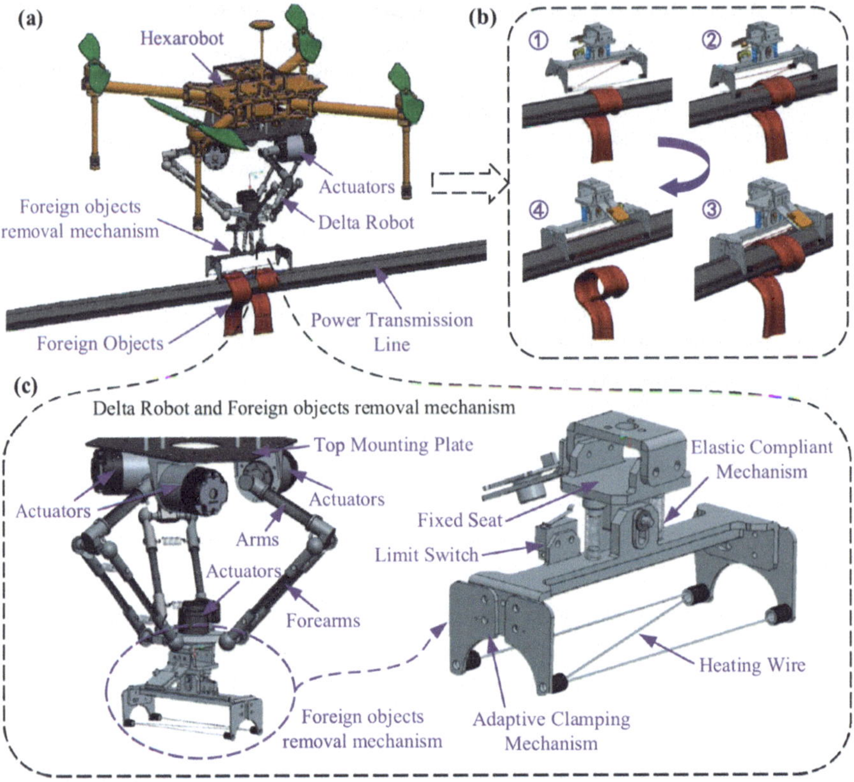

Fig. 7.29 UAM mechanism design overview. (**a**) System overall structure. (**b**) Schematic diagram of the UAM foreign object removal process. (**c**) The design of the delta manipulator and the foreign object removal mechanism

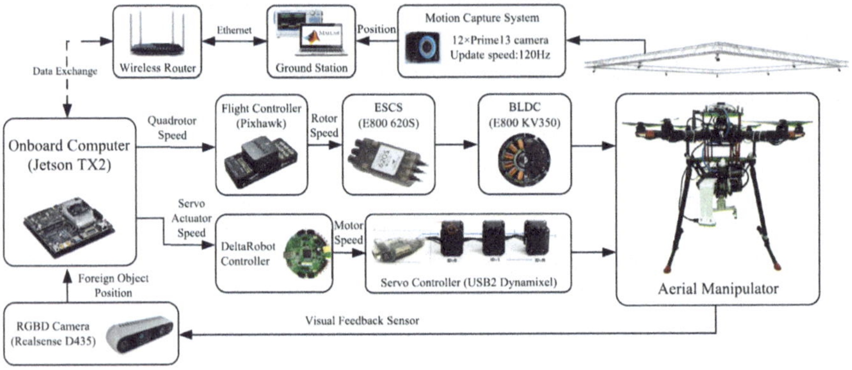

Fig. 7.30 Hardware structure of the experimental system

prove the effectiveness of the method we proposed. First, we will introduce the experimental environment and system configuration. Next, we will introduce the results of the experiment and give some analysis.

The proposed UAM system hardware includes aerial platform, Delta manipulator and its controller, onboard computer, flight controller, and RGB-D camera. The aerial platform is a hexacopter drone, which has the advantages of heavy load and high redundancy and can meet the load requirements of the mechanism. A Pixhawk flight controller with PX4 firmware is used to control the hexacopter drone's altitude and position. At the same time, an onboard computer equipped with an NVIDIA Jetson TX2 is used to process the information of the vision sensor and send control commands to the flight controller. The Delta manipulator controller is a self-made PCB circuit board based on the Arm Cortex-M4 core. It receives control commands from the onboard computer through the Ethernet and controls the actuators through the CAN bus. As an RGB-D camera, an Intel RealSense D435 depth camera was selected. In addition, the actuators adopt an intelligent servo driver (INNFOS QDD Lite series) that integrates the controller, motor, and reducer. It has the advantages of a fast communication cycle (up to 500 Hz), large power density, and high accuracy (19-bit absolute encoder), which provides the possibility for dynamic control.

Furthermore, the visual servoing control algorithm is implemented in ROS (Robot Operating System) Kinetic on the onboard computer and sent the desired velocity to the underlying PX4 controller via MAVROS. The lower embedded software, including flight controller software, and Delta manipulator control board software are based on the embedded real-time operating system (RTOS). The motion capture system and the ground station capture the pose of the aerial manipulator at a frequency of 120 Hz and broadcast it to the network. Meanwhile, we calibrate the camera's external parameters. The results are as follows:

$$
{}^{B}\boldsymbol{T}_{C}=\begin{pmatrix} -0.771 & 0.099 & 0.629 & 0.209 \\ 0.065 & 0.994 & -0.076 & -0.041 \\ -0.633 & -0.017 & -0.773 & -0.141 \\ 0 & 0 & 0 & 1 \end{pmatrix} \tag{7.159}
$$

Finally, the parameters mentioned above are set as follows: $\Lambda_1 = \mathrm{diag}\,(0.2, 0.2, 0.2), \Lambda_2 = \mathrm{diag}\,(0.17, 0.17, 0.17, 0.17), \Lambda_3 = \mathrm{diag}\,(0.065, 0.065)$, and $k_\psi = 0.36$. The initial joint value of the robot manipulator is $(-0.16\pi, -0.16\pi, -0.16\pi, 0)$.

The results of the experiment are shown in Figs. 7.31, 7.32, and 7.33. Some snapshots of the experiment are shown in Fig. 7.31. Firstly, the drone moves above the power transmission line manually. Then UAM starts to patrol the line. In this process, the transmission line will be guaranteed within the camera's view by keeping the joints of the manipulator maintain the initial angle. This progress is shown in Fig. 7.31a–c. Figure 7.31d shows that the drone starts to patrol the line and starts searching for the target. In Fig. 7.31g–i, some details of the removing process

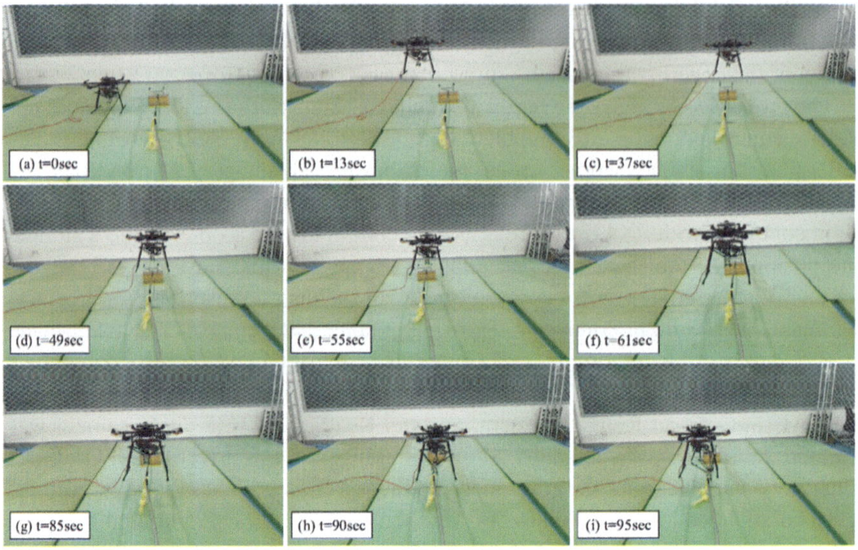

Fig. 7.31 Snapshots of the experiments

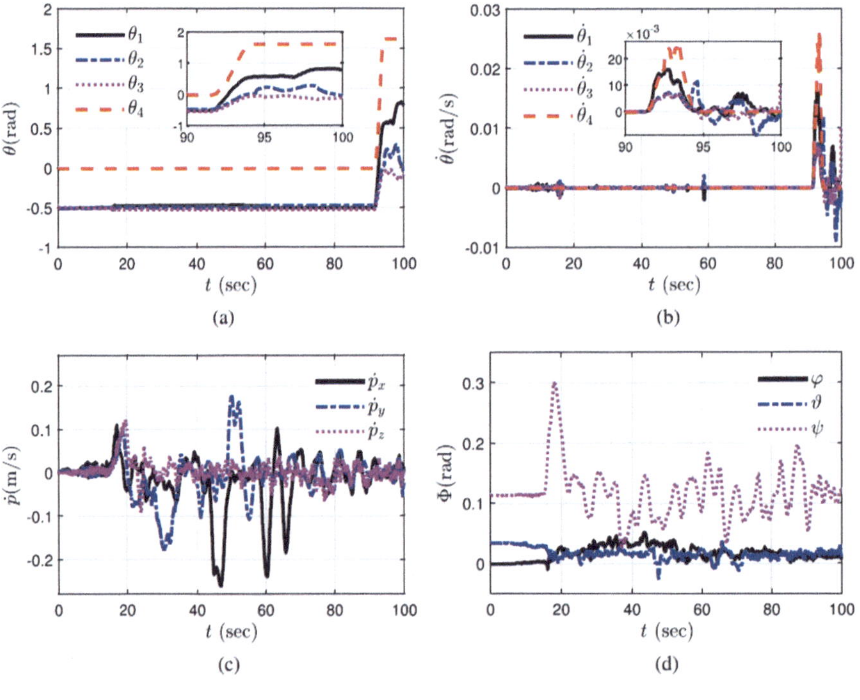

Fig. 7.32 Results of the experiment. (**a**) Joint angles of the manipulator. (**b**) Joint angular velocity of the manipulator. (**c**) Linear velocity of the drone. (**d**) Euler angles of the drone

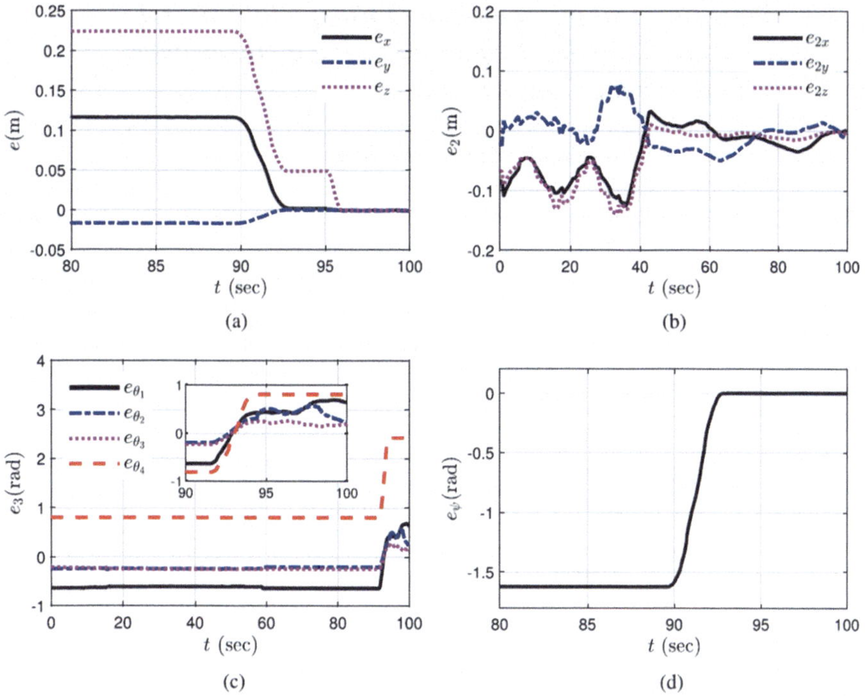

Fig. 7.33 Results of the experiment. (**a**) Time history of the visual error **e**. (**b**) Time history of subtask 1 error $\mathbf{e}_2$. (**c**) Time history of subtask 2 error $\mathbf{e}_3$. (**d**) Time history of the orientation control error e_ψ

are proved. From Fig. 7.31, it can be seen that the UAM can detect the foreign object and can remove the target under the visual servo control.

Figures 7.32 and 7.33 show some details of visual servo process. Figure 7.32a and b shows the joint angles $\boldsymbol{\theta}$ and the angular velocity $\dot{\boldsymbol{\theta}}$ of the manipulator, which begins to grasp from 92 s. Figure 7.32c shows the linear velocity $\dot{\boldsymbol{p}}$ of the drone, and Fig. 7.32d shows the angular velocity $\dot{\boldsymbol{\Phi}}$ of the drone. From the picture, we can learn that the drone only works 92 s before. Therefore, we can learn that the aerial manipulator can achieve hierarchical control of the drone and the manipulator according to the proposed method.

Figure 7.33 shows the trend of visual error e. As can be seen from Fig. 7.33a, e finally gradually becomes zero. This indicates that the end effector grasps the target, and UAM reaches the grab position. Figure 7.33b shows the error $\mathbf{e}_2$ of subtask 1, which finally become zero. This indicates that the camera maintains the desired value. Figure 7.33c shows the error e_3 of subtask 2. Since subtask 2 cannot be satisfied while satisfying the main task and subtask 1, the error e_3 of subtask 2 did not eventually tends to zero. Figure 7.33d shows the error e_ψ of the extra-task, which finally becomes zero. Results show the effectiveness of the task-priority control and the visual control.

7.4.2 Reference Optimization-Based Compliant Control for Aerial Pipeline Inspection Using a Hexacopter with a Robotic Contact Device

7.4.2.1 System Dynamics and Admittance Filter

The aerial contact inspection system is shown in Fig. 7.34, where the system is assumed to be rigid. Let $\mathcal{F}_w\{O_w x_w y_w z_w\}$ be the inertial frame, $\mathcal{F}_b\{O_b x_b y_b z_b\}$ denotes the body frame of the hexacopter, and $\mathcal{F}_e\{O_e x_e y_e z_e\}$ denotes the environment frame. Define the vectors $\boldsymbol{p} = [x, y, z]^\top \in \mathbb{P}^3$ and $\boldsymbol{v} = [v_x, v_y, v_z]^\top \in \mathbb{P}^3$ are the position and the linear velocity of the hexacopter expressed in the inertial frame $\mathcal{F}_w$, respectively. The vector $\boldsymbol{\Phi} = [\phi, \theta, \psi]^\top \in \mathbb{R}^3$ represents the roll-pitch-yaw Euler angle, and $\boldsymbol{R}_{b,\,w} \in SO(3)$ denotes the transformation matrix from the body frame $\mathcal{F}_b$ to the inertial frame $\mathcal{F}_w$.

Denote the system state variable $\boldsymbol{q} = [\boldsymbol{p}^\top, \boldsymbol{\Phi}^\top]^\top \in \mathbb{R}^6$, and then the system dynamics can be modeled by Newton-Euler method as [29, 30]

$$\boldsymbol{H}(\boldsymbol{q})\ddot{\boldsymbol{q}} + \boldsymbol{C}(\boldsymbol{q}, \dot{\boldsymbol{q}})\dot{\boldsymbol{q}} + \boldsymbol{G}(\boldsymbol{q}) = \boldsymbol{u} - \boldsymbol{d}_e \tag{7.160}$$

where $\boldsymbol{H}(\boldsymbol{q}) = \mathrm{diag}\,\{m\boldsymbol{I}_{3\times 3}, \boldsymbol{J}\}$ is the symmetric positive definite inertia matrix with the total mass m and the inertia tensor $\boldsymbol{J} \in \mathbb{P}^{3\times 3}$. The matrix $\boldsymbol{C}(\boldsymbol{q}, \dot{\boldsymbol{q}})$ and the vector $\boldsymbol{G}(\boldsymbol{q})$ represent the Coriolis and gravity terms, respectively. For brevity, $\boldsymbol{H}(\boldsymbol{q})$, $\boldsymbol{C}(\boldsymbol{q}, \dot{\boldsymbol{q}})$, and $\boldsymbol{G}(\boldsymbol{q})$ will be written as $\boldsymbol{H}$, $\boldsymbol{C}$, and $\boldsymbol{G}$ later. The vector $\boldsymbol{d}_e = [\boldsymbol{f}_w^\top, \boldsymbol{\tau}_w^\top]^\top$ denotes the external wrench with the external force $\boldsymbol{f}_w$ and torque $\boldsymbol{\tau}_w$, where $\boldsymbol{f}_w = \boldsymbol{f}_{dis} + \boldsymbol{f}_e$ includes the disturbance force $\boldsymbol{f}_{dis}$ and the interaction force

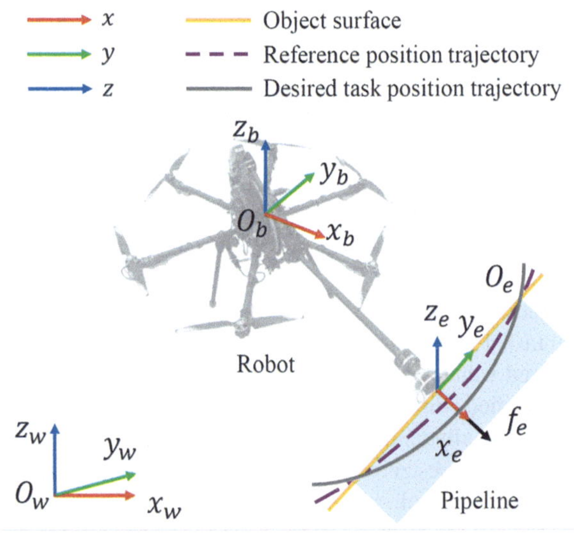

Fig. 7.34 Coordinate definitions of the aerial contact inspection system

$\boldsymbol{f}_e = \left[f_{e,x}, f_{e_y}, f_{e,z}\right]^{\mathrm{T}}$. The vector $\boldsymbol{u}$ is the generalized input of the system, which consists of the total thrust f and input torque $\boldsymbol{\tau}$. The actuator inputs f and $\boldsymbol{\tau}$ can be calculated by utilizing the generalized input $\boldsymbol{u}$ as [31]

$$\dot{s}_1 = \dot{z}_1 - \dot{\chi}_1 = s_2 + \boldsymbol{\alpha}_1 - \dot{\boldsymbol{q}}_c + \boldsymbol{K}_1\chi_1 + \frac{a_1\chi_1}{\sqrt{\|\chi_1\|^2 + \eta_1^2}} \tag{7.161}$$

where $\boldsymbol{T}_b$ represents the transformation matrix between the Euler angle rate and the angular velocity $_b$ expressed in the inertial frame $\mathcal{F}_b$. Further, the first two elements of $\boldsymbol{u}$ and the commanded yaw angle ψ_c are adopted for computing the commanded roll and pitch angles as

$$\begin{bmatrix} \phi_c \\ \theta_c \end{bmatrix} = \begin{bmatrix} \sin^{-1}\left[\frac{1}{f}(\boldsymbol{u}(1)\sin\psi_c - \boldsymbol{u}(2)\cos\psi_c)\right] \\ \tan^{-1}\left[\frac{1}{\mathbf{u}(3)}(\boldsymbol{u}(1)\cos\psi_c + \boldsymbol{u}(2)\sin\psi_c)\right] \end{bmatrix} \tag{7.162}$$

Next, without loss of generality, the dynamic model of a static environment is expressed as [32]

$$\boldsymbol{f}_e = \boldsymbol{K}_e(\boldsymbol{p} - \boldsymbol{p}_e) + \boldsymbol{D}_e\dot{\boldsymbol{p}} + \boldsymbol{M}_e\ddot{\boldsymbol{p}} \tag{7.163}$$

where $\boldsymbol{M}_e$, $\boldsymbol{D}_e$, and $\boldsymbol{K}_e$ denote the unknown inertia, damping, and stiffness matrices of the environment, respectively. $\boldsymbol{p}_e$ is the position of the hexacopter just when the hexacopter interacts with the environment without force exerted. Further, according to the studies in [33], the environment model in (7.163) can be also simplified as

$$\begin{aligned} \boldsymbol{f}_e &= \boldsymbol{K}_e(\boldsymbol{p} - \boldsymbol{p}_e) + \boldsymbol{D}_e\dot{\boldsymbol{p}} \\ \boldsymbol{f}_e &= \boldsymbol{K}_e(\boldsymbol{p} - \boldsymbol{p}_e) \end{aligned} \tag{7.164}$$

To achieve the compliant control for aerial contact inspection, an admittance filter is introduced and driven by the interaction force $\boldsymbol{f}_e$. The target admittance filter is expressed as [34, 35]

$$\boldsymbol{M}_d(\ddot{\boldsymbol{p}}_c - \ddot{\boldsymbol{p}}_r) + \boldsymbol{D}_d(\dot{\boldsymbol{p}}_c - \dot{\boldsymbol{p}}_r) + \boldsymbol{K}_d(\boldsymbol{p}_c - \boldsymbol{p}_r) = -\boldsymbol{f}_e \tag{7.165}$$

where $\boldsymbol{M}_d$, $\boldsymbol{D}_d$, and $\boldsymbol{K}_d$ denote the inertia, damping, and stiffness matrices of the admittance filter. $\boldsymbol{p}_r$ and $\boldsymbol{p}_c$ represent the reference position trajectory and the modified commanded position trajectory, respectively. From the admittance filter (7.165), it shows that the modified commanded position trajectory $\boldsymbol{p}_c$ is generated by the reference position trajectory $\boldsymbol{p}_r$ and the interaction force $\boldsymbol{f}_e$.

In order to tune the response of the system motion to the interaction force $\boldsymbol{f}_e$, the admittance coefficient matrices $\boldsymbol{M}_d$, $\boldsymbol{D}_d$, and $\boldsymbol{K}_d$ and the reference position trajectory $\boldsymbol{p}_r$ are required to be given. Notice that fixed admittance coefficient matrices and a fixed reference trajectory are conservative for interaction, and the environment dynamics should be considered to implement expected interaction performance. To solve this problem, this article focuses on designing the reference trajectory for optimal aerial contact inspection performance.

7.4.2.2 Aerial Contact Inspection Control Design

The control objective of this article is to design a compliant control framework for implementing optimal aerial contact inspection performance, and the overall structure of the compliant control framework is displayed in Fig. 7.35. To implement compliant control, the first step is to investigate the optimal interaction behavior using a reference learning scheme, and then the second step is to design a tracking controller for achieving the ideal tracking performance of the system under interaction disturbances such that $\mathbf{p} \to \mathbf{p}_c$ as $t \to \infty$.

In more detail, a cost function is introduced to describe the interaction performance, and then the reference learning scheme is utilized to investigate the optimization-based interaction behavior. Next, a pose tracking controller is developed for the system to ensure that $\mathbf{p} \to \mathbf{p}_r$ under external wrenches. In particular, the command filter-based adaptive backstepping technique is presented for designing the tracking controller to avoid calculating intermediate quantities, thereby reducing computational complexity.

7.4.2.3 Cost Function

Humans are prone to compensate for the original task trajectory for interaction with compliant environments according to the research results in [36]. If the interactive environment is stiff, humans will adjust trajectories to reduce interaction forces; thereby, the reference trajectory will gradually deform toward the surface of the environment. Typically, this phenomenon can be modeled by maintaining a balance between tracking errors and interaction forces. Thus, the aerial contact inspection performance can be quantified by the following cost function as

$$W(t) = \int_{t_0}^{t_f} \left((\boldsymbol{p} - \boldsymbol{p}_t)^{\mathrm{T}} \mathcal{Q} (\boldsymbol{p} - \boldsymbol{p}_t) + \boldsymbol{f}_e^{\mathrm{T}} \mathcal{R} \boldsymbol{f}_e \right) \mathrm{d}t \tag{7.166}$$

where $\boldsymbol{p}_t$ denotes the desired task trajectory and $\mathcal{Q}$ and $\mathcal{R} \in \mathbb{P}^{3 \times 3}$ denote the positive definite diagonal weight matrices. t_0 and t_f are the times when each motion starts and ends, respectively. $W(t)$ quantifies the expected interaction performance, which is a trade-off between the trajectory tracking error and the interaction force. By minimizing $W(t)$, the optimal interaction performance can be implemented and seen in Fig. 7.35; thus, the generated reference trajectory $\boldsymbol{p}_r$ can be regarded as the system optimal interaction behavior for the interaction force.

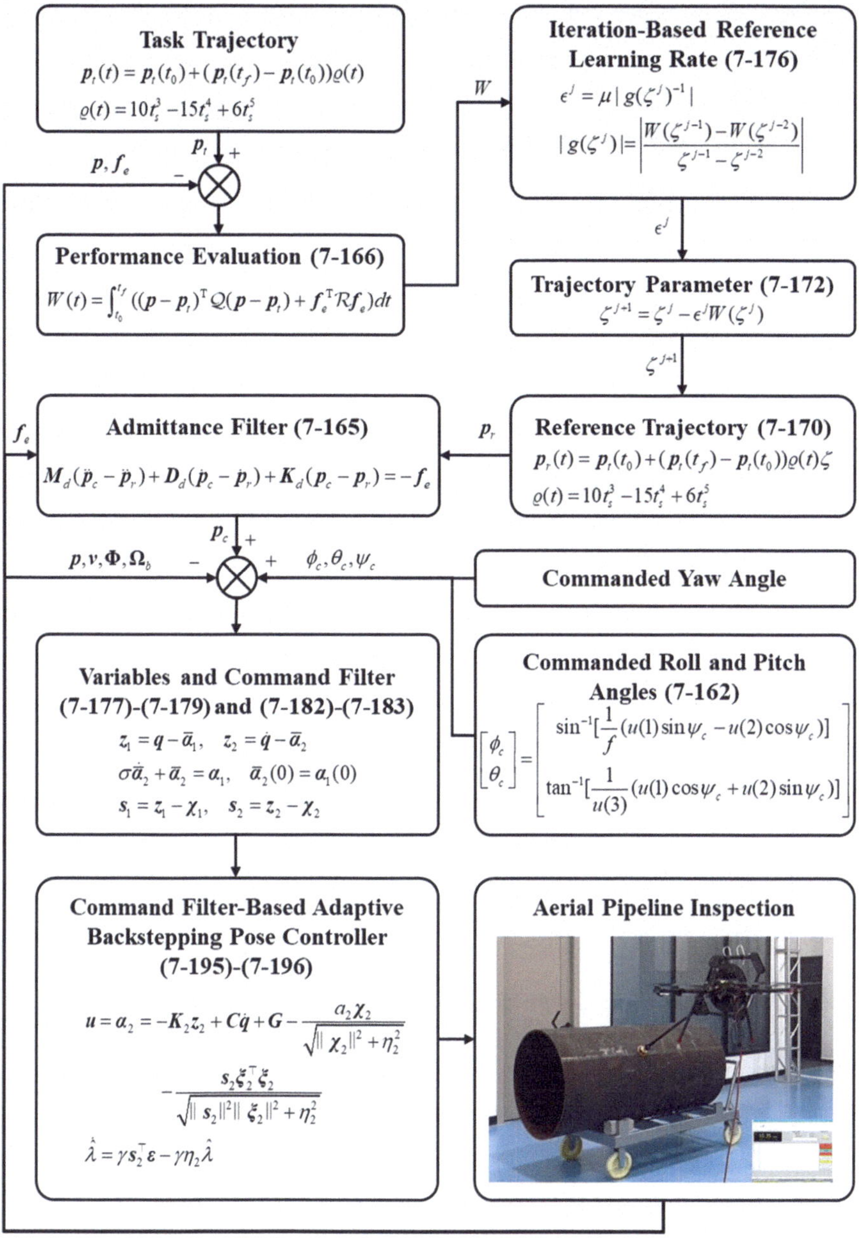

Fig. 7.35 Control framework for aerial compliant interaction

Different from existing interaction methods, the advantage of the cost function (7.166) is that it does not depend on the environment model, i.e., without the knowledge of the environment. When the environment changes, the expected interaction performance can also be guaranteed without the predefined feedback gains.

7.4.2.4 Function Parametrization

The ideal tracking performance of the system gives rise to $\boldsymbol{p} \rightarrow \boldsymbol{p}_c$ as $t \rightarrow \infty$, and then the admittance filter (7.165) can be written as

$$\boldsymbol{M}_d(\ddot{\boldsymbol{p}} - \ddot{\boldsymbol{p}}_r) + \boldsymbol{D}_d(\dot{\boldsymbol{p}} - \dot{\boldsymbol{p}}_r) + \boldsymbol{K}_d(\boldsymbol{p} - \boldsymbol{p}_r) = -\boldsymbol{f}_e \tag{7.167}$$

Combining (7.163) and (7.167) yields

$$(\boldsymbol{M}_d + \boldsymbol{M}_e)\ddot{\boldsymbol{p}} + (\boldsymbol{D}_d + \boldsymbol{D}_e)\dot{\boldsymbol{p}} + (\boldsymbol{K}_d + \boldsymbol{K}_e)\boldsymbol{p} = \boldsymbol{K}_d\boldsymbol{p}_r + \boldsymbol{D}_d\dot{\boldsymbol{p}}_r + \boldsymbol{M}_d\ddot{\boldsymbol{p}}_r\boldsymbol{K}_e\boldsymbol{p}_e \tag{7.168}$$

It shows that the system actual trajectory $\boldsymbol{p}$ is given from the reference position trajectory $\boldsymbol{p}_r$ and the environmental location $\boldsymbol{p}_e$, where $\boldsymbol{p}_e$ is constant owing to the system interacting with the static environment.

Let the reference trajectory $\boldsymbol{p}_r$ parameterize as $\boldsymbol{p}_r(\zeta)$ with a trajectory parameter ζ. From (7.167), the variables $\boldsymbol{p}$ and $\boldsymbol{f}_e$ can be parameterized as $\boldsymbol{p}(\zeta)$ and $\boldsymbol{f}_e(\zeta)$, respectively. According to (7.166), the cost function $W(t)$ can also be parameterized as $W(\zeta)$. Thus, the control objective becomes to find an optimal set ζ^* of ζ such that $W(\zeta)$ is minimized by

$$\zeta^* = \arg\min_{\zeta} W(\zeta) \tag{7.169}$$

It is worth noting that as long as the reference trajectory $\boldsymbol{p}_r$ can be parameterized, it is applicable to the reference learning method in this article. According to [37] and [38], the reference position trajectory $\boldsymbol{p}_r(t)$ is parameterized by ζ with the minimum jerk trajectory form as

$$\begin{aligned} \boldsymbol{p}_r(t) &= \boldsymbol{p}_t(t_0) + (\boldsymbol{p}_t(t_f) - \boldsymbol{p}_t(t_0))\varrho(t)\zeta \\ \varrho(t) &= 10t_s^3 - 15t_s^4 + 6t_s^5 \end{aligned} \tag{7.170}$$

where $t_s = (t - t_0)/(t_f - t_0)$ with the current time t.

7.4.2.5 Iteration-Based Reference Learning and Learning Rate Selection

An iteration-based reference learning scheme is utilized to acquire the optimal set ζ^*. First, the mapping is constructed with the minimum of W^* by

$$W^* - W(\zeta^{j+1}) = \overline{\delta}(W^* - W(\zeta^j)) \tag{7.171}$$

where $W^* = W(\zeta^*)$, $\overline{\delta}$ denotes the convergence rate, and j represents the iteration index. According to [39], when $|\overline{\delta}| < 1$, then $W(\zeta) \to W^*$ as $j \to \infty$.

To avoid the unknown parameter W^* applying for the learning law design, thus a learning law is regulated and designed to implement the mapping (7.171) as

$$\zeta^{j+1} = \zeta^j - \epsilon^j W(\zeta^j) \tag{7.172}$$

where ϵ^j is the learning rate. Thus, the constructed mapping changes by

$$\begin{aligned} W(\zeta^{j+1}) &= W(\zeta^j) + (W(\zeta^{j+1}) - W(\zeta^j)) \\ &= W(\zeta^j) + g(\zeta^j)(\zeta^{j+1} - \zeta^j) \end{aligned} \tag{7.173}$$

Combining (7.172) and (7.173) yields

$$W(\zeta^{j+1}) = (1 - g(\zeta^j)\epsilon^j)W(\zeta^j) \tag{7.174}$$

From (7.174), $W(\zeta)$ converges to the minimum value with $|\overline{\delta}| = |1 - g(\zeta^j)\epsilon^j| < 1$. Thus, the learning rate ϵ^j should be selected to ensure that $W(\zeta)$ converges to the minimum. Note that the selection of ϵ^j relies on the gradient $g(\zeta^j)$ based on (7.174), if both the sign and bounds of $g(\zeta^j)$ are known such that

$$\underline{g} < g(\zeta^j) < \overline{g} \tag{7.175}$$

where $\overline{g}$ and $\underline{g}$ denote the positive upper and lower bounds of $g(\zeta^j)$, respectively. Then, the learning rate ϵ^j can be selected as $\epsilon^j = \mu\overline{g}^{-1}$ with a positive parameter $\mu \in (0, 1]$.

When both the sign and bounds of $g(\zeta^j)$ are unknown, extra method is required to determine ϵ^j. Note that if the sign of ϵ^j is chosen wrongly, then the value of the cost function $W(\zeta)$ will increase. Alternatively, additional trials are performed to determine the proper sign of ϵ^j. Typically, if ϵ^j is k-dimensional, then 2^k sets of trials are required to determine the proper sign of ϵ^j [40].

Meanwhile, the sign of the redefined gradient $g(\zeta^j)$ is quite important, and it is not always possible to obtain the proper sign of $g(\zeta^j)$. Thus, it can be given the amplitude of $g(\zeta^j)$ as

$$|g(\zeta^j)| = \left|\frac{W(\zeta^{j-1}) - W(\zeta^{j-2})}{\zeta^{j-1} - \zeta^{j-2}}\right| \tag{7.176}$$

Then, the learning rate ϵ^j is accordingly regulated to be positive by $\epsilon^j = \mu|g(\zeta^j)^{-1}|$.

7.4.2.6 Pose Tracking Controller Design

A pose tracking controller is developed for the system under external wrenches. Define the commanded state trajectory $\boldsymbol{q}_c = \left[\boldsymbol{p}_c^{\top}, \boldsymbol{\Phi}_c^{\top}\right]^{\top} \in \mathbb{R}^6$, where $\boldsymbol{p}_c$ is obtained according to the admittance filter and $\boldsymbol{\Phi}_c$ is calculated from (7.162). Then, the following variables are defined as

$$z_1 = \boldsymbol{q} - \overline{\boldsymbol{\alpha}}_1 \tag{7.177}$$

$$z_2 = \dot{\boldsymbol{q}} - \overline{\boldsymbol{\alpha}}_2 \tag{7.178}$$

where $\overline{\boldsymbol{\alpha}}_1$ is defined as $\overline{\boldsymbol{\alpha}}_1 = \boldsymbol{q}_c$. $\overline{\boldsymbol{\alpha}}_2$ is the output of the first-order command filter of the virtual input $\boldsymbol{\alpha}_2$, and $\overline{\boldsymbol{\alpha}}_2$ is defined as

$$\sigma\dot{\overline{\boldsymbol{\alpha}}}_2 + \overline{\boldsymbol{\alpha}}_2 = \boldsymbol{\alpha}_1, \quad \overline{\boldsymbol{\alpha}}_2(0) = \boldsymbol{\alpha}_1(0) \tag{7.179}$$

where σ is a positive constant. The command filter-based adaptive backstepping tracking control method is given as follows:

Step 1: Calculating the derivative of z_1 gives

$$\dot{z}_1 = \dot{\boldsymbol{q}} - \dot{\overline{\boldsymbol{\alpha}}}_1 = z_2 + (\overline{\boldsymbol{\alpha}}_2 - \boldsymbol{\alpha}_1) + \boldsymbol{\alpha}_1 - \dot{\boldsymbol{q}}_c \tag{7.180}$$

In order to handle the influence of the error $(\overline{\boldsymbol{\alpha}}_2 - \boldsymbol{\alpha}_1)$, the compensation $\boldsymbol{\chi}_1$ is defined as

$$\dot{\boldsymbol{\chi}}_1 = -\boldsymbol{K}_1\boldsymbol{\chi}_1 + (\overline{\boldsymbol{\alpha}}_2 - \boldsymbol{\alpha}_1) + \boldsymbol{\chi}_2 - \frac{a_1\boldsymbol{\chi}_1}{\sqrt{\|\boldsymbol{\chi}_1\|^2 + \eta_1^2}} \tag{7.181}$$

where $\boldsymbol{K}_1 \in \mathbb{P}^{6 \times 6}$ is a positive definite diagonal gain matrix and a_1 and η_1 are positive constants. $\boldsymbol{\chi}_1(0) = 0_6$, and χ_2 is designed later. Further, define the compensation errors s_1 and s_2 as

$$s_1 = z_1 - \boldsymbol{\chi}_1 \tag{7.182}$$

$$s_2 = z_2 - \boldsymbol{\chi}_2 \tag{7.183}$$

Based on (7.181), the derivative of $\mathbf{s}_1$ can be computed by

$$\dot{s}_1 = \dot{z}_1 - \dot{\chi}_1 = s_2 + \boldsymbol{\alpha}_1 - \dot{q}_c + K_1\chi_1 + \frac{a_1\chi_1}{\sqrt{\|\chi_1\|^2 + \eta_1^2}} \tag{7.184}$$

Design the following Lyapunov function V_1 as

$$V_1 = \frac{1}{2} s_1^\top s_1 \tag{7.185}$$

According to (7.184), the derivative of V_1 is given by

$$\dot{V}_1 = s_1^\top \dot{s}_1 = s_1^\top \left(s_2 + \boldsymbol{\alpha}_1 - \dot{\boldsymbol{q}}_c + \boldsymbol{K}_1 \chi_1 + \frac{a_1 \chi_1}{\sqrt{\|\chi_1\|^2 + \eta_1^2}} \right) \tag{7.186}$$

Thus, the virtual controller $\boldsymbol{\alpha}_1$ is designed as

$$\boldsymbol{\alpha}_1 = -\boldsymbol{K}_1 z_1 - \frac{a_1 \boldsymbol{\chi}_1}{\sqrt{\|\boldsymbol{\chi}_1\|^2 + \eta_1^2}} + \dot{\boldsymbol{q}}_c \tag{7.187}$$

Substituting $\boldsymbol{\alpha}_1$ into (7.186) yields

$$\dot{V}_1 = -\mathbf{s}_1^\top \boldsymbol{K}_1 s_1 + s_1^\top s_2 \tag{7.188}$$

Step 2: From (7.188) and the system dynamics (7.160), the derivative of s_2 is obtained by

$$\dot{z}_2 = \ddot{\boldsymbol{q}} - \dot{\bar{\boldsymbol{\alpha}}}_2 = \boldsymbol{H}^{-1}(\boldsymbol{u} - \boldsymbol{d}_e - \boldsymbol{C}\dot{\boldsymbol{q}} - \boldsymbol{G}) - \dot{\bar{\boldsymbol{\alpha}}}_2 \tag{7.189}$$

Construct the compensation $\boldsymbol{\chi}_2$ as

$$\dot{\chi}_2 = -\boldsymbol{H}^{-1} \boldsymbol{K}_2 \chi_2 - \boldsymbol{H}^{-1} \frac{a_2 \chi_2}{\sqrt{\|\chi_2\|^2 + \eta_2^2}} \tag{7.190}$$

where $\boldsymbol{K}_2 \in \mathbb{P}^{6 \times 6}$ is a positive definite diagonal gain matrix, a_2 and η_2 are positive constants, and $\boldsymbol{\chi}_2(0) = 0_6$. Then, from (7.183), the derivative of $\mathbf{s}_2$ is obtained by

$$\dot{s}_2 = \dot{z}_2 - \dot{\chi}_2 = \boldsymbol{H}^{-1}(\boldsymbol{u} - \boldsymbol{d}_e - \boldsymbol{C}\dot{\boldsymbol{q}} - \boldsymbol{G}) - \dot{\bar{\boldsymbol{\alpha}}}_2 + \boldsymbol{H}^{-1} \boldsymbol{K}_2 \chi_2 + \boldsymbol{H}^{-1} \frac{a_2 \boldsymbol{\chi}_2}{\sqrt{\|\boldsymbol{\chi}_2\|^2 + \eta_2^2}} \tag{7.191}$$

Design the following Lyapunov function V_2 as

$$V_2 = V_1 + \frac{1}{2} s_2^\top \boldsymbol{H} s_2 + \frac{1}{2\gamma} \tilde{\lambda}^2 \tag{7.192}$$

where $\widetilde{\lambda} = \lambda - \widehat{\lambda}$ and γ is a positive constant. Define $\lambda = \sup_{t \geq 0} \|\boldsymbol{\Pi}\|$ with $\boldsymbol{\Pi} = [\boldsymbol{I}_{6\times 6}, \text{diag}\{\mathbf{d}_e\}, \boldsymbol{H}]^\top \in \mathbb{R}^{18\times 6}$. Then, utilizing (7.188) and (7.191), the derivative of V_2 can be derived by

$$\begin{aligned}
\dot{V}_2 &= \dot{V}_1 + \boldsymbol{s}_2^\top \boldsymbol{H}\dot{\boldsymbol{s}}_2 - \frac{1}{\gamma}\widetilde{\lambda}\dot{\widehat{\lambda}} \\
&= -\boldsymbol{s}_1^\top \boldsymbol{K}_1 \boldsymbol{s}_1 + \boldsymbol{s}_1^\top \boldsymbol{s}_2 + \boldsymbol{s}_2^\top \left(\boldsymbol{u} - \boldsymbol{d}_e - \boldsymbol{C}\dot{\boldsymbol{q}} - \boldsymbol{G} - \boldsymbol{H}\dot{\boldsymbol{\alpha}}_2 + \boldsymbol{K}_2\boldsymbol{\chi}_2 + \frac{a_2\boldsymbol{\chi}_2}{\sqrt{\|\boldsymbol{\chi}_2\|^2 + \eta_2^2}} \right) \\
&\quad - \frac{1}{\gamma}\widetilde{\lambda}\dot{\widehat{\lambda}} \\
&= -\boldsymbol{s}_1^\top \boldsymbol{K}_1 \boldsymbol{s}_1 + \boldsymbol{s}_2^\top \left(\boldsymbol{u} - \boldsymbol{C}\dot{\boldsymbol{q}} - \boldsymbol{G} + \frac{a_2\boldsymbol{\chi}_2}{\sqrt{\|\boldsymbol{\chi}_2\|^2 + \eta_2^2}} + \boldsymbol{K}_2\boldsymbol{\chi}_2 \right) + \boldsymbol{s}_2^\top \boldsymbol{\Pi}^\top \boldsymbol{\Psi} - \frac{1}{\gamma}\widetilde{\lambda}\dot{\widehat{\lambda}}
\end{aligned} \tag{7.193}$$

where $\boldsymbol{\Psi} = \left[\boldsymbol{s}_1^\top, -\boldsymbol{I}_6^\top, -\dot{\boldsymbol{\alpha}}_2^\top\right]^\top \in \mathbb{R}^{18}$. Note that there exists a parameter $\boldsymbol{\varepsilon}$ such that $\boldsymbol{s}_2^\top \boldsymbol{\Pi}^\top \boldsymbol{\Psi} \leq \lambda \boldsymbol{s}_2^\top \boldsymbol{\varepsilon} + \eta_2 \lambda$ with $\boldsymbol{\varepsilon} = \frac{\boldsymbol{s}_2 \boldsymbol{\Psi}^\top \boldsymbol{\Psi}}{\sqrt{\|\boldsymbol{s}_2\|^2 \|\boldsymbol{\Psi}\|^2 + \eta_2^2}}$, where the proof is given later. Then, (7.193) can be rewritten as

$$\begin{aligned}
\dot{V}_2 &\leq -\boldsymbol{s}_1^\top \boldsymbol{K}_1 \boldsymbol{s}_1 + \boldsymbol{s}_2^\top \left(\boldsymbol{u} - \boldsymbol{C}\dot{\boldsymbol{q}} - \boldsymbol{G} + \frac{a_2\boldsymbol{\chi}_2}{\sqrt{\|\boldsymbol{\chi}_2\|^2 + \eta_2^2}} + \boldsymbol{K}_2\boldsymbol{\chi}_2 \right) \\
&\quad + \lambda \mathbf{s}_2^\top \boldsymbol{\varepsilon} + \eta_2 \lambda - \frac{1}{\gamma}\widetilde{\lambda}\dot{\widehat{\lambda}} \\
&= -\boldsymbol{s}_1^\top \boldsymbol{K}_1 \boldsymbol{s}_1 + \boldsymbol{s}_2^\top \left(\boldsymbol{u} - \boldsymbol{C}\dot{\boldsymbol{q}} - \boldsymbol{G} + \frac{a_2\boldsymbol{\chi}_2}{\sqrt{\|\boldsymbol{\chi}_2\|^2 + \eta_2^2}} + \boldsymbol{K}_2\boldsymbol{\chi}_2 \right) + \widehat{\lambda} \boldsymbol{s}_2^\top \boldsymbol{\varepsilon} \\
&\quad + \widetilde{\lambda}\left(\boldsymbol{s}_2^\top \boldsymbol{\varepsilon} - \frac{1}{\gamma}\dot{\widehat{\lambda}} \right) + \eta_2 \lambda
\end{aligned} \tag{7.194}$$

Thus, the generalized control input can be designed as

$$\boldsymbol{u} = \boldsymbol{\alpha}_2 = -\boldsymbol{K}_2 \boldsymbol{z}_2 + \boldsymbol{C}\dot{\boldsymbol{q}} + \boldsymbol{G} - \frac{a_2\boldsymbol{\chi}_2}{\sqrt{\|\boldsymbol{\chi}_2\|^2 + \eta_2^2}} - \frac{\boldsymbol{s}_2 \boldsymbol{\xi}_2^\top \boldsymbol{\xi}_2}{\sqrt{\|\boldsymbol{s}_2\|^2 \|\boldsymbol{\xi}_2\|^2 + \eta_2^2}} \tag{7.195}$$

with the adaptive law

$$\dot{\widehat{\lambda}} = \gamma \boldsymbol{s}_2^\top \boldsymbol{\varepsilon} - \gamma \eta_2 \widehat{\lambda} \tag{7.196}$$

where $\xi_2 = \widehat{\lambda} \boldsymbol{\varepsilon} := \widehat{\lambda} \frac{\boldsymbol{s}_2 \boldsymbol{\Psi}^\top \boldsymbol{\Psi}}{\sqrt{\|\boldsymbol{s}_2\|^2 \|\boldsymbol{\Psi}\|^2 + \eta_2^2}}$. Further, substituting (7.195) and (7.196) into (7.194) yields

$$\dot{V}_2 \leq -\boldsymbol{s}_1^\top \boldsymbol{K}_1 \boldsymbol{s}_1 - \boldsymbol{s}_2^\top \boldsymbol{K}_2 \boldsymbol{s}_2 + \eta_2(1 + \lambda) + \eta_2 \widetilde{\lambda} \widehat{\lambda} \tag{7.197}$$

Theorem 7.1 Considering the system dynamics (7.160) under the generalized input (7.195) with the adaptive law (7.196), then for any initial conditions, all the system state variables of the closed-loop system are bounded under external wrenches, and the system state $\boldsymbol{q}$ will converge to the reference state $\boldsymbol{q}_c$.

Proof The following proof process is given:

Step 1: It will be proven that there exists a parameter $\boldsymbol{\varepsilon}$ satisfying $\boldsymbol{s}_2^\top \boldsymbol{\Pi}^\top \boldsymbol{\Psi} \leq \lambda \boldsymbol{s}_2^\top \boldsymbol{\varepsilon} + \eta_2 \lambda$ with $\boldsymbol{\varepsilon} = \frac{\boldsymbol{s}_2 \boldsymbol{\Psi}^\top \boldsymbol{\Psi}}{\sqrt{\|\boldsymbol{s}_2\|^2 \|\boldsymbol{\Psi}\|^2 + \eta_2^2}}$. From the definitions of $\boldsymbol{\Pi}$ and $\boldsymbol{\Psi}$ in (7.193), it gives

$$\boldsymbol{s}_2^\top \boldsymbol{\Pi}^\top \boldsymbol{\Psi} \leq \|\boldsymbol{s}_2\| \|\boldsymbol{\Pi}\| \|\boldsymbol{\Psi}\| = \lambda \|\boldsymbol{s}_2\| \|\boldsymbol{\Psi}\| \tag{7.198}$$

Then, it is concluded as

$$\|\boldsymbol{s}_2\| \|\boldsymbol{\Psi}\| - \frac{\boldsymbol{s}_2^\top \boldsymbol{s}_2 \boldsymbol{\Psi}^\top \boldsymbol{\Psi}}{\|\boldsymbol{s}_2\| \|\boldsymbol{\Psi}\| + \eta_2} = \frac{\eta_2 \|\boldsymbol{s}_2\| \|\boldsymbol{\Psi}\|}{\|\boldsymbol{s}_2\| \|\boldsymbol{\Psi}\| + \eta_2} < \eta_2 \tag{7.199}$$

From (7.199), it yields

$$\|\boldsymbol{s}_2\| \|\boldsymbol{\Psi}\| \leq \frac{\boldsymbol{s}_2^\top \boldsymbol{s}_2 \boldsymbol{\Psi}^\top \boldsymbol{\Psi}}{\|\boldsymbol{s}_2\| \|\boldsymbol{\Psi}\| + \eta_2} < \eta_2 \tag{7.200}$$

Further, it can be easily obtained as

$$\frac{\boldsymbol{s}_2^\top \boldsymbol{s}_2 \boldsymbol{\Psi}^\top \boldsymbol{\Psi}}{\|\boldsymbol{s}_2\| \|\boldsymbol{\Psi}\| + \eta_2} \leq \frac{\boldsymbol{s}_2^\top \boldsymbol{s}_2 \boldsymbol{\Psi}^\top \boldsymbol{\Psi}}{\sqrt{\|\boldsymbol{s}_2\|^2 \|\boldsymbol{\Psi}\|^2 + \eta_2^2}} \tag{7.201}$$

Based on (7.198) and (7.201), it yields

$$s_2^{\top} \boldsymbol{\Pi}^{\top} \boldsymbol{\Psi} \leq \lambda \left(\frac{s_2^{\top} s_2 \boldsymbol{\Psi}^{\top} \boldsymbol{\Psi}}{\sqrt{\|s_2\|^2 \|\boldsymbol{\Psi}\|^2 + \eta_2^2}} + \eta_2 \right) \tag{7.202}$$

Denoting $\boldsymbol{\varepsilon} = \frac{s_2 \boldsymbol{\Psi}^{\top} \boldsymbol{\Psi}}{\sqrt{\|s_2\|^2 \|\boldsymbol{\Psi}\|^2 + \eta_2^2}}$, (7.202) becomes

$$s_2^{\top} \boldsymbol{\Pi}^{\top} \boldsymbol{\Psi} \leq \lambda s_2^{\top} \boldsymbol{\varepsilon} + \eta_2 \lambda \tag{7.203}$$

Step 2: It will be proven that the variables s_1, s_2, and $\tilde{\lambda}$ can converge to the equilibrium point. Note that the following inequality always holds by

$$\tilde{\lambda}\lambda = \tilde{\lambda}\left(\lambda - \tilde{\lambda}\right) \leq -\frac{\tilde{\lambda}^2}{2} + \frac{\lambda^2}{2} \tag{7.204}$$

Then, substituting (7.204) into (7.197) yields

$$\begin{aligned} \dot{V}_2 &\leq -s_1^{\top} \boldsymbol{K}_1 s_1 - s_2^{\top} \boldsymbol{K}_2 s_2 - \frac{\eta_2}{2} \tilde{\lambda}^2 + \eta_2 \left(1 + \lambda + \frac{\lambda^2}{2} \right) \\ &\leq - \min\left\{ \mu_{\min}\{2\boldsymbol{K}_1\}, \mu_{\min}\{2\boldsymbol{H}^{-1}\boldsymbol{K}_2\}, \gamma\eta_2 \right\} V_2 + \eta_2 \left(1 + \lambda + \frac{\lambda^2}{2} \right) \\ &= -k_V V + c_V \end{aligned} \tag{7.205}$$

where $k_V := \min\{\mu_{\min}\{2\boldsymbol{K}_1\}, \mu_{\min}\{2\boldsymbol{H}^{-1}\boldsymbol{K}_2\}, \gamma\eta_2\}$ and $c_V := \eta_2(1 + \lambda + \lambda^2/2)$. $\mu_{\min}\{\cdot\}$ denotes the minimum eigenvalue of $\{\cdot\}$. Similar to the analysis in [30], (7.205) implies

$$V_2 \leq \left(V_2(0) - \frac{c_V}{k_V} \right) \exp^{-k_V t} + \frac{c_V}{k_V}, \quad \forall t \geq 0 \tag{7.206}$$

which indicates the convergence and boundedness of the variables s_1, s_2, and $\tilde{\lambda}$.

Step 3: It will be proven that the compensation terms $\boldsymbol{\chi}_1$ and $\boldsymbol{\chi}_2$ can converge to the equilibrium point, and then the system state variables can also converge to the equilibrium point. Construct the Lyapunov function as

$$V = \frac{1}{2} \boldsymbol{\chi}_1^{\top} \boldsymbol{\chi}_1 + \frac{1}{2} \boldsymbol{\chi}_2^{\top} \boldsymbol{H} \boldsymbol{\chi}_2 \tag{7.207}$$

Utilizing (7.181) and (7.190), the derivative of V is given by

$$\begin{aligned}\dot{V} &= \chi_1^{\top}\dot{\chi}_1 + \chi_2^{\top}\boldsymbol{H}\dot{\chi}_2 \\ &= -\chi_1^{\top}\boldsymbol{K}_1\chi_1 + \chi_1^{\top}(\overline{\boldsymbol{\alpha}}_2 - \boldsymbol{\alpha}_1) + \chi_1^{\top}\chi_2 - \frac{a_1\chi_1^{\top}\chi_1}{\sqrt{\|\chi_1\|^2 + \eta_1^2}} \\ &\quad - \chi_2^{\top}\boldsymbol{K}_2\chi_2 - \frac{a_2\chi_2^{\top}\chi_2}{\sqrt{\|\chi_2\|^2 + \eta_2^2}}\end{aligned} \tag{7.208}$$

Similar to the proof in Step 1 in this subsection, it has

$$\|\chi_k\| - \frac{\chi_k^{\top}\chi_k}{\sqrt{\|\chi_k\|^2 + \eta_k^2}} \le \|\chi_k\| - \frac{\chi_k^{\top}\chi_k}{\|\chi_k\| + \eta_k} \le \eta_k \tag{7.209}$$

for $k = 1, 2$. Then, (7.208) can be rewritten as

$$\dot{V} \le -\chi_1^{\top}\boldsymbol{K}_1\chi_1 - \chi_2^{\top}\boldsymbol{K}_2\chi_2 + \chi_1^{\top}(\overline{\boldsymbol{\alpha}}_2 - \boldsymbol{\alpha}_1) - \boldsymbol{a}_1\|\chi_1\| - \boldsymbol{a}_2\|\chi_2\| + a_1\eta_1 + a_2\eta_2 \tag{7.210}$$

Considering $\|\overline{\boldsymbol{\alpha}}_2 - \boldsymbol{\alpha}_1\| \le a_0$, if $a_1$is selected such that $a_1 - a_0 > 0$, then (7.210) yields

$$\begin{aligned}\dot{V} &\le -\chi_1^{\top}\boldsymbol{K}_1\chi_1 - \chi_2^{\top}\boldsymbol{K}_2\chi_2 + \chi_1^{\top}\chi_2 - (a_1 - a_0)\|\chi_1\| - a_2\|\chi_2\| + a_1\eta_1 + a_2\eta_2 \\ &\le -\mu_{\min}\{\boldsymbol{K}_1\}\|\chi_1\|^2 - \mu_{\min}\{\boldsymbol{K}_2\}\|\chi_2\|^2 + \|\chi_1\|\|\chi_2\| + a_1\eta_1 + a_2\eta_2 \\ &\le -\mu_{\min}\{\boldsymbol{N}\} \times \left(\|\chi_1\|^2 + \|\chi_2\|^2\right) + a_1\eta_1 + a_2\eta_2\end{aligned} \tag{7.211}$$

where the matrix $\boldsymbol{N} \in \mathbb{P}^{2 \times 2}$ is defined as $N_{1,1} = \mu_{\min}\{\boldsymbol{K}_1\}$, $N_{1,2} = N_{2,1} = -1/2$, and $N_{2,2} = \mu_{\min}\{\boldsymbol{K}_2\}$, where $\boldsymbol{N}$ is always positive definite by choosing $\mu_{\min}\{\boldsymbol{K}_1\}\mu_{\min}\{\boldsymbol{K}_2\} > 1/4$.

Similar to the proof in Step 2 in this subsection, it can be concluded that the compensation terms χ_1 and χ_2 can converge to the equilibrium point. From the expressions of (7.182) and (7.183), it gives the boundedness of the variables z_1 and z_2. Thus, the system state variables can also converge to the equilibrium point. Theorem 7.1 is proven.

7.4.2.7 Simulation Results

The numerical simulations are executed to verify the effectiveness of the proposed framework, including the iteration-based reference learning scheme and the pose tracking controller. In the simulations, the model parameters are given by $m = 2.5$ kg, $\boldsymbol{J} = \text{diag}\{0.0820, 0.0845, 0.1377\}$ kg·m^2, and $g = 9.81$ kg·m/s^2. The parameters

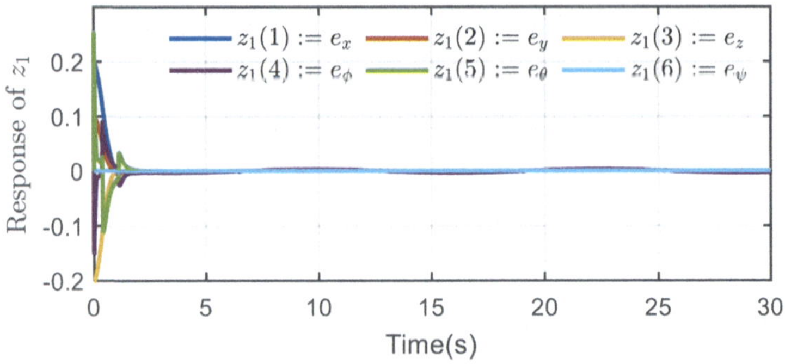

Fig. 7.36 Simulation results of the controller performance evaluation

of the admittance filter are set as $\boldsymbol{M}_d =$ diag $\{2.5, 2.5, 2.5\}, \boldsymbol{D}_d =$ diag $\{25, 25, 25\}$, and $\boldsymbol{K}_d =$ diag $\{15, 15, 15\}$. The parameters of the system tracking controller are selected as $\boldsymbol{K}_1 =$ diag $\{2.5, 3, 3, 2.5, 3, 3\}$, $\boldsymbol{K}_2 =$ diag $\{1, 1, 1, 1, 1, 1\}$, $a_1 = a_2 = 0.01$, and $\eta_1 = \eta_2 = 0.005$. The parameters of the first-order command filter and the adaptive law in the tracking controller are chosen as $\sigma = 0.05$ and $\gamma = 0.1$. The commanded yaw angle ψ_c is set as zero throughout the simulations, where the commanded roll angle ϕ_c and pitch angle θ_c are computed according to (7.162).

The simulation in this subsection illustrates the feasibility of the system tracking controller in (7.195) under external wrenches. The initial state of the system is $\boldsymbol{q} = [0.2, 0.1, 0.8, 0, 0, 0]^\top$, and the commanded position is $\boldsymbol{p}_r = [0, 0, 1]^\top$. The external wrench is set as $\boldsymbol{d}_e = [0.5 \sin 0.5t\boldsymbol{I}_3; 0.02 \sin 0.5t\boldsymbol{I}_3]$. The results of the controller performance evaluation are shown in Fig. 7.36. Figure 7.36 depicts the response of the error variable z_1, where $z_1(1:3)$ represents the position tracking error and $\mathbf{z}_1(4:6)$ represents the attitude error. It shows that the state errors can converge to zero in nearly 2 s. This simulation demonstrates the feasibility of the proposed tracking controller with favorable tracking performance under external wrenches.

The optimization-based interaction behavior is investigated in this subsection while considering the reference learning scheme and the pose tracking controller in (7.195). Without loss of generality, the interaction in the x direction is considered, and the interactions in the y and z directions are omitted in simulations for brevity. The desired task position trajectory is $\boldsymbol{p}_t(t) = \boldsymbol{p}_t(t_0) + (\boldsymbol{p}_t(t_f) - \boldsymbol{p}_t(t_0)) \times (10t_s^3 - 15t_s^4 + 6t_s^5)$ with $t_s = (t - t_0)/(t_f - t_0)$, and the reference position trajectory is set in (7.170), where $t_0 = 0$, $t_f = 30$ and $\boldsymbol{p}_t(t_0) = [0, 0, 1]^\top$, $p_t(t_f) = [0.2, 0, 1]^\top$ are set. The initial trajectory parameter is chosen as $\zeta = 1$. The parameters of the cost function in (7.166) are selected as $\mathcal{Q} = \mathcal{R} = \text{diag}\{0.1, 0.1, 0.1\}$. The environment model is chosen as $\boldsymbol{f}_e = \boldsymbol{K}_e(\boldsymbol{p} - \boldsymbol{p}_e)$ with $\boldsymbol{K}_e =$ diag $\{300, 300, 300\}$ N/m and $\boldsymbol{p}_e(1) = 0.1$. The initial position trajectory of the system is $\boldsymbol{p} = [0, 0, 1]^\top$.

The simulation results of interactions are illustrated in Fig. 7.37. In Fig. 7.37a–d, the lines with square and circle markers denote the 1st to 3rd iterations and the 10th to 12th iterations, respectively. Specifically, Fig. 7.37a plots the responses of the

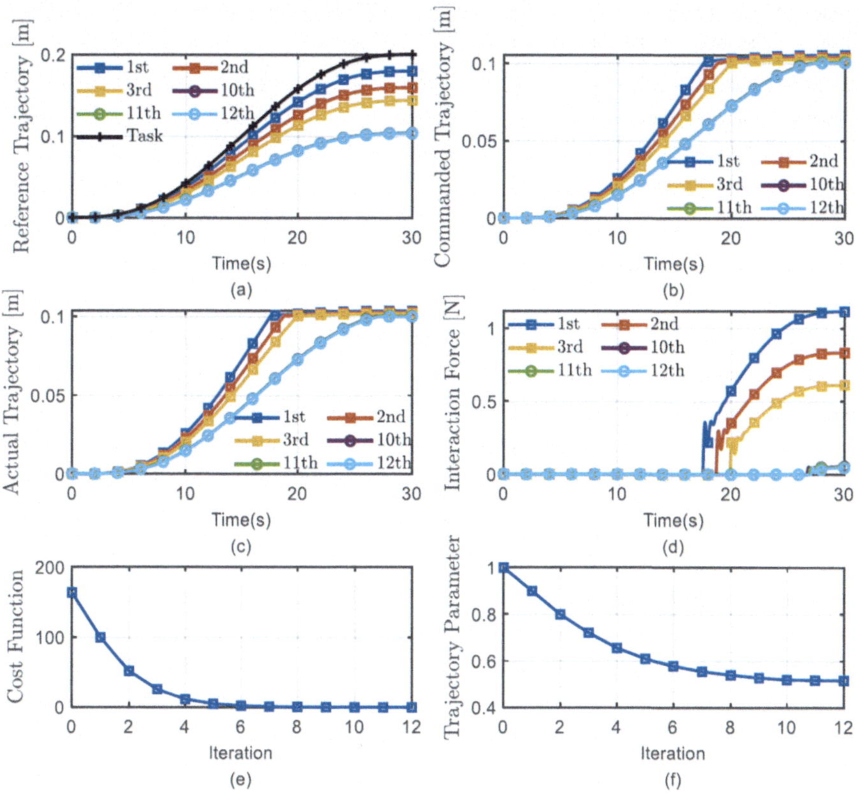

Fig. 7.37 Simulation results of interactions with the environmental stiffness $\mathbf{K}_e =$ diag $\{300, 300, 300\}$ N/m and $\mathcal{Q} = \mathcal{R} = \text{diag}\{0.1, 0.1, 0.1\}$ under the proposed framework. In (**a**–**d**), the lines with square and circle markers represent the first three (1st to 3rd) and the last three (10th to 12th) iterations, respectively. (**a**) Reference position trajectory $\mathbf{p}_r(1)$ during iterations and the desired task position trajectory $\mathbf{p}_t(1)$ with the black line and plus sign marker. (**b**) Modified commanded position trajectory $\mathbf{p}_c(1)$. (**c**) Actual position trajectory $\mathbf{p}(1)$. (**d**) Interaction force $\mathbf{f}_e(1)$ in the x direction. (**e**) Cost function W. (**f**) Trajectory parameter ζ

desired task trajectory $\boldsymbol{p}_t(1)$ and the reference position trajectory $\boldsymbol{p}_r(1)$ during iterations, respectively. Figure 7.37b depicts the response of the modified commanded position trajectory $\boldsymbol{p}_c$, which is calculated by $\boldsymbol{p}_r$ and the interaction force $\mathbf{f}_e(1)$ according to (7.165). The actual motion position trajectory $\boldsymbol{p}(1)$ is shown in Fig. 7.37c, and it is found that the actual trajectory $\boldsymbol{p}(1)$ can well track the commanded trajectory $\boldsymbol{p}_c(1)$. Figure 7.37d displays the trajectories of the interaction force during iterations. From Fig. 7.37e, it shows that the cost function W gradually decreases as the number of iterations increases, and the trajectory parameter ζ converges to the minimum, shown in Fig. 7.37f. This case demonstrates the effectiveness of the interaction using reference learning scheme, and the optimization-based interaction behavior described by a trade-off between the trajectory tracking and interaction force can be accomplished.

Table 7.3 Simulation comparison results of optimal interactions with different environmental stiffness at the moment of stable interaction

Environmental stiffness $\mathbf{K}_e(1)$	300	100	500
Reference trajectory $\mathbf{p}_r(1)$ [m]	0.1036	0.1039	0.1034
Commanded trajectory $\mathbf{p}_c(1)$ [m]	0.1005	0.1008	0.1004
Actual trajectory $\mathbf{p}(1)$ [m]	0.1002	0.1005	0.1001
Interaction force $\mathbf{f}_e(1)$ [N]	0.04563	0.04636	0.04344

To further explain the performance of the proposed framework, the interactions with different stiffness environments are studied, where the environmental stiffness $\boldsymbol{K}_e = \text{diag}\{100, 100, 100\}$ N/m and $\boldsymbol{K}_e = \text{diag}\{500, 500, 500\}$ N/m are considered. The other settings are the same as the previous simulation. The simulation comparison results of optimal interactions with different environmental stiffness are shown in Table 7.3, where the optimal comparison values are selected at the final time of 30 s. From Table 4.6, the comparison results are slightly different when interacting with environments of different stiffness. Although the environmental stiffness changes, the optimization-based interaction behavior can also be implemented under the proposed method.

Further, the different cost functions in (7.166) are studied to realize the desired optimization-based interaction behavior. The weights in the cost function are changed into $\boldsymbol{Q} = \text{diag}\{1, 1, 1\}$ and $\mathcal{R} = \text{diag}\{0.1, 0.1, 0.1\}$. The other settings are the same as the previous simulation. The comparison results of different weight cost functions are shown in Fig. 7.38. When the weight of Q is relatively greater, the trade-off is more in favor of trajectory tracking, and the corresponding tracking performance is better, as shown in Fig. 7.38a. From Fig. 7.38b, the corresponding interaction force $f_e(1)$ becomes larger. Conversely, the trade-off is more in favor of the minimum interaction force if the weight of $\mathcal{R}$ is relatively greater, and this proposal is omitted in this article for brevity. Figure 7.38c and d depicts the trajectories of the cost function W and trajectory parameter ζ, respectively. This simulation case further demonstrates the feasibility of the proposed framework with different weight cost functions.

7.4.2.8 Application: Aerial Pipeline Inspection

Real-world experiments are performed on the hexacopter with the robotic contact device system as shown in Fig. 7.39 to verify the proposed method in this section.

The system's total mass is about 4.422 kg. A hexacopter is adopted as the flying platform with six KV330 BLDC rotors and 15×5.5 propellers, and the vehicle's wheelbase is approximately 0.96 m. The hexacopter is equipped with a flight control module Pixhawk for low-level processing and an onboard computer Intel NUC-i7 for high-level processing. All software is developed on Ubuntu 20.04 with ROS Noetic system, and the communication between the flight controller and the onboard computer is realized via MAVLink protocol. The control frequency is set as 100 Hz.

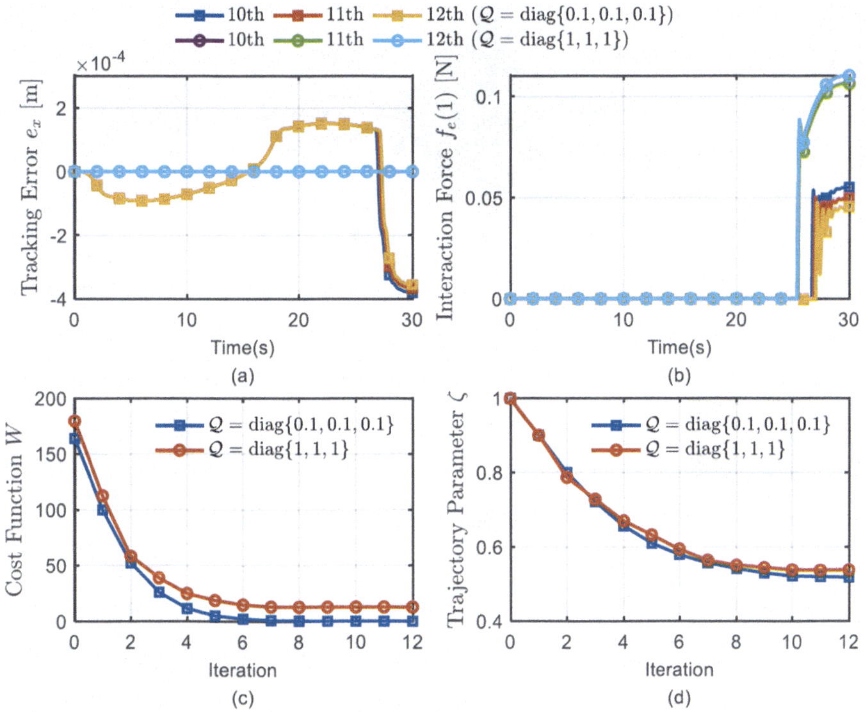

Fig. 7.38 Simulation comparison results of interactions with the different weight matrices. In (**a** and **b**), the lines with square and circle markers denote the 10th to 12th iterations with $\mathcal{Q} = \text{diag}\{0.1, 0.1, 0.1\}$ and $\mathcal{Q} = \text{diag}\{1, 1, 1\}$, respectively. (**a**) x direction tracking error e_x. (**b**) Interaction force $\mathbf{f}_e(1)$. (**c**) Cost function W. (**d**) Trajectory parameter ζ

A robotic contact device is rigidly installed in the vehicle, and the designed device is manufactured using custom carbon fiber components, 3-D printed connection parts with a lightweight plastic material, and aluminum alloys. The contact device is equipped with a six-axis force/torque sensor SRI-M3713A and ultrasonic inspection equipment, where the force/torque sensor is filtered for further usage, and a spring is mounted to protect the ultrasonic inspection equipment from large impacts. Further, the inspection value can be displayed on the PC software through radio transmission, as shown in the bottom right corner of Fig. 7.39b.

In addition, the contact device can achieve circular motion in the x-z plane with respect to the vehicle through the gear transmission mechanism. More information, the main body of the contact device is a large transmission gear with a 360° gear chute embedded inside, and the chute is connected to a small gear on the Dynamixel XH540-W270-R motor, where the transmission ratio is approximately 1:4.9. When the rotation direction of the contact device coincides with the positive direction of the x axis of the body, this situation is defined as zero. Then, the clockwise direction is defined as the positive direction, and the counterclockwise direction is the negative direction, i.e., the rotation angle of the contact device in Fig. 7.39b is 30°.

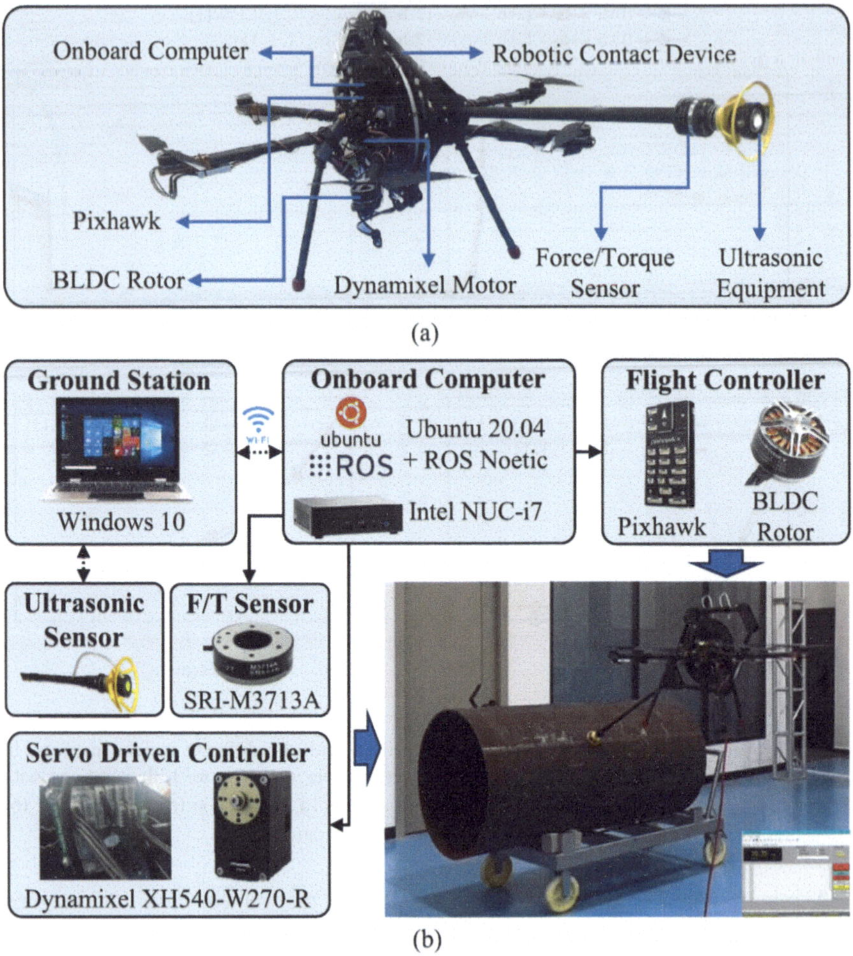

Fig. 7.39 (**a**) System hardware diagram. (**b**) Control flowchart of aerial pipeline inspection, where the robotic contact device is tilted at 30°

This subsection presents the real-world aerial contact inspections with a pipeline, whose size is ∅0.76×1.50 m. The OptiTrack motion capture system is adopted to offer the external linear and angular pose information. The parameters of the admittance filter are set as $\boldsymbol{M}_d = m\boldsymbol{I}_{3 \times 3}$, $\boldsymbol{D}_d =$ diag $\{30, 30, 30\}$, and $\boldsymbol{K}_d =$ diag $\{10, 10, 10\}$.

To verify the feasibility of the proposed tracking controller, fixed-point tracking and circle trajectory tracking are executed under external 3.5 m/s wind disturbances, where the reference point is $\boldsymbol{p}_r = [0, -0.34, 1.62]^\top$ and the reference circle trajectory is $\boldsymbol{p}_r = [0.8 \cos(0.1t), -0.34 + 0.8 \sin(0.1t), 1.62]^\top$. The experimental results of fixed-point tracking and circle trajectory tracking are shown in Figs. 7.40 and 7.41, respectively. Figure 7.40a and b represents the position tracking and the linear

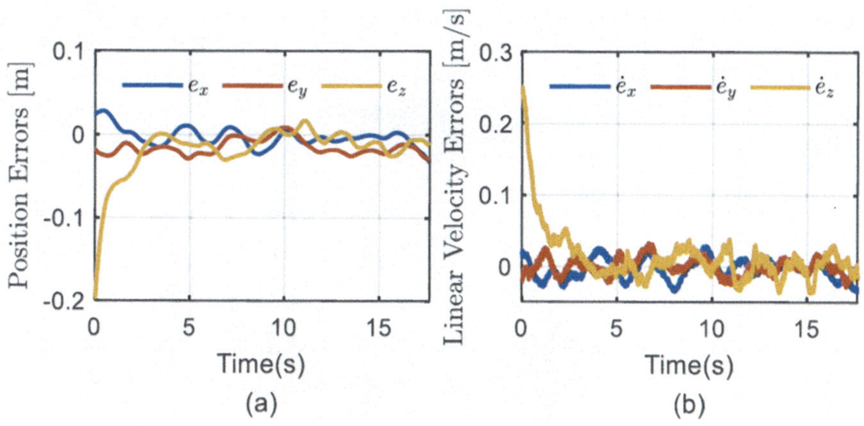

Fig. 7.40 Experimental results of fixed-point tracking

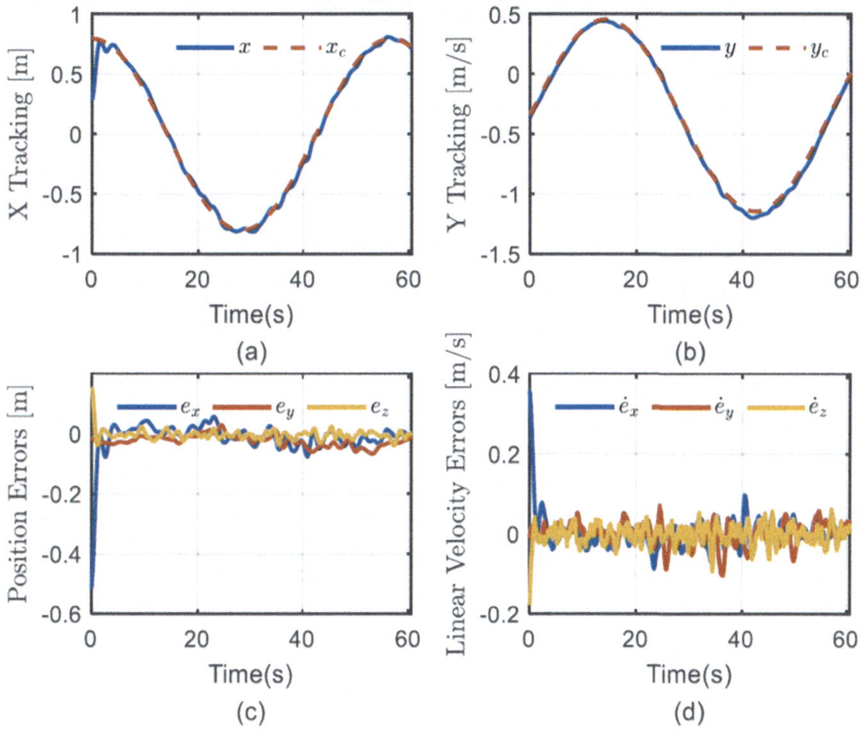

Fig. 7.41 Experimental results of circle trajectory tracking

velocity errors, respectively. Figure 7.41a and b shows the tracking performances in the x and y directions. Then, Fig. 7.41c and d depicts the position and linear velocity errors under circle trajectory tracking. These cases illustrate that the proposed tracking controller performs well tracking performance under external wrenches.

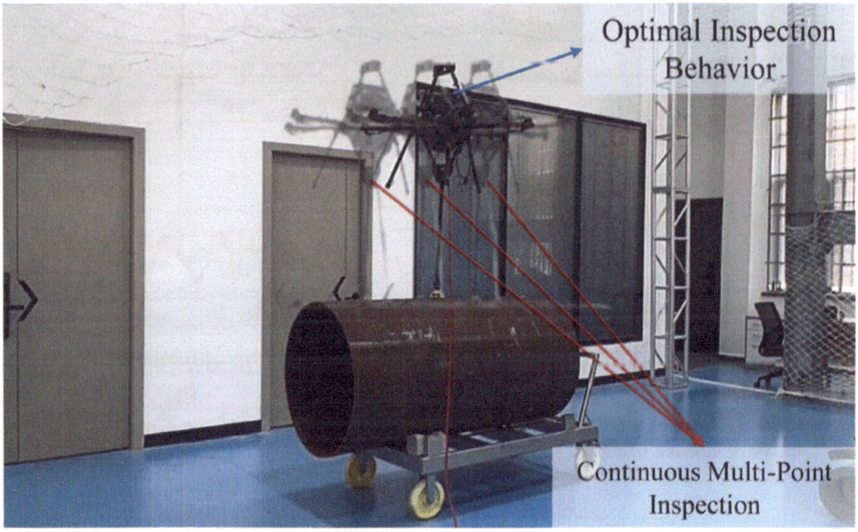

Fig. 7.42 Screenshots of the aerial inspection with the contact device tilted at 90°

The built hexarotor with the robotic contact device is applied for optimization-based aerial inspection with a pipeline, as shown in the real image of Fig. 7.42. The contact device is tilted at 90°, which means that only z direction contact is considered. The reference position trajectory is similar to simulations, where $t_f - t_0 = 30$, $\boldsymbol{p}_t(t_0) = [0.918, 0, 2.398]^\top$, and $p_t(t_f) = [0.918, 0, 1.998]^\mathrm{T}$. The parameters in [35] are set as $\mathcal{Q} = \mathcal{R} = \text{diag}\{1, 1, 1\}$. Figure 7.43 displays the experimental results. Figure 7.43a plots the reference trajectory z_r during iterations, and Fig. 7.43b and c depicts the commanded trajectory z_c according to the admittance filter and the actual motion trajectory z, respectively. The tracking errors are shown in Fig. 7.43d, and it means that the actual trajectory can successfully track the commanded trajectory. Figure 7.43e shows the interaction forces, and Fig. 7.43f draws the curves of the cost function W and the trajectory parameter ζ, where the eighth iteration exhibits optimal interaction behavior. Further, the system is applied for continuous multipoint contact inspection shown in the blurred images of Fig. 7.42.

To further verify the performance of the proposed framework, the comparisons are conducted with the compliant interaction control method in [35]. To ensure the fairness of comparisons, the parameters in [35] are selected to perform favorable performance as $\boldsymbol{K}_{p1} =$ diag $(7.92, 7.92, 12.96)$, $\boldsymbol{K}_{p2} =$ diag $(3.65, 3.65, 6.28)$, $\boldsymbol{K}_{R1} =$ diag $(0.42, 0.42, 0.60)$, $\boldsymbol{K}_{R2} =$ diag $(0.078, 0.078, 0.086)$, $\boldsymbol{L} = \boldsymbol{I}_{3 \times 3}$, and $\boldsymbol{K}_\mathrm{I} = 0.5\boldsymbol{I}_{3 \times 3}$. The admittance filter parameters and the reference trajectory in [37] are set the same as in this section. The comparison results are shown in Fig. 7.43b–e. The proposed framework shows better performances with smaller tracking errors and interaction forces, which are implemented by the iterative optimization-based reference learning scheme.

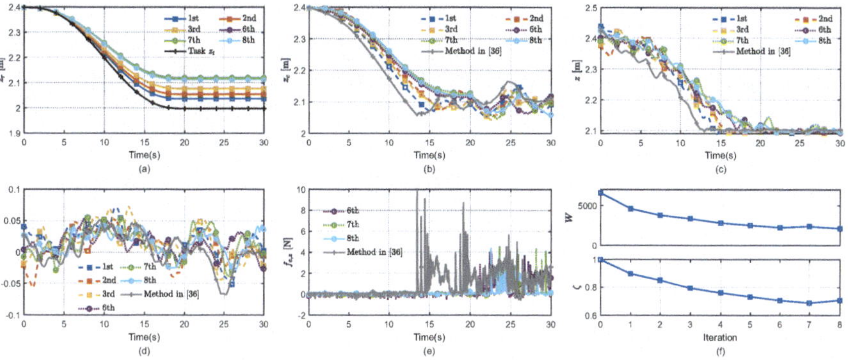

Fig. 7.43 Experimental results of optimization-based aerial inspection with the contact device tilted at 90°, where the comparisons are conducted with the compliant interaction control method in [35]. (**a**) Reference trajectory z_r. (**b**) Commanded trajectory z_c. (**c**) Actual trajectory z. (**d**) Tracking error e_z. (**e**) Interaction force $f_{e,\ z}$. (**f**) Cost function W and trajectory parameter ζ

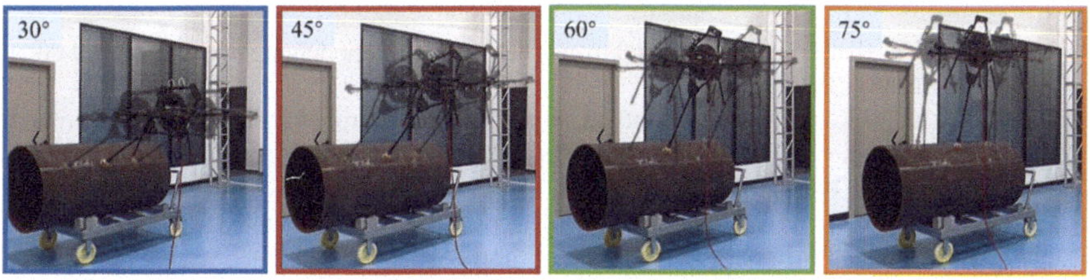

Fig. 7.44 Screenshots of optimal inspection behavior (real images) and continuous multipoint inspection (blurred images) with the robotic contact device tilted at different angles (30°, 45°, 60°, and 75°)

Furthermore, the contact device is tilted at multiple angles for different arc surface inspections of the pipeline, as shown in the real images of Fig. 7.44. Thus, in this case, the x and z directions should be both considered for aerial interaction. From 0 to 20 s, the task trajectory is similar to the previous simulations and experiments according to (7.170). From 20 to 30 s, the task trajectory is set at a fixed position, which is offset by $0.1 \cos \sigma$ m and $-0.1 \sin \sigma$ m with respect to the x and z directions of the contact point, respectively, and σ is the tilted angle of the contact device. The experimental results of optimal inspection behavior with different tilted angles are drawn in Fig. 7.45. The blue, red, yellow, and purple curves present the contact device tilted at 30°, 45°, 60°, and 75°, respectively. Figure 7.45a and b shows the curves of the x and z directions, respectively, including the modified commanded trajectories x_c and z_c, actual trajectories x and z, tracking errors e_x and e_z, and interaction force $f_{e,\ x}$ and $f_{e,\ z}$. Further, the system is applied for continuous multipoint contact inspection with different tilted angles shown in the blurred images of Fig. 7.44.

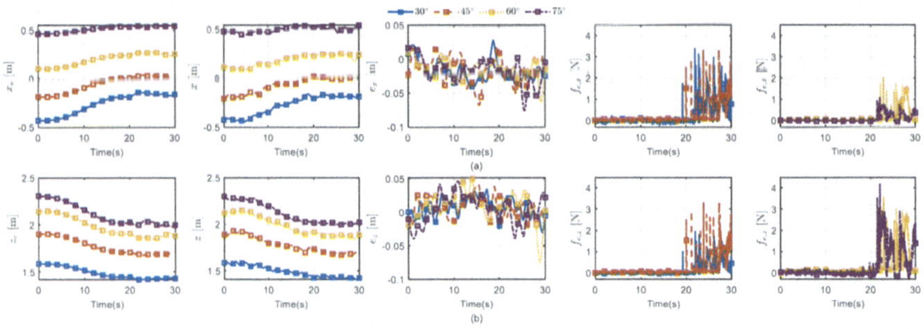

Fig. 7.45 Experimental results of optimal inspection behavior with the robotic contact device tilted at different angles: 30° (blue), 45° (red), 60° (yellow), and 75° (purple)

References

1. Zhong H, Miao Z, Wang Y, et al. A practical visual servo control for aerial manipulation using a spherical projection model. IEEE Transactions on Industrial Electronics, 2019, 67(12): 10564–10574.
2. Meng X, He Y, Han J. Survey on aerial manipulator: system, modeling, and control. Robotica. 2020;38(7):1288–1317.
3. Zhong H, Liang J, Chen Y, et al. Prototype, modeling, and control of aerial robots with physical interaction: A review. IEEE Transactions on Automation Science and Engineering, 2024, https://doi.org/10.1109/TASE.2024.3395658.
4. Gupte S, Mohandas P I T, Conrad J M. A survey of quadrotor unmanned aerial vehicles. In: 2012 Proceedings of IEEE Southeastcon. IEEE, 2012, 1–6.
5. Santos V J, Valero-Cuevas F J. Reported anatomical variability naturally leads to multimodal distributions of Denavit-Hartenberg parameters for the human thumb. IEEE Transactions on Biomedical Engineering, 2006, 53(2): 155–163.
6. Zhang XD. Matrix analysis and applications. Cambridge: Cambridge University Press; 2017.
7. Yoo D W, Oh H D, Won D Y, et al. Dynamic modeling and stabilization techniques for tri-rotor unmanned aerial vehicles. International Journal of Aeronautical and Space Sciences, 2010, 11 (3): 167–174.
8. Kemper M, Fatikow S. Impact of center of gravity in quadrotor helicopter controller design. IFAC Proceedings Volumes, 2006, 39(16): 157–162.
9. Chen Y, Liang J, Wu Y, et al. Adaptive sliding-mode disturbance observer-based finite-time control for unmanned aerial manipulator with prescribed performance[J]. IEEE Transactions on Cybernetics, 2022, 53(5): 3263–3276.
10. Hamel T, Mahony R. Visual servoing of an under-actuated dynamic rigid-body system: An image-based approach. IEEE Transactions on Robotics and Automation, 2002, 18(2): 187–198.
11. Bourquardez O, Mahony R, Hamel T, et al. Stability and performance of image based visual servo control using first order spherical image moments. In: 2006 IEEE/RSJ International Conference on Intelligent Robots and Systems. IEEE, 2006: 4304–4309.
12. Guenard N, Hamel T, Mahony R. A practical visual servo control for an unmanned aerial vehicle. IEEE Transactions on Robotics, 2008, 24(2): 331–340.
13. Mebarki R, Lippiello V, Siciliano B. Nonlinear visual control of unmanned aerial vehicles in GPS-denied environments. IEEE Transactions on Robotics, 2015, 31(4): 1004–1017.
14. Peters J, Mistry M, Udwadia F, et al. A unifying framework for robot control with redundant DOFs. Autonomous Robots. 2008;24:1–12.

15. Robotics P C. Vision and control: Fundamental algorithms in MATLAB. Springer, 2011: 251–262.
16. Santamaria-Navarro A, Grosch P, Lippiello V, et al. Uncalibrated visual servo for unmanned aerial manipulation. IEEE/ASME Transactions on Mechatronics, 2017, 22(4): 1610–1621.
17. Chaumette F, Hutchinson S. Visual servo control. I. Basic approaches. IEEE Robotics & Automation Magazine, 2006, 13(4): 82–90.
18. Kemper M. Control system for unmanned 4-rotor helicopter. European Patent Application. EP. 2008;1901153:A1.
19. Cao Z, Chen X, Yu Y, et al. Image dynamics-based visual servoing for quadrotors tracking a target with a nonlinear trajectory observer. IEEE Transactions on Systems, Man, and Cybernetics: Systems, 2017, 50(1): 376–384.
20. Jabbari H, Oriolo G, Bolandi H. An adaptive scheme for image-based visual servoing of an underactuated UAV. International Journal of Robotics and Automation, 2014, 29(1): 92–104.
21. Zheng D, Wang H, Wang J, et al. Toward visibility guaranteed visual servoing control of quadrotor UAVs. IEEE/ASME Transactions on Mechatronics, 2019, 24(3): 1087–1095.
22. Zhang G, He Y, Dai B, et al. Robust control of an aerial manipulator based on a variable inertia parameters model. IEEE Transactions on Industrial Electronics, 2019, 67(11): 9515–9525.
23. Mebarki R, Lippiello V. Image-based control for aerial manipulation. Asian Journal of Control, 2014, 16(3): 646–656.
24. Sobie E A. An introduction to MATLAB. Science signaling, 2011, 4(191): tr7.
25. Wang L, Su J. Robust disturbance rejection control for attitude tracking of an aircraft. IEEE Transactions on Control Systems Technology, 2015, 23(6): 2361–2368.
26. Mekki H, Letaief M. Path planning for 3D visual servoing: For a wheeled mobile robot. In: 2013 International Conference on Individual and Collective Behaviors in Robotics (ICBR). IEEE, 2013: 86–91.
27. Abadianzadeh F, Derhami V, Rezaeian M. Visual servoing control of robot manipulator in 3D space using fuzzy hybrid controller. In: 2016 4th international conference on robotics and mechatronics (ICROM). IEEE, 2016: 61–65.
28. Chaumette F, Hutchinson S. Visual servo control. II. Advanced approaches [Tutorial]. IEEE Robotics & Automation Magazine, 2007, 14(1): 109–118.
29. Bodie K, Brunner M, Pantic M, et al. Active interaction force control for contact-based inspection with a fully actuated aerial vehicle. IEEE Transactions on Robotics, 2020, 37(3): 709–722.
30. Liang J, Chen Y, Wu Y, et al. Active physical interaction control for aerial manipulator based on external wrench estimation. IEEE/ASME Transactions on Mechatronics, 2023, 28(5): 2774–2785.
31. Liang J, Chen Y, Lai N, et al. Low-complexity prescribed performance control for unmanned aerial manipulator robot system under model uncertainty and unknown disturbances. IEEE Transactions on Industrial Informatics, 2021, 18(7): 4632–4641.
32. Komati B, Clévy C, Lutz P. Force tracking impedance control with unknown environment at the microscale. In: 2014 IEEE International Conference on Robotics and Automation (ICRA). IEEE, 2014: 5203–5208.
33. Lin Y, Chen Z, Yao B. Unified motion/force/impedance control for manipulators in unknown contact environments based on robust model-reaching approach. IEEE/ASME Transactions on Mechatronics, 2021, 26(4): 1905–1913.
34. Liang J, Wang Y, Zhong H, et al. Robust variable impedance control for aerial compliant interaction with stability guarantee. IEEE Transactions on Industrial Informatics, 2024, 20(3): 3351–3360.
35. Ryll M, Muscio G, Pierri F, et al. 6D interaction control with aerial robots: The flying end-effector paradigm. The International Journal of Robotics Research, 2019, 38(9): 1045–1062.
36. Burdet E, Osu R, Franklin D W, et al. The central nervous system stabilizes unstable dynamics by learning optimal impedance. Nature, 2001, 414(6862): 446–449.

37. Amirabdollahian F, Loureiro R, Harwin W. Minimum jerk trajectory control for rehabilitation and haptic applications. In: Proceedings 2002 IEEE International Conference on Robotics and Automation. IEEE, 2002, 4: 3380–3385.
38. Huang P, Xu Y, Liang B. Global minimum-jerk trajectory planning of space manipulator. International Journal of Control, Automation, and Systems, 2006, 4(4): 405–413.
39. Arimoto S, Kawamura S, Miyazaki F. Bettering operation of robots by learning. Journal of Robotic systems, 1984, 1(2): 123–140.
40. Xu J X, Huang D, Pindi S. Optimal tuning of PID parameters using iterative learning approach. SICE Journal of Control, Measurement, and System Integration, 2008, 1(2): 143–154.

Chapter 8
Conclusions and Future Directions

This chapter presents the conclusions of the book and future directions of electric power robots. The conclusions are related to the development of condenser cleaning mobile manipulator robots, transmission line robot manipulators, and aerial manipulator robots with applications. Faced challenges involve novel environmental perception, decisional autonomy and planning, intelligent control, human-robot interaction, mechanism design, and related technologies. Future directions of electric power robots are fully autonomous with multifunctional operation modes, collaborative autonomy, and novel morphology. Finally, the four major driving forces are summarized for accelerating the applications of electric power robots.

8.1 Conclusions

The application of electric power robots is growing rapidly due to the huge power requirements. The book introduces the typical operational robots used in the power industry, including condenser cleaning mobile manipulator robots, transmission line de-icing robot manipulators, and aerial manipulator robots.

The condenser cleaning mobile manipulator robot system is presented in the second chapter and third chapter of the book. The structure design and operating principle provide a reference for researchers and practitioners engaged in the related field. The presented adaptive trajectory tracking control and adaptive force/motion control approaches can be potentially suitable for other ground mobile manipulator robot systems.

The transmission line de-icing robot manipulator system is presented in the fourth chapter and fifth chapter of the book. The inverse kinematic solution algorithms determine the corresponding joint variables based on the known pose of the end effector, and the inverse kinematic solution is one of the key technologies in robot trajectory planning and motion control. Further, the transmission line robot

Y. Wang et al., *Electric Power Robots*,
https://doi.org/10.1007/978-981-96-2512-3_8

manipulator system can be employed for complex bolt tightening and insulator installation tasks.

The aerial manipulator robot system is presented in the sixth chapter and seventh chapter of the book. Compared with other types of combined mobile operating robots, combined aerial manipulator robot systems are highly coupled and sensitive to environmental interference, which brings huge challenges to system stability and safety. The adaptive control approaches are presented to stabilize aerial vehicles under disturbances and input saturation. Regarding accurate aerial manipulation, onboard cameras are mounted on aerial manipulator robots to implement various manipulation tasks. It has been shown in the seventh chapter that both kinematic-level and dynamic-level image-based visual servo control methods can be employed for aerial grasping. The relevance of aerial manipulator robotics for autonomous removing foreign objects for power transmission lines and aerial power pipeline inspection has been also shown in the book.

8.2 Challenges

The key characteristics and technologies for electric power robots include environmental perception, decisional autonomy and planning, intelligent control, human-robot interaction, etc. Regarding robots used in the power industry, technical and application challenges arise accordingly.

8.2.1 Technical Challenges

Perception and recognition technology is the basis for decision-making, planning, and control of robot systems. Most of the existing operational robots are researched in laboratory environments and rely on indoor positioning systems. However, in real-world applications, electric power robots need to have the ability to perceive the environment and adapt to changes in the surrounding environment. The available solutions include simultaneous localization and mapping (SLAM) and object detection, which have been widely investigated based on the camera, combined IMU and camera, and LIDAR.

Decisional autonomy and planning technology is a specific plan or strategy for electric power robots to complete given operation tasks. The challenges generally involve environmental dynamics, task complexity, perception errors, and limited computing power. Thus, electric power robots are required to quickly, efficiently, and autonomously complete operation tasks with the smallest task cost, the most reasonable target ratio, and the optimal motion trajectory. Also, electric power robots should take account into the highly reliable fault tolerance and robustness.

Control execution technology is the basis for electric power robots to complete various complex tasks. Typically, electric power robots are equipped with rigid operating mechanisms, which form a strongly coupled time-varying system with the environment. Especially in aerial manipulator robot systems, the time lag of the aerodynamics system is large, which can easily cause several problems such as difficulty in flight stability and weak adaptability to dynamic environments. These problems bring great challenges to existing theoretical techniques. The existing solutions mostly design compliance control algorithms from the algorithm level, which are difficult and low in safety. Therefore, it is necessary to study the coupled modeling and autonomous control of electric power robots from the perspective of mechanism-algorithm-task integration to improve the stability and safety of the system in contact operation mode.

8.2.2 *Application Challenges*

The design of the operating mechanism is the basis for autonomous electric power robots to complete tasks efficiently and reliably, and the performance of the operating mechanism will directly affect the overall capabilities of the robot. Although prototypes of autonomous electric power robots have been developed and verified, practical applications still face a series of challenges, such as insufficient operation time and load and complex operating mechanisms with single capabilities. On the one hand, new energy suppliers can be explored and developed for electric power robots. Also, highly adaptable electric power robots can be designed and integrated while taking into account reasonable mass distribution and lightweight overall mechanism. On the other hand, the environment faced by electric power robots is becoming extremely complex, and the operational tasks are diverse in the power industry; thus, traditional and single operating mechanisms are different to meet the actual requirements. It is necessary to study the design and integration of task-driven and bionic operating mechanisms.

8.3 Further Directions

1. Trend: With the development of mechatronics and artificial intelligence, autonomous electric power robots are developing from teleoperation and semiautonomous to fully autonomous. Further, the complexity of power operational tasks has also evolved from a single function to a multifunctional mode that integrates sensing and execution.
2. Collaborative Autonomy: As the complexity of power operational tasks increases, it is difficult for a single power operational robot to complete APhI tasks effectively. Thus, the development of human-robot and multi-robot collaborative autonomous operations is the future trend.
3. Novel Morphology: With the rapid development of new materials, it is the future development trend to combine soft materials with aerial robots to achieve

rigid-soft integrated power operation and manipulation. Relying on the variable stiffness capability of the soft mechanism, the rigid-soft integrated electric power robots can implement active compliance and dexterous operations, thereby enhancing the safety of operations in the power industry.

Overall, the four major driving forces can be summarized for accelerating the application of electric power robots:

1. Manipulation ability, which is the basis for the application of electric power robots and determines the application prospects of robots.
2. Autonomy, which is the key to enabling electric power robots with autonomous manipulation, and it will reduce manual intervention and effectively release manual labor.
3. Learning ability, which is the prerequisite and guarantees for electric power robots to perform complex tasks.
4. Human-robot collaboration, which is the basis for the large-scale promotion and application of electric power robots in various scenarios.